FRICTION DYNAMICS

Related title

Handbook of Friction-Vibration Interactions
Xiandong Liu, Gang Sheng Chen
(ISBN: 978-0-85709-458-2)

FRICTION DYNAMICS
Principles and Applications

GANG SHENG CHEN

XIANDONG LIU

ELSEVIER

Amsterdam • Boston • Cambridge • Heidelberg
London • New York • Oxford • Paris • San Diego
San Francisco • Singapore • Sydney • Tokyo
Woodhead Publishing is an imprint of Elsevier

WP
WOODHEAD
PUBLISHING

Woodhead Publishing is an imprint of Elsevier
The Officers' Mess Business Centre, Royston Road, Duxford, CB22 4QH, United Kingdom
50 Hampshire Street, 5th Floor, Cambridge, MA 02139, United States
The Boulevard, Langford Lane, Kidlington, OX5 1GB, United Kingdom

Library of Congress Cataloging-in-Publication Data
A catalog record for this book is available from the Library of Congress

British Library Cataloguing-in-Publication Data
A catalogue record for this book is available from the British Library

ISBN: 978-0-08-100285-8 (print)
ISBN: 978-0-08-100283-4 (online)

For information on all Woodhead Publishing publications visit our website at https://www.elsevier.com/

Publisher: Matthew Deans
Acquisition Editor: Glyn Jones
Editorial Project Manager: Harriet Clayton
Production Project Manager: Debasish Ghosh
Designer: Greg Harris

Typeset by TNQ Books and Journals

CONTENTS

ABOUT THE AUTHORS

Gang Sheng Chen is the J. Robert Fletcher Associate Professor of Engineering at Marshall University, West Virginia, USA. He received his B.S. and M.S. from Shanghai JiaoTong University, P.R., China, and his Ph.D. degree from Nanyang Technical University, Singapore. He is a fellow of SAE and ASME. He has written over 100 journal papers, 7 book chapters and 3 books. Gang has had many years' industry experience as a principal engineer and research scientist at DSI Singapore, Sony Singapore Research Lab, IBM Corp. Technical Division, California and Gates Corp. Technological Center in Michigan, USA.

Xiandong Liu is a Professor of Automotive Engineering at the School of Transportation Science and Engineering, Beihang University, Beijing, P.R. China. He received his B.S. and M.S. from Jilin University and his Ph.D. from Beihang University, P.R. China. He teaches noise and vibration control of vehicles and his research interests include mechanical system dynamics, noise and vibration control, strength analysis and lightweight design of structure. He has published 104 journal papers and 1 book, and is a recipient of 5 patents.

PREFACE

Due to its significance in engineering applications, the friction dynamics of complex systems has been widely investigated by using modeling, analysis and experiments. This book introduces basic principles of friction dynamics. The friction dynamics are presented in a unified theoretical framework with focusing on some most important engineering applications. The first three chapters in the book introduce basic concepts and analytical methods of friction and dynamics. These are designed for the readers who may need to recap basic knowledge in the relevant areas. The fourth chapter presents vehicle brake friction dynamics. In the fifth chapter the friction dynamics of oil drilling and lifting systems is described. The chapter six presents the friction dynamics of rotor systems and also touches on the fault diagnosis. This book provides the information necessary to understand and deal with friction dynamics by covering the fundamental principles and the most critical engineering applications which represent the state-of-the-art in the study of friction dynamics.

This book gives a comprehensive picture of friction dynamics, intended for practicing engineers, designers, researchers, and educators to use as a reference book. It is also a special text of engineering for upper-level undergraduate and graduate students as an independent study text. Research students will also find this book useful.

This book has the following key features: widely interested and concerned topic, from mechanical engineering, automotive engineering to oil & gas engineering; the first reference of this kind to introduce the framework of friction dynamics principles; the analysis, experiments and technical treatment of varied the real systems; broad coverage of contemporary perspectives of complex friction dynamics including condition/health monitoring and diagnosis; presenting comprehensive and new view of friction dynamics using modern tools; very useful for users who want to control friction dynamics or make use of it.

We sincerely hope that this book will help students, teachers, and researchers in developing an appreciation for the topic of friction dynamics. Any errors, oversights, omissions, or other comments to improve the book can be communicated to authors and will be greatly appreciated.

We wish to acknowledge the contributions to this book by many colleagues and students. We here acknowledge those individuals who helped us to develop understanding: Wade Bray, Prof. Mohammed Qatu, Prof. Jonah Lee, Dr. Bill Chen, Dr. Feng Xiao and Prof. Leroy Hulsey. We would like to thank Dr. T. He, Dr. Y. C. Shan and Dr. H. X. Wang, Beihang Univeristy, for their help in the preparation of the draft, their reviews of portions of the manuscript and helpful comments. We also are grateful to the graduate students, D. Shang, K. Shang, D. Qu, W. H. Chai, H. Dong, Q. Gao and J. J. Yi, Beihang Univeristy, for their help in creating figures and proofreading of the manuscript.

CHAPTER 1

Introduction

1.1 DEFINITIONS OF FRICTION, DYNAMICS, AND FRICTION DYNAMICS

When the surfaces of two objects are placed in contact and allowed to slide against one another, there is a resistance to this motion. This resistance is known as *friction*. Friction is one of the most important problems in tribology, the science and technology of interacting surfaces in relative motion concerned with friction, wear, and lubrication in interfaces. *Dynamics* deals with the motion of rigid or deformable bodies and the forces associated with it. Dynamical systems with friction often give rise to diverse forms of motion.

The contact and friction events between two contact surfaces constitute an energy transition and dissipation process. The kinetic energy of driving bodies is usually transferred to the driven body through friction and is also transferred into the dissipative energy of irregular microscopic motion of the interface asperities, particles, and atoms for deformable bodies. The friction usually gives rise to motion, oscillatory vibrations, or vibrations propagating in media.

Up to now, there has been no universally accepted friction model or theory to cover general friction phenomena due to its multiphysics nature. Different models have been developed for individual conditions [1−4]. Part of the reason for this is that friction is a complex process in which forces are transmitted, mechanical energy is converted, surface topography is altered, interface material can be removed or formed, and physical and even chemical changes can occur. Actually, friction can be assumed to be a variable in a dynamical system with a sliding interface, which is complicated by the fact that this system's boundary conditions are not stationary and deterministic due to interfacial changes in geometrical, mechanical, material, physical, and chemical aspects. As such, the coefficients of friction are not intrinsic properties of materials. Rather, they depend on the properties of the contact surfaces, their operational conditions, their time history, environmental conditions, and even their interactions.

Friction Dynamics
ISBN 978-0-08-100285-8
http://dx.doi.org/10.1016/B978-0-08-100285-8.00001-8

Tribology and dynamics used to be two distinct fields. However, with the recent rapid developments in the related area of friction dynamics, especially in the application of varied engineering disciplines, engineers are turning to combining tribology and dynamics for efficient methods to handle and analyze the vast amounts of practical cases.

This book, *Friction Dynamics,* offers a combined treatment of modeling, analyzing, and testing of many problems that application engineers and scientists are trying to solve. After delineating these mathematical characterizations, it presents several applications currently in use for analyzing friction dynamics. The emphasis of the text is on the fundamental aspects and the contemporary knowledge in the area of friction dynamics.

1.2 SIGNIFICANCES AND CHALLENGES OF STUDIES ON FRICTION DYNAMICS

Friction dynamics exists in numerous engineering systems, including brake squealing of automobiles and motorcycles, chattering of machine tools, rubbing impact in rotor systems, and friction whirl of drilling systems, just to name a few. These examples of friction dynamics cover numerous phenomena in science and engineering. Even in the interior of a car, from time to time we may perceive clutch shuddering, wiper squealing, belt drive chirping, instrument panel squeaking, tire road friction noise, and other phenomena. This book considers friction dynamics in its broader meaning, but with a concentration on the fundamentals and engineering applications.

To give some examples of the problems treated in the book, let's consider the immense effort that is being put into dealing with friction dynamics in the mechanical and automotive industries. The need for rotatory systems is tremendously great and continues to increase. There is a variety of mechanical systems with varying rotor systems, from washing machines to airplane engines. In these contexts, interface friction and contact-induced friction dynamics and instability have been the most challenging and critical problems faced by the product designers.

Our lives are highly dependent on vehicles of various types. Most vehicles, like automobiles, motorcycles, trains, and aircraft, use some kind of brake system for stopping. In these brake systems, friction is a principal functional/safety performance factor and is also a potential cause of undesirable noise and vibration. The structures and principles of the varied brake systems in vehicles are analogous and similar, but noise has been an unsolved problem for the last half-century. Friction dynamics affects the reliability and quality of brake systems in many ways. For instance, brake squealing, a complex vibration, has been the most challenging issue in automotive brake systems, as it has been equal to the quality of products perceived by customers. It has been estimated that the warranty work of noise and vibration of automotive brake systems costs approximately $1 billion a year in the Detroit area alone. Even in aircraft braking systems, friction dynamics has been a critical issue. Friction-induced torque oscillations can lead to excessive high loads in the landing gear and brake structures, and this results in passenger discomfort, component failure, or both, and thereby warranty claims.

Understanding the nature of friction dynamics and solving the technological problems associated with it are the essence of these fields. Modeling of friction dynamics in mechanical and other systems requires an accurate definition and description of friction. Unfortunately, there is no universally accepted friction model or theory to cover general friction phenomena due to the reasons that have already been mentioned. On the other hand, the resultant dynamics often exhibit various nonlinear, transient/nonstationary, stochastic and uncertain properties [5–17]. Moreover, small changes in interfacial parameters could have significant effects on the resultant dynamics, and thus the scales of influencing factor range from the macrometer, micrometer, and nanometer levels. The boundary conditions of the problems are not fixed; rather, they depend on environmental conditions, operation conditions, system interactions, and time. Because of the complexity of friction dynamics, it has been considered to be an unsolved problem in many engineering applications. Since the modeling and the predictions are not very reliable, the trial-and-error approach has been extensively used.

Recent extensive efforts on modeling, analytical, and experimental investigations have made much substantial progress in many practical applications. Many techniques, such as advanced testing, advanced signal processing, and spectrum analysis and contemporary nonlinear stochastic dynamics techniques have been used as efficient means to address the nonlinear, nonstationary, stochastic and uncertain dynamics; it enables friction dynamics to be quantified efficiently. The emergence and application of scanning tunneling microscopy and atomic force microscopy have allowed for systematic investigation of interfacial problems with high resolution, which have led to the development of the insight of friction at the micro-, nano-, molecular-, and even atomic-scale levels. There are several purposes of research in the area of friction dynamics, such as to develop a fundamental understanding of friction dynamics, to control the motion of dynamical systems with friction, to realize some physical processes for products, and to reduce and eliminate instability in engineering systems caused by friction.

1.3 ORGANIZATION OF THE BOOK

The book has two purposes. The first is to give a general introduction to the theory of friction and dynamics by offering a physical view of the fundamental theory. The second is to give examples of the applications of these theoretical approaches. The author will provide contemporary coverage of the primary concepts and techniques in the treatment of friction dynamics.

There are six chapters. The basic principles of vibrations and friction are discussed in the first three chapters. This chapter gives a brief introduction to the whole book. Chapter 2 provides a comprehensive analysis of vibrations and advanced dynamics. The vibration segment of this chapter covers vibrations of linear systems, random excited systems, and nonlinear systems, and discusses most of the major areas and applications, including the Duffing equation, Van der Pol equation, and Mathieu equation. The advanced dynamics portion presents rigid body dynamics including Euler equation.

Chapter 3 describes contact and friction principles. Chapter 4 presents the friction dynamics of vehicle brake systems, covering various vibrations and instability issues of brake systems and numerical and analysis approaches. Chapter 5 presents the friction dynamics of oil-well drill-strings and sucker rods. Finally, Chapter 6 presents the friction dynamics of rotor systems.

Reference sections in the book provide a comprehensive perspective on developments in friction dynamics, as well as various applications. For didactic reasons, the text flow is not interrupted by the inclusion of references. However, at the end of each chapter, the relevant literatures published are cited.

REFERENCES

[1] Armstrong-Helouvry, B., Dupont, P., Canudas De Wit, C., 1994. A survey of models, analysis tools, and compensation methods for the control of machines with friction. Automatica 30 (7), 1083–1138.

[2] Berger, E.J., 2002. Friction modeling for dynamic system simulation. Appl. Mech. Rev. 55 (6), 535–576.

[3] Bhushan, B., 1999. Principles and Applications of Tribology. John Wiley & Sons, New York.
[4] Chen, G., 2014. Handbook of Friction-Vibration Interactions. Elsevier.
[5] Chen, G., Sheng, G., 2012. Vehicle Noise, Vibrations, and Sound Quality. SAE International, Warrendale, PA.
[6] Choy, F.K., Padovan, J., 1987. Non-linear transient analysis of rotor-casing rub event. J. Sound Vib. 13 (3), 529–545.
[7] Guran, A., Pfeiffer, F., Popp, K., 1996. Dynamics with Friction. World Scientific Publishing, New Jersey.
[8] Ibrahim, R.A., 1994. Friction-induced vibration, chatter, squeal, and chaos. Part 1: Mechanics of contact and friction. Appl. Mech. Rev. 47 (7), 209–226.
[9] Ibrahim, R.A., 1994. Friction-induced vibration, chatter, squeal, and chaos. Part 2: Dynamics and modeling. Appl. Mech. Rev. 47 (7), 227–253.
[10] Liu, X.D., Li, Q.H., 1998. Impact-rub fault characteristics and holospectrum analysis of rotating machinery with imbalance. J. Aerosp. Power 13 (4), 428–430.
[11] Liu, X.D., Li, Q.H., 1998. Impact-rub model and chaotic feature of partial rub of eccentric rotor. J. Aerosp. Power 13 (4), 361–365.
[12] Nayfeh, A.H., Mook, D.T., 1979. Nonlinear Oscillations. Wiley, New York.
[13] Oden, J.T., Martins, J.A.C., 1985. Models and computational methods for dynamic friction phenomena. Comp. Meth. Appl. Mech. Eng 52 (1–3), 527–634.
[14] Sheng, G., Wang, Q., 2000. Brake NVH Technology, AMC report, 2002–82.
[15] Sheng, G., Liu, B., Zhu, Y.L., 1998. Vibrations in contact magnetic recording systems. In: Bhushan, B., Ono, K. (Eds.), Advances in Information Storage System. World Scientific Publishing, New Jersey.
[16] Shi, H.M., Sheng, G., Wu, Y., 1991. Mechanical Vibration System: Analysis, Measurement, Modeling, and Control. HUST Press, Wuhan, China.
[17] Stewart, D.E., 2000. Rigid-body dynamics with friction and impact. Siam Rev. 42 (1), 3–39.

CHAPTER 2

Vibrations and Advanced Dynamics

2.1 INTRODUCTION

This chapter presents information on vibrations and advanced dynamics. After the introduction, the sections 2.2-2.5 of this chapter are devoted to the analysis of linear vibration systems. The section 2.6 is dedicated to the random vibrations. The section 2.7 presents the fundamentals of nonlinear vibrations. The final section describes the advanced dynamics.

2.2 LINEAR VIBRATIONS UNDER DETERMINISTIC EXCITATIONS

2.2.1 Vibration of Linear Discrete and Continuous Systems

Vibration is the oscillatory motion of a body or structure. It takes place when a body is displaced from its stable equilibrium position by a restoring force.

A vibration system with a finite number of unknown variables is said to be *discrete,* while a system whose variables are functions of location as well as time is called *continuous*. Real systems are continuous, and their parameters are distributed. In many situations, it is possible to approximate a continuous system by discrete ones.

The analytical description of the vibrations of a discrete case is a set of ordinary differential equations, while for the continuous case, it is a set of partial differential equations. If the dependent variables in the differential equation are to the first power, then the system is linear. If there are fractional or higher powers, then the system is nonlinear. The superposition principle holds only for linear systems.

The independent coordinates required to quantify the configuration of a vibration system are called *generalized coordinates*. The number of generalized coordinates is defined as the number of degrees of freedom of the system. A discrete model of a dynamic system possesses a finite number of

Friction Dynamics
ISBN 978-0-08-100285-8
http://dx.doi.org/10.1016/B978-0-08-100285-8.00002-X

degrees of freedom, whereas a continuous model has an infinite number of degrees of freedom.

The excitation of a vibration system is usually a function of time. The vibratory motion of the system caused by excitations is referred to as the *response*. If the vibratory motion is periodic, then the system repeats its motion at equal time intervals. The minimum time required for the system to repeat its motion is called a *period;* this is the time needed to complete one cycle of motion. *Frequency* is defined as the number of times that the motion repeats itself per unit of time. Free vibrations describe the natural behavior of vibration of a system. Many systems need to be treated as damped systems due to the dissipation of motion energy.

The excitation may be either deterministic or a random function of time. In deterministic vibrations, the response at any designated future time can be completely predicted from past history; random forced vibrations are defined statistically, and only the probability of occurrence of designated magnitudes and frequencies can be predicted [1–5].

2.2.2 Vibration of Linear Discrete Systems: SDOF System

Consider a sinusoidal periodic motion:

$$x(t) = X \sin\left(2\pi f_0\, t + \theta\right), \qquad [2.1]$$

where X is amplitude, f_0 is cyclical frequency in cycles per unit of time, θ is initial phase angle with respect to the time origin in radians, and $x(t)$ is the instantaneous value at time t.

The time interval required for one full cycle of sinusoidal motion is called the period (T_P). The number of cycles per unit time is called the frequency (f_0). The frequency and period are related to each other as follows:

$$T_p = {}^{1}\!/_{f_0}$$

Complex periodic motion can be defined mathematically by a time-varying function whose waveform exactly repeats itself at regular intervals such that

$$x(t) = x(t \pm nT_P),\ \mathrm{n} = 1, 2, 3, \ldots$$

The complex periodic motion can be expanded into a Fourier series as

$$x(t) = \frac{a_0}{2} + \sum_{n=1}^{\infty}\left(a_n \cos 2\pi n f_1 t + b_n \sin 2\pi n f_1 t\right), \qquad [2.2]$$

where $f_1 = 1/T_P$:

$$a_n = \frac{2}{T_p}\int_0^{T_p} x(t)\cos 2\pi n f_1 t dt \quad n = 1, 2, \ldots$$

$$b_n = \frac{2}{T_p}\int_0^{T_p} x(t)\sin 2\pi n f_1 t dt \quad n = 1, 2, \ldots.$$

Another way of expressing the Fourier series for complex periodic data is

$$x(t) = X_0 + \sum_{n=1}^{\infty} X_n \sin(2\pi n f_1 t - \theta_n), \qquad [2.3]$$

where $X_0 = a_0/2$, $X_n = \sqrt{a_n^2 + b_n^2}$, $\theta_n = tan^{-1}(a_n/b_n)$, $n = 1, 2, \ldots$. Eq. [2.3] implies that complex periodic data consist of a static component, X_0, and an infinite number of sinusoidal components called *harmonics,* which have amplitude X_n and phase θ_n. The frequencies of the harmonic components are all integral multiples of f_1. The phase angles are often ignored when periodic data is analyzed in practice. In this case, Eq. [2.3] can be characterized by a discrete spectrum.

Transient motion is defined as all nonperiodic motion other than the almost-periodic data discussed previously. The concept also includes all motion that can be described by some suitable time-varying function. Physical phenomena that produce transient data are numerous and diverse. The important characteristic of transient motion is its continuous spectral representation, which can be obtained in most cases from a Fourier integral given by

$$X(f) = \int_{-\infty}^{\infty} x(t)e^{-j2\pi f t} dt. \qquad [2.4]$$

The Fourier spectrum $X(f)$ is generally a complex number that can be expressed in complex polar notation as $X(f) = |X(f)|e^{-j\theta(f)}$, where $|X(f)|$ is the magnitude of $X(f)$ and $\theta(f)$ is the argument.

2.3 SDOF SYSTEM

Here, we discuss the vibration of a single-degree-of-freedom (SDOF) model, as shown in Figure 2.1. From Newton's law, we obtain

$$F(t) - F_s(t) - F_d(t) = m\,\ddot{x}(t), \qquad [2.5]$$

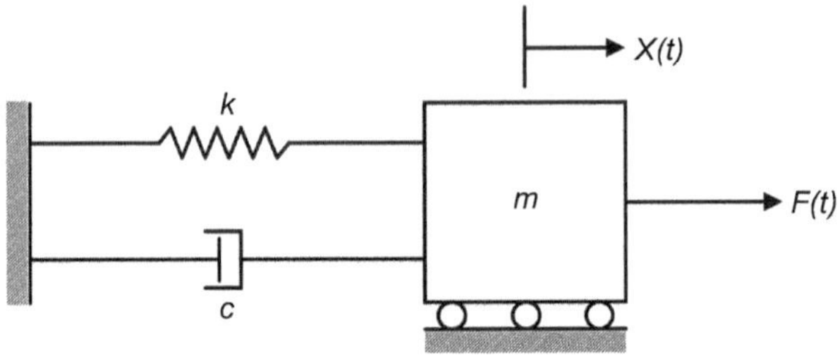

Figure 2.1 SDOF system

where $F(t)$, $F_s(t)$, and $F_d(t)$ are the exciting, spring, and damping forces, respectively, m denotes the mass of the body, and $\ddot{x}(t)$ its acceleration. Because $F_s(t) = kx(t)$ and $F_d(t) = c\,\dot{x}(t)$, Eq. [2.5] becomes

$$m\ddot{x}(t) + c\dot{x}(t) + kx(t) = F(t), \tag{2.6}$$

where c and k are the viscous damping and stiffness coefficients, respectively.

Eq. [2.6] is the equation of motion of the linear SDOF system and is a second-order linear ordinary differential equation with constant coefficients.

In the case of the free vibration of an SDOF system, the exciting force $F(t) = 0$ and the equation of motion is

$$m\ddot{x}(t) + c\dot{x}(t) + kx(t) = 0. \tag{2.7}$$

If we define $\omega_n^2 = k/m$ and $\xi = c/2m\omega_n$, Eq. [2.7] can be written as

$$\ddot{x}(t) + 2\xi\omega_n\dot{x}(t) + \omega_n^2 x(t) = 0. \tag{2.8}$$

To solve Eq. [2.8], assume that

$$x(t) = A\,e^{st}, \tag{2.9}$$

where A is a constant and s is a parameter to be determined. By substituting Eq. [2.9] into Eq. [2.8], one obtains

$$\left(s^2 + 2\xi\omega_n s + \omega_n^2\right) A\,e^{st} = 0. \tag{2.10}$$

Since $A\,e^{st} \neq 0$, then

$$s^2 + 2\xi\omega_n s + \omega_n^2 = 0. \tag{2.11}$$

Eq. [2.11] is known as the characteristic equation of the system. This equation has the following two roots:

$$s_1, s_2 = \left(-\xi \pm \sqrt{\xi^2 - 1}\right)\omega_n. \tag{2.12}$$

For case a, $\xi < 1$ (underdamped conditions):

$$s_1, s_2 = \left(-\xi \pm i\sqrt{1-\xi^2}\right)\omega_n$$
$$x(t) = A\exp(-\xi\omega_n t)\cos\left(\omega_n\sqrt{1-\xi^2}t - \varphi\right) \quad [2.13]$$

$$x(t) = A\exp(-\xi\omega_n t)\cos(\omega_d t - \varphi), \quad [2.14]$$

where ω_n is the natural circular frequency, ξ is the damping factor, and $\omega_d = \omega_n\sqrt{1-\xi^2}$, the damped frequency of the system. Constants A and φ are determined from the initial conditions.

For case b, $\xi > 1$ (overdamped conditions):

$$s_1, s_2 = \left(-\xi \pm \sqrt{\xi^2 - 1}\right)\omega_n$$
$$x(t) = A_1\exp\left(-\xi + \sqrt{\xi^2-1}\right)\omega_n t + A_2\exp\left(-\xi - \sqrt{\xi^2-1}\right)\omega_n t. \quad [2.15]$$

The motion is aperiodic and decays exponentially with time. Constants A_1 and A_2 are determined from the initial conditions.

For case c, $\xi = 1$ (critically damped conditions):

$$s_1 = s_2 = -\omega_n$$
$$x(t) = (A_1 + A_2)\exp(-\omega_n t). \quad [2.16]$$

Eq. [2.16] represents an exponentially decaying response. The constants A_1 and A_2 depend on the initial conditions. For this case, the coefficient of viscous damping has the value $c_c = 2m\omega_n = 2\sqrt{km}$.

$$\xi = c/c_c. \quad [2.17]$$

We consider the undamped condition, in which t_1 and t_2 denote the times corresponding to the consecutive displacements x_1 and x_2, respectively, measured one cycle apart. By using Eq. [2.14], we can write

$$\frac{x_1}{x_2} = \frac{A\exp(-\xi\omega_n t_1)\cos(\omega_d t_1 - \varphi)}{A\exp(-\xi\omega_n t_2)\cos(\omega_d t_2 - \varphi)}. \quad [2.18]$$

Since $t_2 = t_1 + T = t_1 + 2\pi/\omega_d, cos(\omega_d t_1 - \varphi) = cos(\omega_d t_2 - \varphi)$. Eq. [2.18] then reduces to

$$x_1/x_2 = \exp(\xi\omega_n T).$$

We define logarithmic decrement as

$$\delta = \ln (x_1/x_2) = \xi\omega_n T = 2\pi\xi \Big/ \sqrt{1-\xi^2}. \qquad [2.19]$$

To determine the amount of damping in the system, it is sufficient to measure any two consecutive displacements x_1 and x_2 and obtain ξ from the following:

$$\xi = \delta \Big/ \sqrt{(2\pi)^2 + \delta^2}. \qquad [2.20]$$

We now consider the response of an SDOF system to a harmonic excitation, for which the equation of motion is

$$m\ddot{x}(t) + c\dot{x}(t) + kx(t) = F_0 \cos \omega t, \qquad [2.21]$$

where F_0 is the amplitude and ω the frequency of the excitation. Eq. [2.21] can be simplified as

$$\ddot{x}(t) + 2\xi\omega_n\dot{x}(t) + \omega_n^2 x(t) = (F_0/k)\omega_n^2 \cos \omega t. \qquad [2.22]$$

The solution of Eq. [2.22] consists of two parts: the complementary function, which is the solution of the homogeneous equation, and the particular integral. The complementary function dies out with time for $\xi > 0$ and is often called the *transient solution,* whereas the particular solution does not vanish for a large t and is referred to as the *steady-state solution* to the harmonic excitation. We assume a solution that takes the form

$$x(t) = X \cos(\omega t - \varphi), \qquad [2.23]$$

where X and φ are the amplitude and phase angle of response, respectively.

By substituting Eq. [2.23] into Eq. [2.22], we obtain

$$X\left[\left(\omega_n^2 - \omega^2\right) \cos(\omega t - \varphi) - 2\xi\omega_n\omega \sin(\omega t - \varphi)\right] = (F_0/k)\omega_n^2 \cos \omega t. \qquad [2.24]$$

By developing the terms in Eq. [2.24], and equating the coefficients of $\cos\omega t$ and $\sin\omega t$ on both sides of the equation, we obtain

$$X\left[\left(\omega_n^2 - \omega^2\right) \cos \varphi + 2\xi\omega_n\omega \sin \varphi\right] = (F_0/k)\omega_n^2 \qquad [2.25a]$$

$$X\left[\left(\omega_n^2 - \omega^2\right) \sin \varphi - 2\xi\omega_n\omega \cos \varphi\right] = 0. \qquad [2.25b]$$

By solving Eq. [2.25], we get

$$X/(F_0/k) = \left\{\left[1-(\omega/\omega_n)^2\right]^2 + \left[2\xi(\omega/\omega_n)\right]^2\right\}^{-1/2} \quad [2.26]$$

and

$$\varphi = tan^{-1}\left\{[2\xi(\omega/\omega_n)]/\left[1-(\omega/\omega_n)^2\right]\right\}. \quad [2.27]$$

Eqs. [2.26] and [2.27] indicate that the nondimensional amplitude $X/(F_0/k)$ and the phase angle φ are functions of the frequency ratio ω/ω_n and the damping ratio ξ. For ω/ω_n much less than 1, both the inertia and damping forces are small, and this results in a small phase angle φ, with $X/(F_0/k) \cong 1$. However, for ω/ω_n much greater than 1, the phase angle $\varphi \rightarrow 180°$ and $X/(F_0/k) \rightarrow 0$. For $\omega/\omega_n = 1$, the phase angle $\varphi = 90°$ and $X/(F_0/k) = 1/2\xi$. In summary, the complete solution of Eq. [2.22] is given as

$$x(t) = A_1 \exp(-\xi\omega_n t)\cos(\omega_d t + \varphi_1) + \frac{F_0}{k}\frac{\cos(\omega\, t - \varphi)}{\sqrt{\left[1-(\omega/\omega_n)^2\right]^2 + \left[2\xi(\omega/\omega_n)\right]^2}}, \quad [2.28]$$

where the constants A_1 and φ_1 are determined by the initial conditions.

Let us reconsider Eq. [2.22] and represent the excitation by the complex form

$$(F_0/k)\omega_n^2 e^{-i\omega t} = X_s\omega_n^2 e^{-i\omega t}, \quad [2.29]$$

where $X_s = F_0/k$ and is referred to as a static response. We assume a solution in the form

$$x(t) = Xe^{-i\omega t}. \quad [2.30]$$

By substituting Eq. [2.30] into Eq. [2.22], we get

$$\left[\omega_n^2 - \omega^2 - 2i\xi\omega_n\omega\right]Xe^{-i\omega t} = X_s\omega_n^2 e^{-i\omega t} \quad [2.31]$$

$$X/X_s = \left[1-(\omega/\omega_n)^2 - 2i\xi(\omega/\omega_n)\right]^{-1} = H(\omega), \quad [2.32]$$

where $H(\omega)$ is known as the complex frequency response function. Its magnitude $|H(\omega)|$ refers to a magnification factor and is given as

$$|H(\omega)| = \left\{\left[1-(\omega/\omega_n)^2\right]^2 + \left[2\xi(\omega/\omega_n)\right]^2\right\}^{-1/2}. \quad [2.33]$$

The phase angle φ will be

$$\varphi = \tan^{-1}\left[\frac{2\xi(\omega/\omega_n)}{1-(\omega/\omega_n)^2}\right]. \quad [2.34]$$

The excitation considered thus far has been a simple harmonic force. We can generalize the results when the exciting force is periodic because periodic force can be expanded in terms of the Fourier series, as follows:

$$F(t) = a_1 \sin \omega t + b_1 \cos \omega t + a_2 \sin 2\omega t + b_2 \cos 2\omega t + \cdots + a_n \sin n\omega t + b_n \cos n\omega t, \quad [2.35]$$

where a_n and b_n are the coefficients of the Fourier series expansion, and it is assumed that the constant $b_0 = 0$. Because

$$a_n \sin n\omega t + b_n \cos n\omega t = f_n \sin (n\omega t + \alpha_n), \quad [2.36]$$

where $f_n = \sqrt{a_n^2 + b_n^2}$ and $\alpha_n = \tan^{-1}(b_n/a_n)$, it follows that

$$F(t) = f_1 \sin (\omega t + \alpha_1) + f_2 \sin (2\omega t + \alpha_2) + \cdots + f_n \sin (n\omega t + \alpha_n). \quad [2.37]$$

Because superposition is valid, we can consider each term on the right side of Eq. [2.37] as a separate forcing function and obtain the steady-state response by adding individual responses due to each forcing function acting separately. Hence, it follows that

$$x(t) = X_1 \cos (\omega t + \alpha_1 - \varphi_1) + X_2 \cos (2\omega t + \alpha_2 - \varphi_2) + \cdots + X_n \cos(n\omega t + \alpha_n - \varphi_n), \quad [2.38]$$

where

$$X_n = \frac{f_n/k}{\left\{\left[1 - (n\omega/\omega_n)^2\right]^2 + \left[2\xi(n\omega/\omega_n)\right]^2\right\}^{1/2}}$$

and

$$\varphi_n = \tan^{-1}\frac{2\xi(n\omega/\omega_n)}{1 - (n\omega/\omega_n)^2} \qquad n = 1, 2, \ldots$$

Hence, the steady-state response is also periodic, with the same period as the forcing function but with a different amplitude and an associated phase lag.

For arbitrary excitation *F*(*t*), the steady-state response is

$$x(t) = \frac{1}{m\omega_d} \int_0^t F(\tau)\exp[-\xi\omega_n(t - \tau)]\sin \omega_d(t - \tau)d\tau. \quad [2.39]$$

2.4 LINEAR MDOF SYSTEM

Next, we consider the multiple-degree-of-freedom (MDOF) discrete system shown in Figure 2.2. Its general equations of motion are written as

$$[m]\{\ddot{x}(t)\} + [c]\{\dot{x}(t)\} + [k]\{x(t)\} = \{F(t)\}, \qquad [2.40]$$

where $\{F(t)\}$ denotes the externally applied force. In Eq. [2.40], $[m]$, $[c]$, and $[k]$ are the $n \times n$ mass, damping, and stiffness matrices, respectively. For linear systems, these matrices are constant, whereas for nonlinear systems, the elements of these matrices are functions of generalized displacements and velocities that are time dependent. The response $\{x(t)\}$ of Eq. [2.40] consists of two parts: first, $\{x_h(t)\}$, the homogeneous solution which is the transient response, and second, $\{x_p(t)\}$, the particular solution which is the steady state or forced response.

2.4.1 Eigenvalues and Eigenvectors

Next, we discuss the simplest case, the equations of motion for the free vibration of an undamped MDOF system, focusing on its eigenvalues and eigenvectors. By setting $[c]$ and $\{F(t)\}$ in Eq. [2.40] to zero, it follows that

$$[m]\{\ddot{x}(t)\} + [k]\{x(t)\} = \{0\}. \qquad [2.41]$$

We use a linear transformation to replace $\{x\}$:

$$\{x\} = [\varphi]\,\{y\}, \qquad [2.42]$$

where $[\varphi]$ is a constant nonsingular square matrix to be specified in the following analysis. It is referred to as a transformation matrix:

$$\{\ddot{x}(t)\} = [\varphi]\{\ddot{y}(t)\}. \qquad [2.43]$$

Substituting Eqs. [2.42] and [2.43] into Eq. [2.41], we obtain

$$[m][\varphi]\{\ddot{y}(t)\} + [k][\varphi]\{y(t)\} = \{0\}. \qquad [2.44]$$

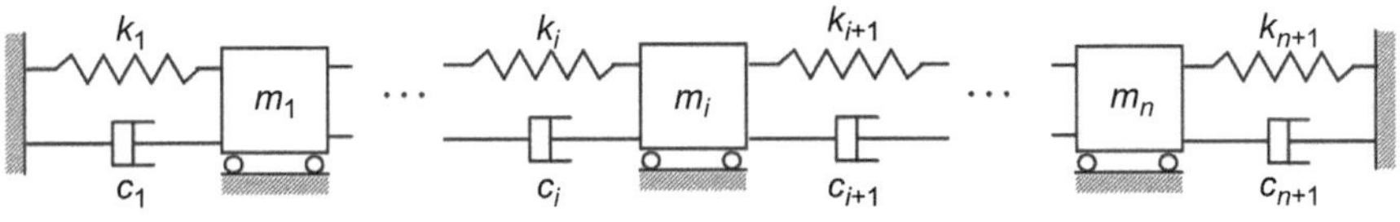

Figure 2.2 Linear MDOF system

Pre-multiplying both sides of Eq. [2.44] by $[\varphi]^T$ to yield

$$[\varphi]^T[m][\varphi]\{\ddot{y}(t)\} + [\varphi]^T[k][\varphi]\{y(t)\} = \{0\}. \qquad [2.45]$$

From Eq. [2.45], it follows that

$$[m^*]\{\ddot{y}(t)\} + [k^*]\{y(t)\} = \{0\}, \qquad [2.46]$$

where $[m^*]$, $[k^*]$ are diagonal matrices known as the generalized mass and the stiffness matrix, respectively. Eq. [2.46] refers to the uncoupled homogeneous equations of motion of the system. It follows that the uncoupled equation of motion for the *i*th degree of freedom is

$$\ddot{y}_i + \omega_i^2 y_i = 0, \qquad [2.47]$$

where ω_i is the frequency corresponding to the *i*th mode of vibration. The solution of Eq. [2.47] is given as

$$y_i(t) = A_i \sin \omega_i t + A_i^* \cos \omega_i t, \qquad [2.48]$$

where arbitrary constants A_i and A_i^* are determined by the initial conditions $x_i(0)$ and $\dot{x}_i(0)$.

We now consider Eq. [2.41] and premultiply both sides by $[m]^{-1}$ to yield

$$[m]^{-1}[m]\{\ddot{x}(t)\} + [m]^{-1}[k]\{x(t)\} = \{0\}. \qquad [2.49]$$

Eq. [2.49] can be written as

$$[I]\{\ddot{x}(t)\} + [D]x(t)\} = \{0\}, \qquad [2.50]$$

where $[I]$ is the unit matrix and $[D] = [m]^{-1}[k]$, which is known as the dynamic matrix.

Let us assume a harmonic motion such that

$$\{x\} = \{A\}e^{i\omega t}. \qquad [2.51]$$

Eq. [2.51] yields

$$\{\ddot{x}(t)\} = -\omega^2\{A\}e^{i\omega t} = -\lambda\{x(t)\}, \qquad [2.52]$$

where $\lambda = \omega^2$. Substituting Eqs. [2.51] and [2.52] into Eq. [2.49] results in

$$[[D] - \lambda[I]]\ \{x\}\ = \{0\}. \qquad [2.53]$$

The characteristic equation of the system, then, is as follows, with the determinant being zero:

$$|[D] - \lambda[I]| = 0. \qquad [2.54]$$

The roots λ_i of the characteristic equation are called *eigenvalues*. The natural frequencies of the system are determined from

$$\lambda_i = \omega_i^2. \qquad [2.55]$$

By substituting λ_i into the matrix [Eq. [2.53]], we obtain the corresponding mode shapes, which are called *eigenvectors*. Thus, for an n-degree-of-freedom system, there are n eigenvalues and n eigenvectors.

Let us now consider two distinct solutions corresponding to the rth and the sth modes of the eigenvalue problem, ω_r^2, $\{\varphi^{(r)}\}$ and ω_s^2, $\{\varphi^{(s)}\}$ of the eigenvalue problem, respectively. Because these solutions satisfy Eq. [2.41], it follows that

$$[k]\{\varphi^{(r)}\} = \omega_r^2[m]\{\varphi^{(r)}\} \qquad [2.56]$$

and

$$[k]\{\varphi^{(s)}\} = \omega_s^2[m]\{\varphi^{(s)}\}. \qquad [2.57]$$

We premultiply both sides of Eq. [2.56] by $\{\varphi^{(s)}\}^T$ and both sides of Eq. [2.57] by $\{\varphi^{(r)}\}^T$ to obtain

$$\{\varphi^{(s)}\}^T[k]\{\varphi^{(r)}\} = \omega_r^2\{\varphi^{(s)}\}^T[m]\{\varphi^{(r)}\} \qquad [2.58a]$$

$$\{\varphi^{(r)}\}^T[k]\{\varphi^{(s)}\} = \omega_s^2\{\varphi^{(r)}\}^T[m]\{\varphi^{(s)}\}. \qquad [2.58b]$$

Now we take the transposition of Eq. [2.58b] to get

$$\{\varphi^{(s)}\}^T[k]\{\varphi^{(r)}\} = \omega_s^2\{\varphi^{(s)}\}^T[m]\{\varphi^{(r)}\}. \qquad [2.59]$$

By subtracting Eq. [2.59] from Eq. [2.58a], we obtain

$$(\omega_r^2 - \omega_s^2)\{\varphi^{(s)}\}^T[m]\{\varphi^{(r)}\} = 0. \qquad [2.60]$$

Because $\omega_r \neq \omega_s$, we conclude that

$$\{\varphi^{(s)}\}^T[m]\{\varphi^{(r)}\} = 0 \quad r \neq s. \qquad [2.61]$$

Eq. [2.61] represents the orthogonality condition of modal vectors. It can also be shown that

$$\{\varphi^{(s)}\}^T[k]\{\varphi^{(r)}\} = 0 \quad r \neq s. \qquad [2.62]$$

Thus, $[\varphi]$ is composed of $\{\varphi^{(i)}\}, i = 1, 2, \ldots, n$. If each column of the modal matrix $[\varphi]$ is divided by the square root of the generalized mass M_i^*, the new matrix $[\overline{\varphi}]$ is called the *weighted modal matrix*. It can be seen that

$$[\overline{\varphi}]^{\mathrm{T}}[m][\overline{\varphi}] = [I] \qquad [2.63]$$

and

$$[k][\overline{\varphi}] = [m][\overline{\varphi}][\omega^2]. \qquad [2.64]$$

Premultiplying Eq. [2.64] by $[\overline{\varphi}]^{\mathrm{T}}$ results in

$$[\overline{\varphi}]^{\mathrm{T}}[k][\overline{\varphi}] = [\overline{\varphi}]^{\mathrm{T}}[m][\overline{\varphi}][\omega^2] = [\omega^2]. \qquad [2.65]$$

2.4.2 Forced Vibration Solution of an MDOF System

We consider Eq. [2.40] and first solve the undamped free-vibration problem to obtain the eigenvalues and eigenvectors, which describe the normal modes of the system and the weighted modal matrix $[\overline{\varphi}]$. Let

$$\{x\} = [\overline{\varphi}]\{y\}. \qquad [2.66]$$

Substituting Eq. [2.66] into Eq. [2.40] yields

$$[\mathrm{m}][\overline{\varphi}]\{\ddot{y}\} + [\mathrm{c}][\overline{\varphi}]\{\dot{y}\} + [\mathrm{k}][\overline{\varphi}]\{y\} = \{F(t)\}. \qquad [2.67]$$

Premultiplying both sides of Eq.[2.67] by $[\overline{\varphi}]^{\mathrm{T}}$, we obtain

$$[\overline{\varphi}]^{\mathrm{T}}[m][\overline{\varphi}]\{\ddot{y}\} + [\overline{\varphi}]^{\mathrm{T}}[c][\overline{\varphi}]\{\dot{y}\} + [\overline{\varphi}]^{\mathrm{T}}[k][\overline{\varphi}]\{y\} = [\overline{\varphi}]^{\mathrm{T}}\{F(t)\}. \qquad [2.68]$$

Notice that the matrices $[\overline{\varphi}]^{\mathrm{T}}[m][\overline{\varphi}]$ and $[\overline{\varphi}]^{\mathrm{T}}[k][\overline{\varphi}]$ on the left side of Eq. [2.68] are diagonal matrices that correspond to the matrices $[I]$ and $[\omega^2]$, respectively. However, the matrix $[\overline{\varphi}]^{\mathrm{T}}[c][\overline{\varphi}]$ is not diagonal. If $[c]$ is proportional to $[m]$ or $[k]$ or both, $[\overline{\varphi}]^{\mathrm{T}}[c][\overline{\varphi}]$ becomes diagonal, in which case it can be said that the system has proportional damping. The equations of motion are then completely uncoupled, and the ith equation will be

$$\ddot{y}_i + 2\xi_i\omega_i\dot{y} + \omega_i^2 y_i = \overline{f}_i(t) \quad i = 1, 2, \ldots, n \qquad [2.69]$$

where $\overline{f}_i(t) = \{\overline{\varphi}^{(i)}\}\{F(t)\}$. Thus, instead of n-coupled equations, we will have n-uncoupled equations.

Let $[c] = \alpha[m] + \beta[k]$, in which α and β are proportionality constants. Then we have

$$[\overline{\varphi}]^{\mathrm{T}}[c][\overline{\varphi}] = [\overline{\varphi}]^{\mathrm{T}}(\alpha[m] + \beta[k])[\overline{\varphi}] = \alpha[I] + \beta[\omega^2]. \qquad [2.70]$$

This will yield the uncoupled *i*th equation of motion as

$$\ddot{y}_i + \left(\alpha + \beta\omega_i^2\right)\dot{y}_i + \omega_i^2 y_i = \bar{f}_i(t), \qquad [2.71]$$

and the modal damping can be defined as

$$2\xi_i\omega_i = \alpha + \beta\omega_i^2. \qquad [2.72]$$

The solution of Eq. [2.69] is obtained by using Eq. [2.39] with initial conditions $y_i(0)$ and $\dot{y}_i(0)$:

$$\begin{aligned} y_i(t) = {} & \frac{1}{\omega_d}\int_0^t \bar{f}_i(\tau)\exp[-\xi_i\omega_i(t-\tau)]\sin\omega_{di}(t-\tau)d\tau \\ & + \frac{y_i(0)\exp(-\xi_i\omega_i t)}{\left(1-\xi_i^2\right)^{1/2}}\cos(\omega_{di}t - \psi_i) + \frac{\dot{y}_i(0)\exp(-\xi_i\omega_i t)}{\omega_{di}}\sin\omega_{di}t, \end{aligned} \qquad [2.73]$$

where $\omega_{di} = \left(1-\xi_i^2\right)^{1/2}\omega_i$ and $\psi_i = \tan^{-1}\left[\xi_i\Big/\left(1-\xi_i^2\right)^{1/2}\right]$.

Similarly, the contribution from each normal mode is calculated and substituted into Eq. [2.66] to obtain the complete response of the system. This is known as the *normal mode summation method*. The contributions of the higher vibration modes to the system response are often quite small, and for all practical purposes, they may be ignored in the summation procedure by considering fewer modes of vibration.

2.5 VIBRATION OF CONTINUOUS SYSTEMS

So far in this chapter, we have discussed discrete systems where elasticity and mass are modeled as discrete properties. Discrete systems have a finite number of degrees of freedom that specify system finite configuration. Continuous systems are distributed systems such as strings, cables, rods (bars), and beams, as well as plates where elasticity and mass are distributed parameters. We consider the continuous distribution of elasticity, mass, and damping and assume that each of the infinite number of elements of the system can vibrate.

The displacement of these elements is described by a continuous function of position and time. The governing equations of motion for discrete systems are ordinary differential equations, whereas the governing equations are partial differential equations for the continuous systems, and exact solutions can be obtained for only a few special configurations. For the vibration analysis of systems with distributed elasticity and mass, it must be assumed that the material is homogeneous and isotropic and follows Hooke's law.

2.5.1 Transverse Vibrations of String and Wave Equation

Consider a stretched flexible string of mass ρ per unit length with its end points attached to fixed surfaces. The string is free to vibrate in a vertical (x–y) plane, as shown in Figure 2.3(a). The coordinate of y is a function of both positions along string x and time t, or

$$y = y(x, t). \qquad [2.74]$$

The equilibrium position of the string is shown by the thick black line in Figure 2.3(a), and its differential element in any possible position of motion is shown in Figure 2.3(b). To develop the governing equation of motion for the string, the following assumptions are made: resistances of air and internal friction and gravitational forces are neglected in comparison with the tension in the string, which is quite large. The displacement of any point in the string is very small and occurs only in the x–y plane.

Denote the tension in the string as T and the change in slope as $\frac{\partial^2 y}{\partial x^2}dx$. Considering the vertical motion of the differential element in Figure 2.3(b), we can write Newton's second law as

$$-T\sin\frac{\partial y}{\partial x} + T\sin\left(\frac{\partial y}{\partial x} + \frac{\partial^2 y}{\partial x^2}dx\right) = \rho dx\frac{\partial^2 y}{\partial t^2}. \qquad [2.75]$$

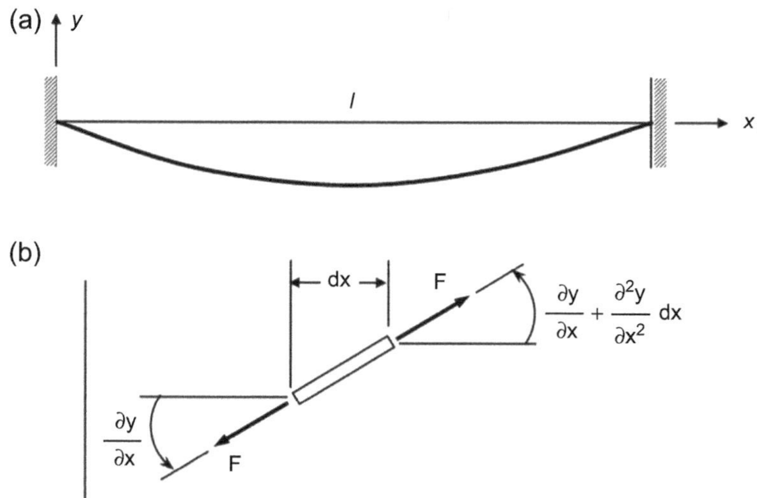

Figure 2.3 Schematic of a string and the differential element

Since the slopes are very small, we use the approximation $\sin\theta \approx \theta$, and Eq. [2.75] becomes

$$-T\frac{\partial y}{\partial x} + T\left(\frac{\partial y}{\partial x} + \frac{\partial^2 y}{\partial x^2}dx\right) = \rho dx\frac{\partial^2 y}{\partial t^2} \quad [2.76]$$

or

$$T\frac{\partial^2 y}{\partial x^2} = \rho\frac{\partial^2 y}{\partial t^2}. \quad [2.77]$$

Eq. [2.77] is a linear, second-order, partial differential equation with constant coefficients and represents the governing equation for transverse motion of the string. Rewrite Eq. [2.77] as

$$c^2\frac{\partial^2 y}{\partial x^2} = \frac{\partial^2 y}{\partial t^2} \quad [2.78]$$

where $c = \sqrt{T/\rho}$. Eq. [2.78] is called the *one-dimensional (1D) wave equation,* and the constant c is called the *wave speed.* We assume that the solution is

$$y(x,t) = F(x)\,G(t). \quad [2.79]$$

Then substituting Eq. [2.79] into Eq. [2.78], we have

$$c^2\frac{\partial^2 F}{\partial x^2}\Big/F = \frac{\partial^2 G}{\partial t^2}\Big/G = \mu, \quad [2.80]$$

where μ is a constant. The initial conditions are

$$y(x,0) = F(x)\,G(0) = F(x)\,G_0 \quad [2.81a]$$

$$\frac{\partial y(x,0)}{\partial t} = F(x)\dot{G}(0) = F(x)\dot{G}_0. \quad [2.81b]$$

The boundary conditions are

$$y(0,t) = F(0)\,G(t) = 0 \quad [2.82]$$

$$y(l,t) = F(l)\,G(t) = 0. \quad [2.83]$$

The solution of Eq. [2.80] depends on the value of μ. It can easily be verified that $\mu < 0$ is the only possibility that satisfies both the differential equations and the boundary conditions. Hence, for $\mu < 0$, we assume $\mu = -\omega^2$, and Eq. [2.80] can be rewritten as

$$\ddot{G} + \omega^2 G = 0 \quad [2.84]$$

$$F'' + (\omega/c)^2 F = 0 \quad [2.85]$$

which have the solutions

$$G(t) = A \sin \omega t + B \cos \omega t \quad [2.86]$$

$$F(x) = C \sin(\omega/c)x + D \cos(\omega/c)x. \quad [2.87]$$

The displacements are then given by

$$y(x, t) = (A \sin \omega t + B \cos \omega t) \left[C \sin(\omega/c)x + D \cos(\omega/c)x\right]. \quad [2.88]$$

Now, applying the boundary conditions given by Eqs. [2.82]–[2.83] will give

$$\sin(\omega/c)l = 0 \quad [2.89]$$

or

$$(\omega/c)l = \pi, 2\pi, \ldots, n\pi, \ldots. \quad [2.90]$$

Eq. [2.90] will give an infinite number of natural frequencies:

$$\omega_n = \frac{n\pi c}{l} = \frac{n\pi}{l}\sqrt{\frac{T}{\rho}} \quad n = 1, 2, 3, \ldots. \quad [2.91]$$

In Eq. [2.91], each ω_n corresponds to a principal mode with the harmonic mode shape sin $(n\pi/l)$. The general solution can be written as

$$y(x, t) = \sum_{n=1}^{\infty}(A_n \sin \omega_n t + B_n \cos \omega_n t)\sin\frac{\omega_n}{c}x, \quad [2.92]$$

where $A_n = AC$ and $B_n = BC$ for each mode.

Now consider the case of initially deforming the string into the shape of a half-sine curve and releasing it from rest. The initial conditions are then

$$y(x, 0) = y_0 \sin\frac{\pi}{l}x \quad [2.93]$$

and

$$\frac{\partial y(x, 0)}{\partial t} = 0. \quad [2.94]$$

Now, applying the initial conditions to Eq. [2.92] gives

$$\begin{aligned} A_n &= 0 \quad \text{for all } n \\ B_1 &= y_0 \\ B_n &= 0 \quad n \neq 1. \end{aligned} \quad [2.95]$$

The displacement of every point on the string is then given by

$$y(x,t) = y_0 \cos\frac{\pi c}{l} t \sin\frac{\pi}{l} x. \qquad [2.96]$$

Eq. [2.96] represents the vibration of the string in its fundamental mode, and in this mode, every point moves harmonically with the amplitude $y_0 \sin(\pi x/l)$. The shape of the string is a half-sine wave at all times. The ratio of the amplitude of any two points is always the same. The entire string goes through the equilibrium position simultaneously. There is no apparent progress in the axial direction and the string appears to move up and down. Such motion is referred to as a *standing wave*. The wavelength λ is defined as the length of one complete sine wave [namely, $\sin(n\pi x/l)$]. That is, for

$$\frac{n\pi\lambda}{l} = 2\pi \qquad [2.97]$$

or

$$\lambda = \frac{2l}{n},$$

where λ is the wavelength. Figure 2.4 shows the standing waves for the first three principal modes.

It can be shown that the wave equation [Eq. [2.78]] has an explicit solution given by

$$y(x,t) = yl(x - ct) + yr(x + ct), \qquad [2.98]$$

where yl and yr are interpreted as left-going and right-going traveling waves, respectively. Therefore, it is possible to consider the solution to the wave equation as a sum of traveling waves, through which a single traveling wave is investigated and simulated using another approach instead of modal analysis.

Traveling waves can be described by a simple harmonic solution at each point on the wave. At each point, waves still have a frequency or period, which describes their periodicity in time. However, waves also have a wavelength, which determines how often they repeat in space. We therefore need to introduce two new quantities to describe a wave: the above-mentioned wave velocity c, and the wavelength λ:

$$\lambda = c/f = 2\pi c/\omega. \qquad [2.99]$$

This is the most important equation in wave motion. There are other quantities related to the wavelength and containing the same physics; the wave number $k = 2\pi/\lambda$ is often used instead of the wavelength.

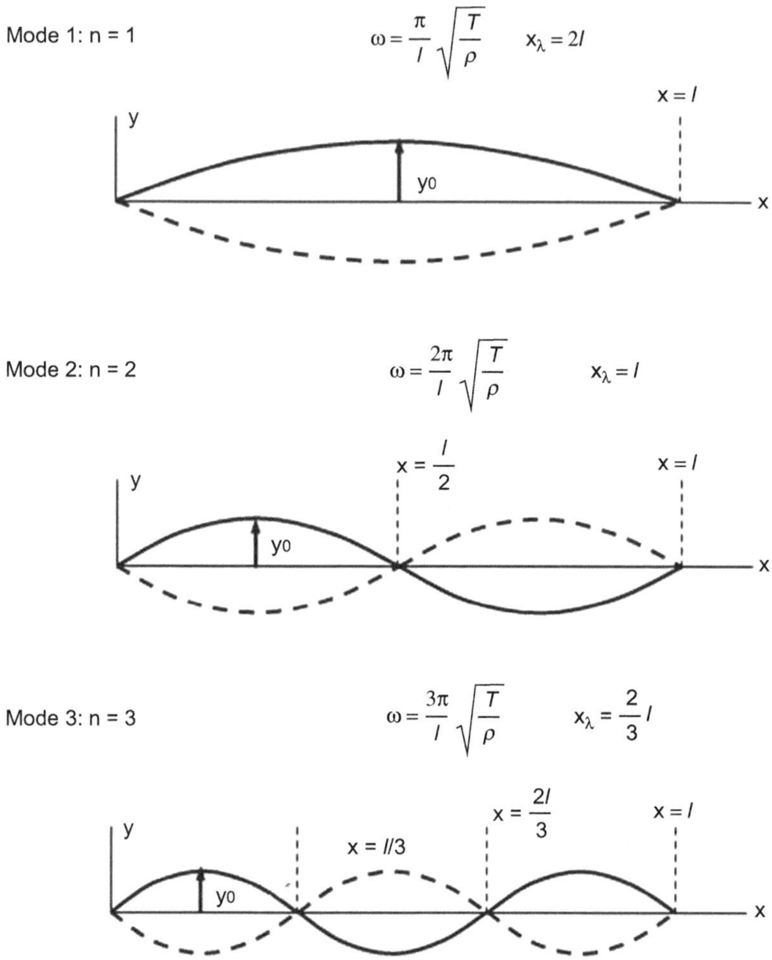

Figure 2.4 Natural modes of string

There are two broad classes of waves: transverse waves and longitudinal waves. The most common wave is the transverse wave, such as the waves on a string as discussed previously and the waves at the surface of liquids such as water. They are called *transverse waves* because the displacements are perpendicular or transverse to the direction of wave motion. The height of a transverse traveling wave is given by

$$y(x, t) = A\cos(kx - \omega t). \qquad [2.100]$$

For each situation where a wave occurs, the frequency, wavelength, and velocity are related to the properties of the material. An alternative type of wave is a longitudinal wave. The most notable example of this type of wave is a sound wave. Sound waves are really oscillations in the pressure inside the material.

When two waves meet, they interfere, and the following two cases illustrate what can happen. If two waves have crests at the same places at the same time, they add; this is called *constructive interference,* and the waves are said to be in phase. If two waves are out of phase, so that one wave has a maximum just where the other has a minimum, the two waves interfere destructively. In fact, if the two waves have exactly the same amplitude, they annihilate each other. The property of interference is important, and many scientific and engineering devices rely on understanding and controlling wave interference, with methods such as active noise control. When a wave hits a barrier, it can reflect. If the barrier is hard, then the wave inverts upon reflection.

2.5.2 Longitudinal Vibration of Rods and Torsional Vibration of Shafts

Consider the vibration of a uniform slender rod. The rod can execute axial or longitudinal vibrations when the equilibrium condition is disturbed axially due to the axial forces. Figure 2.5 shows the free-body diagram of a differential element of this rod of length dx. The equilibrium position of the element is denoted by x and the deformed position is u. That is, if u is the displacement at x, then the displacement at $x + dx$ is $u + \frac{\partial u}{\partial x} dx$. In other words, the deformed length of the element $\left(1 + \frac{\partial u}{\partial x}\right) dx$ is greater than the

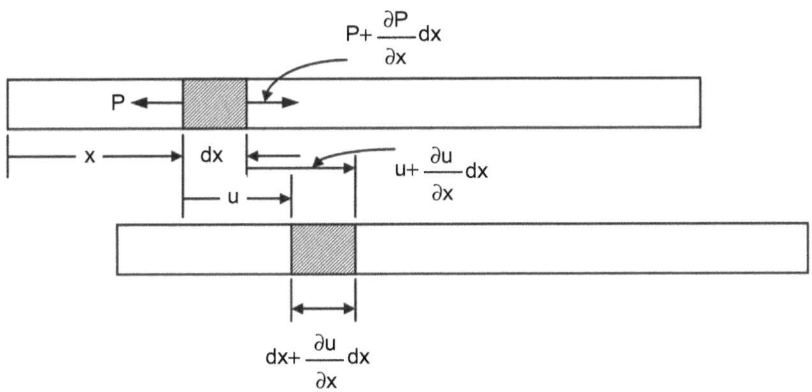

Figure 2.5 Displacement of an element of a rod

original length. Application of Newton's second law to the differential element gives

$$-\sigma A+\left(\sigma+\frac{\partial\sigma}{\partial x}dx\right)A=\rho A dx\frac{\partial^2 u}{\partial t^2}, \qquad [2.101]$$

where A is the cross-sectional area of the rod, ρ is the mass density of the rod material, and σ is stress.

For elastic deformations, Hooke's law gives the ratio of the unit stress to unit strain, which is equal to the modulus of elasticity E of the material of the rod. Thus,

$$\sigma=E\varepsilon=E\frac{\partial u}{\partial x}. \qquad [2.102]$$

Now, combining Eqs. [2.101] and [2.102] gives

$$E\frac{\partial^2 u}{\partial x^2}=\rho\frac{\partial^2 u}{\partial t^2}. \qquad [2.103]$$

Eq. [2.103] can be written as

$$c^2\frac{\partial^2 u}{\partial x^2}=\frac{\partial^2 u}{\partial t^2}, \qquad [2.104]$$

where $c=\sqrt{E/\rho}$ is the velocity of propagation of the displacement or stress wave in the rod.

Next, we consider the torsional vibrations of the elastic circular shaft (Figure 2.6a). Consider the differential shaft elements of length dx, as shown in Figure 2.6b, where I_p is the mass polar moment of inertia of a shaft with outside radius r and density ρ:

$$I_p=\frac{\pi r^4}{2}\rho\, dx. \qquad [2.105]$$

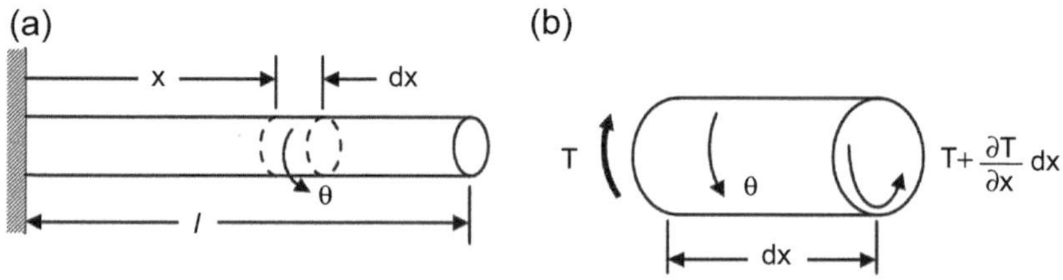

Figure 2.6 Torsional vibration of a long shaft

Applying Newton's second law to an element dx, as shown in Figure 2.6b, we obtain

$$T + \frac{\partial T}{\partial x}dx - T = \frac{\pi r^4 \rho}{2} dx \frac{\partial^2 \theta}{\partial t^2} \qquad [2.106]$$

or

$$\frac{\partial T}{\partial x} = \frac{\pi r^4 \rho}{2} \frac{\partial^2 \theta}{\partial t^2}. \qquad [2.107]$$

Considering the analysis to elastic deformations only, we use the basic relation

$$\tau = G\gamma = Gr\frac{\partial \theta}{\partial x} = \frac{Tr}{J_0} \qquad [2.108]$$

to obtain torque:

$$T = GJ_0\frac{\partial \theta}{\partial x}. \qquad [2.109]$$

In Eq. [2.108], τ is the maximum shear stress, $\gamma = r(\partial\theta/\partial x)$ is the corresponding shear strain, G is the shear modulus of the material, and $J_0 = \pi r^4/2$ is the polar moment of inertia of the cross-sectional area of the shaft. Eq. [2.107] can now be written as

$$G\frac{\partial^2 \theta}{\partial x^2} = \rho\frac{\partial^2 \theta}{\partial t^2} \qquad [2.110]$$

or, as the 1D wave equation,

$$c^2\frac{\partial^2 \theta}{\partial x^2} = \frac{\partial^2 \theta}{\partial t^2}. \qquad [2.111]$$

where $c = \sqrt{G/\rho}$ is the wave speed. Eq. [2.111] takes the same form as that of the longitudinal vibration of rods [Eq. [2.104]], except that θ and G/ρ replace u and E/ρ, respectively.

2.5.3 Transverse Vibration of Beams

If an elastic beam such as the fixed-hinged beam shown in Figure 2.7a is deformed elastically and then released, transverse or lateral oscillation occurs. Assuming that only elastic deflections occur, the displacement of any point on the beam is small and motion occurs only in a direction that is normal to the axis of the beam. We disregard the inertia effects of rotation and shear effect of any section of the beam in this analysis.

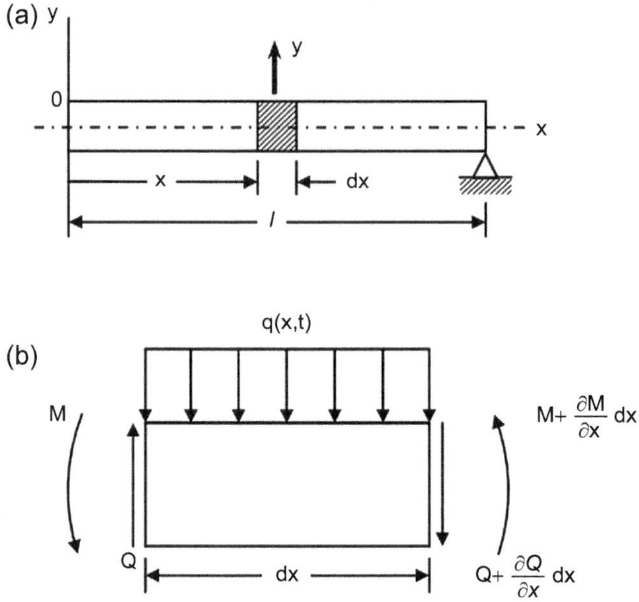

Figure 2.7 Transverse vibration of a beam

The lateral displacement at any point of the beam is represented as

$$y = y(x, t).$$

From the strength of materials, the beam curvature and the bending moment M are related by

$$M = EI\frac{\partial^2 y}{\partial x^2}, \quad [2.112]$$

where EI is the flexural stiffness of the beam, M is the bending moment at any transverse section, E is the modulus of elasticity of the beam material, and I is the moment of inertia of the cross-sectional area of the beam about the axis of bending. Figure 2.7b shows an isolated beam section with bending moment M, shear force Q, and external load per unit length $q(x, t)$.

Taking the sum of the moments about the left end of the section in Figure 2.7b, we get

$$\frac{\partial M}{\partial x}dx - Qdx - \frac{\partial Q}{\partial x}(dx)^2 - q\frac{(dx)^2}{2} = 0. \quad [2.113]$$

Neglecting the higher-order terms containing $(dx)^2$, we obtain

$$Q = \frac{\partial M}{\partial x}. \quad [2.114]$$

If the mass of the beam per unit length is denoted as ρ, then the equation of motion in the vertical direction as given by Newton's second law is

$$-\frac{\partial Q}{\partial x}dx + q dx = \rho dx \frac{\partial^2 y}{\partial t^2}. \quad [2.115]$$

Using Eqs. [2.112] and [2.114], Eq. [2.115] becomes

$$-\frac{\partial^2}{\partial x^2}\left(EI\frac{\partial^2 y}{\partial x^2}\right) + q = \rho\frac{\partial^2 y}{\partial t^2} \quad [2.116]$$

or

$$\frac{\partial^2}{\partial x^2}\left(EI\frac{\partial^2 y}{\partial x^2}\right) + \rho\frac{\partial^2 y}{\partial t^2} = q. \quad [2.117]$$

Assuming that the properties of the beam are constant along its length, Eq. [2.117] becomes

$$EI\frac{\partial^4 y}{\partial x^4} + \rho\frac{\partial^2 y}{\partial t^2} = q. \quad [2.118]$$

The case of free vibration is obtained by setting $q(x, t) = 0$. Hence,

$$EI\frac{\partial^4 y}{\partial x^4} + \rho\frac{\partial^2 y}{\partial t^2} = 0. \quad [2.119]$$

Eq. [2.119] can be rewritten as

$$c^2\frac{\partial^4 y}{\partial x^4} + \frac{\partial^2 y}{\partial t^2} = 0, \quad [2.120]$$

where $c = \sqrt{EI/\rho}$. We consider a separable solution of the form

$$y(x,t) = F(x)\,G(t). \quad [2.121]$$

Then Eq. [2.119] is equivalent to the two ordinary differential equations:

$$\frac{d^2 G}{dt^2} + \omega^2 G = 0 \quad [2.122]$$

and

$$\frac{d^4F}{dx^4} - (\omega/c)^2 F = 0, \qquad [2.123]$$

where ω is to be determined. The solution for Eq. [2.122] can be written by inspection as

$$G(t) = A_1 \sin \omega t + A_2 \cos \omega t. \qquad [2.124]$$

The solution for Eq. [2.123] is assumed as

$$F(x) = Ae^{sx}, \qquad [2.125]$$

where A and s are constants. Substituting the assumed solution into the governing equation gives

$$\left[s^4 - (\omega/c)^2\right]Ae^{st} = 0, \qquad [2.126]$$

from which we obtain the four roots. The solution is then

$$F(x) = C_1e^{\lambda x} + C_2e^{i\lambda x} + C_3e^{-\lambda x} + C_4e^{-i\lambda x}. \qquad [2.127]$$

Eq. [2.127] can be written as

$$F(x) = A_3 \sinh \lambda x + A_4 \cosh \lambda x + A_5 \sin \lambda x + A_6 \cos \lambda x, \qquad [2.128]$$

where we have redefined the constants as

$$\begin{aligned} C_1 &= (A_3 + A_4)/2, C_2 = (A_6 - iA_5)/2,\ C_3 = (A_4 - A_3)/2, \\ &\text{and } C_4 = (A_6 + iA_5)/2. \end{aligned} \qquad [2.129]$$

The solution to Eq. [2.119] is, then,

$$\begin{aligned} y(x,t) = {} & (A_1 \sin \omega t + A_2 \cos \omega t)\,(A_3 \sinh \lambda x + A_4 \cosh \lambda x + A_5 \sin \lambda x \\ & + A_6 \cos \lambda x), \end{aligned} \qquad [2.130]$$

with

$$\omega = \lambda^2 c. \qquad [2.131]$$

The constants in the solution and the natural frequencies are determined by applying the boundary conditions for the beam and the initial conditions of the motion.

For a simply supported beam, the boundary conditions are

$$y(0,t) = y(l,t) = 0 \qquad [2.132]$$

and

$$\left.\frac{\partial^2 y}{\partial x^2}\right|_{x=0} = \left.\frac{\partial^2 y}{\partial x^2}\right|_{x=l} = 0. \qquad [2.133]$$

Eq. [2.133] expresses the absence of bending moment at each end of the beam. Applying the boundary conditions to Eq. [2.130] gives

$$\begin{gathered} A_4 + A_6 = 0 \\ A_3 \sinh \lambda l + A_4 \cosh \lambda l + A_5 \sin \lambda l + A_6 \cos \lambda l = 0 \end{gathered} \qquad [2.134a]$$

$$\begin{gathered} A_4 - A_6 = 0 \\ A_3 \sinh \lambda l + A_4 \cosh \lambda l - A_5 \sin \lambda l - A_6 \cos \lambda l = 0. \end{gathered} \qquad [2.134b]$$

Eqs. [2.134] will be satisfied if

$$A_3 = A_4 = A_6 = 0 \qquad [2.135]$$

and $A_5 \sin \lambda l = 0$, which is the frequency equation. The frequency equation will be satisfied and nontrivial solutions obtained if

$$\lambda l = n\pi \quad n = 1, 2, 3, \ldots. \qquad [2.136]$$

Combining Eqs. [2.136] and [2.131], we obtain the natural frequencies

$$\omega_n = \frac{n^2\pi^2}{l^2} c = n^2\pi^2 \sqrt{\frac{EI}{\rho l^4}}. \qquad [2.137]$$

Each of the frequencies in Eq. [2.137] corresponds to a principal mode of vibration.

The free response of the beam is given by the superposition of the principal modes. Hence,

$$y(x,t) = \sum_{n=1}^{\infty} (A_n \sin \omega_n t + B_n \cos \omega_n t) \sin \frac{n\pi}{l} x \qquad [2.138]$$

By applying the initial conditions, we obtain the two constants A_n and B_n in Eq. [2.138].

Theoretically, the beam has infinite modes. However, the higher the modes, the lower the accuracy of the mode with respect to a real system due to the basic assumption associated with the theory.

At high speeds during the lateral motion of beams, a considerable rotary acceleration associated with a rotary inertia force must be resisted. Also, for shorter beams bearing lateral forces, there is a considerable shear deformation. The combined effect of these two factors, rotary inertia and shear deformation, cannot be neglected, and the thick beam theory or Timoshenko beam theory should be used.

2.6 RANDOM VIBRATIONS

If a vibration motion or the associated force does not exhibit any obvious pattern, the vibration is called a *random vibration*. This is different from what we usually see in the deterministic process, where the records obtained are always alike if an identical experiment is performed several times. On the contrary, when all conditions in the experiment remain unchanged but the records are continually changing, we then consider the process to be random, or nondeterministic. In this situation, a single record is not sufficient to provide a statistical description of the totality of possible records.

In a random process, instead of one time history, a whole family or an ensemble of possible time histories is described, as shown in Figure 2.8. Any

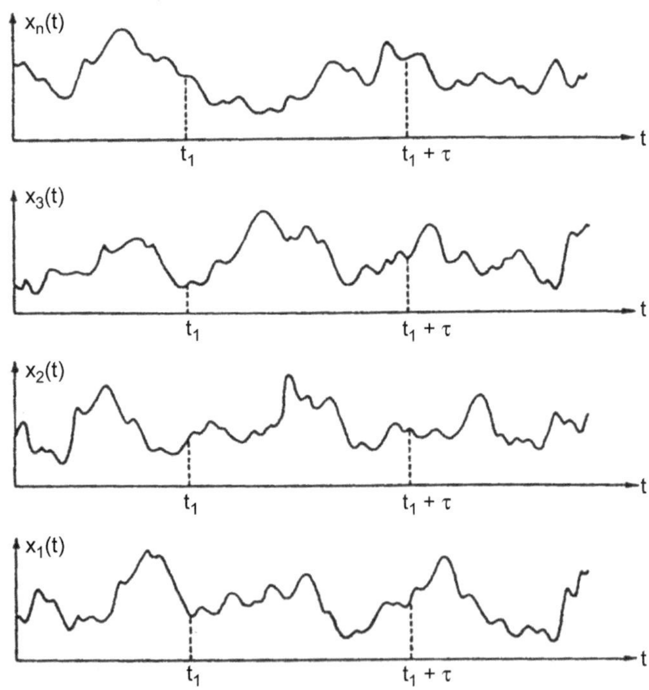

Figure 2.8 Ensemble of sample functions forming a random process

single individual time history that belongs to the ensemble is referred to as a *sample record.*

Let $x^k(t_1)$ be the value of random variable $x(t)$ at time t_1, obtained from the kth record. Then the expected value (average or mean) $E[x(t_1)]$ of $x^k(t_1)$ for a fixed time t_1, obtained from all records (i.e., $k = 1, 2, 3, \ldots, n$) is

$$E[x(t_1)] = \lim_{n \to \infty} \frac{1}{n} \sum_{n=1}^{\infty} x^k(t_1). \qquad [2.139]$$

If $E[x(t_1)]$ is independent of t [i.e., $E[x(t_1)] = E[x(t_1 + t)]$, for all t], then the random process is called *stationary*. If in addition to this property, each record is statistically equivalent to any other record such that $E[x(t_1)]$ in Eq. [2.139] can be replaced by a time average of a sample representative record $x(t)$:

$$\overline{x}(t) = E(x) = \lim_{T \to \infty} \frac{1}{T} \int_0^T x(t)dt, \qquad [2.140]$$

then the stationary process is ergodic. For many applications, this assumption is fairly reasonable. The variance σ_x^2 of $x(t)$ is given by

$$E[x - E(x)]^2 = \lim_{T \to \infty} \frac{1}{T} \int_0^T [x - E(x)]^2 dt. \qquad [2.141]$$

For the special case with $E(x) = 0$, the variance σ_x^2 of x becomes its mean-square value and is given by $\overline{x}^2(t)$, where

$$\overline{x}^2(t) = E[x^2(t)] = \lim_{T \to \infty} \frac{1}{T} \int_0^T x^2(t)dt. \qquad [2.142]$$

2.6.1 Probability Density Function and Autocorrelation Function

The probability density function of random data is the probability that the data will assume a value within some defined range at any instant of time. Here, we consider a sample time history, as shown in Figure 2.9. The probability that $x(t)$ will occur within x and $x + \Delta x$ can be obtained from the ratio T_x/T, where T_x indicates the total amount of time for which $x(t)$ falls within the range of x and $x + \Delta x$ (i.e., $T_x = \sum_{i=1}^{k} \Delta t_1$ and T is observed time). We define

$$\text{Prob}\,[x < x(t) \leq x + \Delta x] = P(x) = \lim_{T \to \infty} (T_x/T). \qquad [2.143]$$

For small Δx, a probability density function $p(x)$ can be defined as

$$\text{Prob}\left[x < x(t) \leq x + \Delta x\right] = P(x) = p(x)\Delta x. \qquad [2.144]$$

More precisely,

$$p(x) = \lim_{\Delta x \to \infty} \frac{P(x)}{\Delta x} = \lim_{\Delta x \to \infty} \frac{1}{\Delta x}\left[\lim_{T \to \infty} (T_x/T)\right]. \qquad [2.145]$$

It is evident from Eq. [2.145] that $p(x)$ is the slope of the cumulative probability distribution $P(x)$. The area under the probability density curve between any two values of x represents the probability of the variable being in this interval. Also, the probability of $x(t)$ being between $x = \pm\infty$ is

$$P(\infty) = \int_{-\infty}^{\infty} p(x)dx = 1. \qquad [2.146]$$

The mean value $\overline{x}(t)$ coincides with the center of the area under the probability density curve $p(x)$, as shown in Figure 2.10. Therefore, in terms of the probability density $p(x)$, the mean value is given by

$$\overline{x}(t) = \int_{-\infty}^{\infty} xp(x)dx. \qquad [2.147]$$

Likewise, the mean-square value $\overline{x}^2(t)$ is determined from the second moment to be

$$\overline{x}^2(t) = \int_{-\infty}^{\infty} x^2p(x)dx. \qquad [2.148]$$

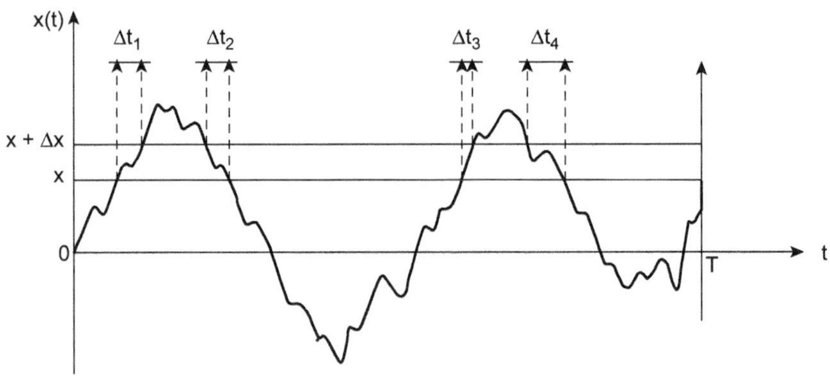

Figure 2.9 Probability measurement

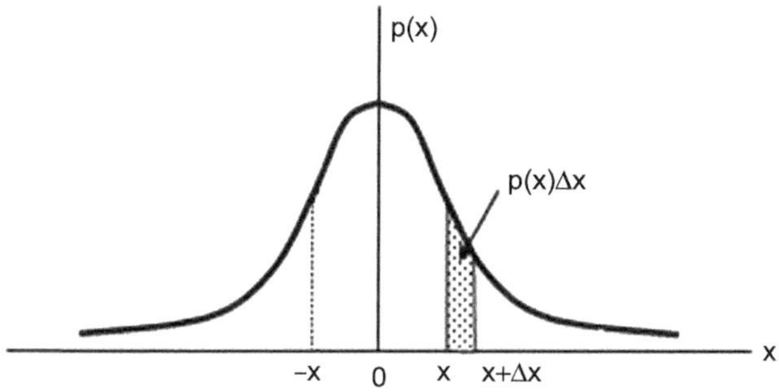

Figure 2.10 Probability density curve

The variance σ_x^2, previously defined as the mean-square value about the mean, is

$$\begin{aligned}\sigma_x^2 &= \int_{-\infty}^{\infty} (x-\overline{x})^2 p(x)dx \\ &= \int_{-\infty}^{\infty} x^2 p(x)dx - 2\overline{x}\int_{-\infty}^{\infty} xp(x)dx + (\overline{x})^2\int_{-\infty}^{\infty} p(x)dx \\ &= \overline{x^2} - 2(\overline{x})^2 + (\overline{x})^2 \\ &= \overline{x^2} - (\overline{x})^2.\end{aligned} \quad [2.149]$$

The standard deviation σ_x is the positive square root of σ_x^2.

The most widely used statistical distribution for modeling random processes is the Gaussian, or normal, random process. The probability density function $x(t)$ of a Gaussian random process is given by

$$p(x) = \frac{1}{\sqrt{2\pi}\sigma_x} e^{-\frac{1}{2}\left[\frac{x-\overline{x}}{\sigma_x}\right]^2}, \quad [2.150]$$

where $\overline{x}$ and σ_x are the mean value and standard deviation of x, respectively. By defining a standard normal variable z as

$$z = (x-\overline{x})/\sigma_x,$$

Eq. [2.150] becomes

$$p(x) = \frac{1}{\sqrt{2\pi}\sigma_x} e^{-\frac{1}{2}z^2}. \quad [2.151]$$

The probability of $x(t)$ in the interval from $-k\sigma$ to $+k\sigma$, assuming $\overline{x} = 0$, is

$$\text{Prob} = [-k\sigma \le x(t) \le k\sigma] = \int_{-k\sigma}^{k\sigma} \frac{1}{\sqrt{2\pi}\sigma} e^{-(1/2)\left(x^2/\sigma^2\right)} dx \quad [2.152]$$

where k is the positive number. Figure 2.11 shows the Gaussian probability density function, which is a bell-shaped curve, symmetric about the mean value.

The autocorrelation function of a stationary random process is defined as the average value of the product $x(t)$ and $x(t + \tau)$. The process is sampled at time t and then again at time $t + \tau$, as shown in Figure 2.12:

$$R_x(\tau) = \lim_{T \to \infty} \frac{1}{T} \int_0^T x(t)x(t + \tau)dt. \qquad [2.153]$$

The quantity $R_x(\tau)$ is always a real-valued even function with a maximum occurring at $\tau = 0$; i.e.,

$$R_x(0) = \lim_{T \to \infty} \frac{1}{T} \int_0^T x^2(t)dt = \overline{x}^2(t). \qquad [2.154]$$

For very large time intervals, with $\tau \to \infty$, the random process will be uncorrelated, and in this case,

$$R_x(\infty) = [\overline{x}(t)]^2;$$

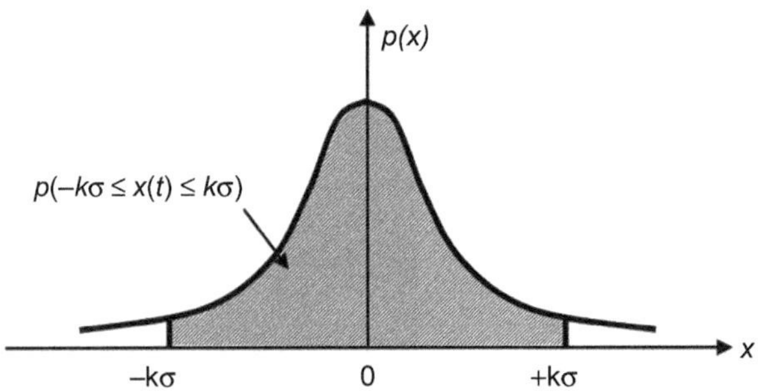

Figure 2.11 Gaussian probability distribution

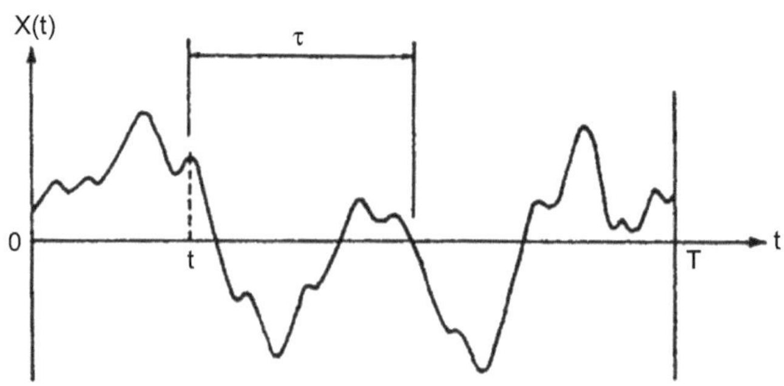

Figure 2.12 Autocorrelation measurement

i.e.,

$$\overline{x}(t) = \sqrt{R_x(\infty)}. \quad [2.155]$$

The mean value of $x(t)$ is equal to the positive square root of the autocorrelation, as the time displacement becomes very long.

2.6.2 Response of an SDOF System to an Arbitrary Function Input

Next, we consider the response of an SDOF vibration system to arbitrary function input.

The equation of motion is given by

$$\ddot{y}(t) + 2\xi\omega_n\dot{y}(t) + \omega_n^2 y(t) = x(t), \quad [2.156]$$

where $x(t) = F(t)/m$, $\omega_n = \sqrt{k/m}$, $\xi = c/c_c$ and $c_c = 2\sqrt{km}$. The arbitrary function $x(t)$ is plotted in Figure 2.13. There are two different ways to obtain the solution of Eq. [2.156]. The first is the impulse response method.

In this method, we assume that the forcing function $x(t)$ is made up of a series of impulses with varying magnitudes, as shown in Figure 2.13, and the amplitude applied at time τ is $x(\tau)\mathrm{d}\tau$. Also, if $y(t) = h(t - \tau)$ denotes the response to the unit impulse excitation $\delta(t - \tau)$, it is called the *impulse response function*. Note that the unit impulse applied at $t = \tau$ is denoted by

$$x(t) = \delta(t{-}\tau),$$

where $\delta(t - \tau)$ is called the *Dirac delta function,* with

$$\delta(t{-}\tau) \rightarrow \infty \text{ as } t \rightarrow \tau \quad [2.157]$$

with $\delta(t - \tau) = 0$ for all t except at $t = \tau$ and $\int_{-\infty}^{\infty} \delta(t - \tau)dt = 1$.

This is shown in Figure 2.14. The total response of the system can be obtained by superimposing the responses to the impulses of $x(\tau)d\tau$ applied

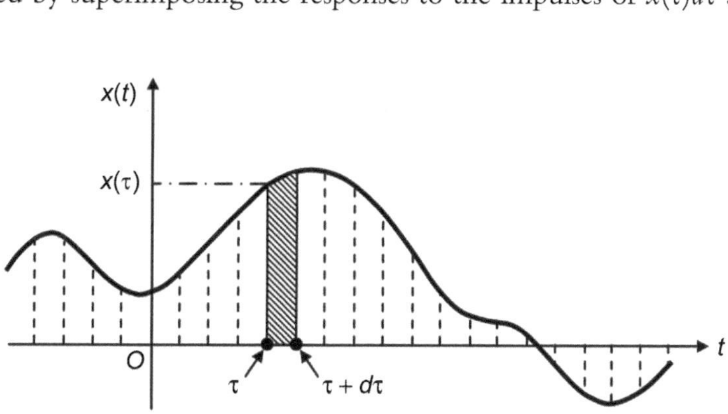

Figure 2.13 Forcing function in the form of a series of impulses

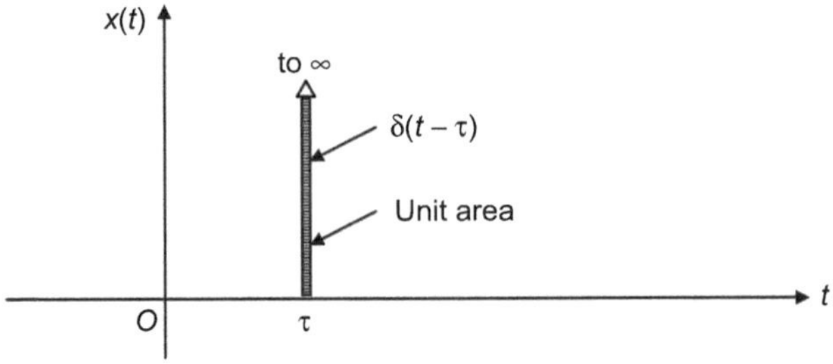

Figure 2.14 Unit impulse excitation at $t = \tau$

at different values of $t = \tau$. Hence, the response to the total excitation is given by the superposition or convolution integral:

$$y(t) = \int_{-\infty}^{t} x(\tau)h(t-\tau)d\tau. \qquad [2.158]$$

where $x(\tau)h(t - \tau)$ is the response to the excitation $x(\tau)d\tau$.

The typical impulse response function is shown in Figure 2.15.

It should be noted here that since $h(t-\tau) = 0$ when $t < \tau$ or $t > \tau$, the upper limit of integration of Eq. [2.158] can be replaced by ∞. That is,

$$y(t) = \int_{-\infty}^{\infty} x(\tau)h(t-\tau)d\tau. \qquad [2.159]$$

Eq. [2.159] can be written as

$$y(t) = \int_{-\infty}^{\infty} x(t-\theta)h(\theta)d\theta, \qquad [2.160]$$

where we have changed the variable from τ to $\theta = t - \tau$ in Eq. [2.159]. Another approach is the frequency response method. The Fourier transform gives,

$$x(t) = \frac{1}{2\pi}\int_{-\infty}^{\infty} X(\omega)e^{i\omega t}dt. \qquad [2.161]$$

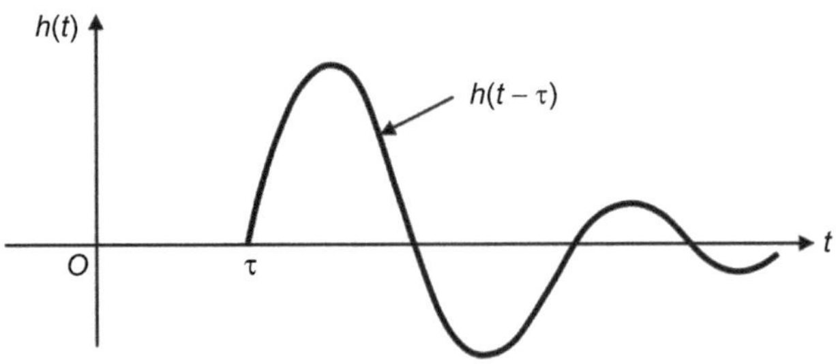

Figure 2.15 Impulse response function

Eq. [2.161] indicates the superposition of components of different frequency ω. If the forcing function of unit modulus is

$$\widetilde{x}(t) = e^{i\omega t}, \tag{2.162}$$

then its response is given by

$$\widetilde{y}(t) = H(\omega)e^{i\omega t}, \tag{2.163}$$

where $H(\omega)$ is the complex frequency response function.

The total response of the system by the superposition principle gives

$$\begin{aligned} y(t) = H(\omega)x(t) &= \int_{-\infty}^{\infty} H(\omega)\frac{1}{2\pi}x(\omega)e^{i\omega t}d\omega \\ &= \frac{1}{2\pi}\int_{-\infty}^{\infty} H(\omega)X(\omega)e^{i\omega t}d\omega \end{aligned} \tag{2.164}$$

or

$$y(t) = \frac{1}{2\pi}\int_{-\infty}^{\infty} Y(\omega)e^{i\omega t}d\omega, \tag{2.165}$$

where $Y(\omega)$ is the Fourier transform of the response function $y(t)$. Note from Eqs. [2.164] and [2.165] that

$$Y(\omega) = H(\omega)X(\omega). \tag{2.166}$$

Eq. [2.164] can be written as

$$y(t) = h(t) = \frac{1}{2\pi}\int_{-\infty}^{\infty} H(\omega)X(\omega)e^{i\omega t}d\omega, \tag{2.167}$$

where $X(\omega)$ is the Fourier transform of $x(t) = \delta(t)$.

$$X(\omega) = \int_{-\infty}^{\infty} x(t)e^{-i\omega t}dt = \int_{-\infty}^{\infty} \delta(t)e^{-i\omega t}dt = 1 \tag{2.168}$$

Now, since $\delta(t) = 0$ everywhere except at $t = 0$, where it has a unit area and $e^{-i\omega t} = 1$ at $t = 0$, Eqs. [2.167] and [2.168] give

$$h(t) = \frac{1}{2\pi}\int_{-\infty}^{\infty} H(\omega)e^{i\omega t}d\omega \tag{2.169}$$

or

$$H(\omega) = \int_{-\infty}^{\infty} h(t)e^{-i\omega t}dt. \tag{2.170}$$

2.6.3 PSD Function

The power spectral density (PSD) of a random process provides the frequency composition of the data in terms of the spectral density of its mean-square value. The mean-square value of a sample time-history record, in a frequency range between ω and $\omega + \Delta\omega$, can be obtained by passing the sample record through a bandpass filter with sharp cutoff features and computing the average of the squared output from the filter. The average square value will approach an exact mean-square value as $T \rightarrow \infty$; i.e.,

$$\Psi_x^2(\omega, \Delta\omega) = \lim_{T\rightarrow\infty} \frac{1}{T}\int_0^T x^2(t, \omega, \Delta\omega)dt, \qquad [2.171]$$

where $x(t, \omega, \Delta\omega)$ is the portion of $x(t)$ in the frequency range ω and $\omega + \Delta\omega$. For a small value of $\Delta\omega$, a PSD function $S_x(\omega)$ is defined as

$$\Psi_x^2(\omega, \Delta\omega) \cong S_x(\omega)\Delta\omega; \qquad [2.172]$$

i.e.,

$$S_x(\omega) = \lim_{\Delta\omega\rightarrow 0} \frac{\Psi_x^2(\omega, \Delta\omega)}{\Delta\omega} = \lim_{\Delta\omega\rightarrow 0} \frac{1}{\Delta\omega} \lim_{T\rightarrow\infty} \frac{1}{T}\int_0^T x^2(t, \omega, \Delta\omega)dt. \qquad [2.173]$$

The quantity $S_x(\omega)$ is always a real-valued, nonnegative function.

In experimental work, a different unit of PSD is often used. The experimental spectral density is defined by $W(f)$, where f denotes the frequency in cycles per unit time. The relation between $S(\omega)$ and $W(f)$ is

$$W(f) = 4\pi S(\omega), \qquad [2.174]$$

where $\omega = 2\pi f$. For a stationary random process, the autocorrelation and PSD functions are related by a Fourier transform as

$$R(\tau) = \int_{-\infty}^{\infty} S(\omega)e^{i\omega\tau}d\omega \qquad [2.175]$$

$$S(\omega) = \frac{1}{2\pi}\int_{-\infty}^{\infty} R(\tau)e^{-i\omega\tau}d\tau. \qquad [2.176]$$

In a limiting case, where $\tau = 0$,

$$R(0) = E[x^2(t)] = \int_{-\infty}^{\infty} S(\omega)d\omega. \qquad [2.177]$$

In other words, the mean-square value is equal to the sum over all frequencies of $S(\omega)d\omega$; therefore, $S(\omega)$ may be interpreted as a mean-square spectral density. Probability density functions, autocorrelation functions, and PSDs for four sample time-history records are shown in Figure 2.16.

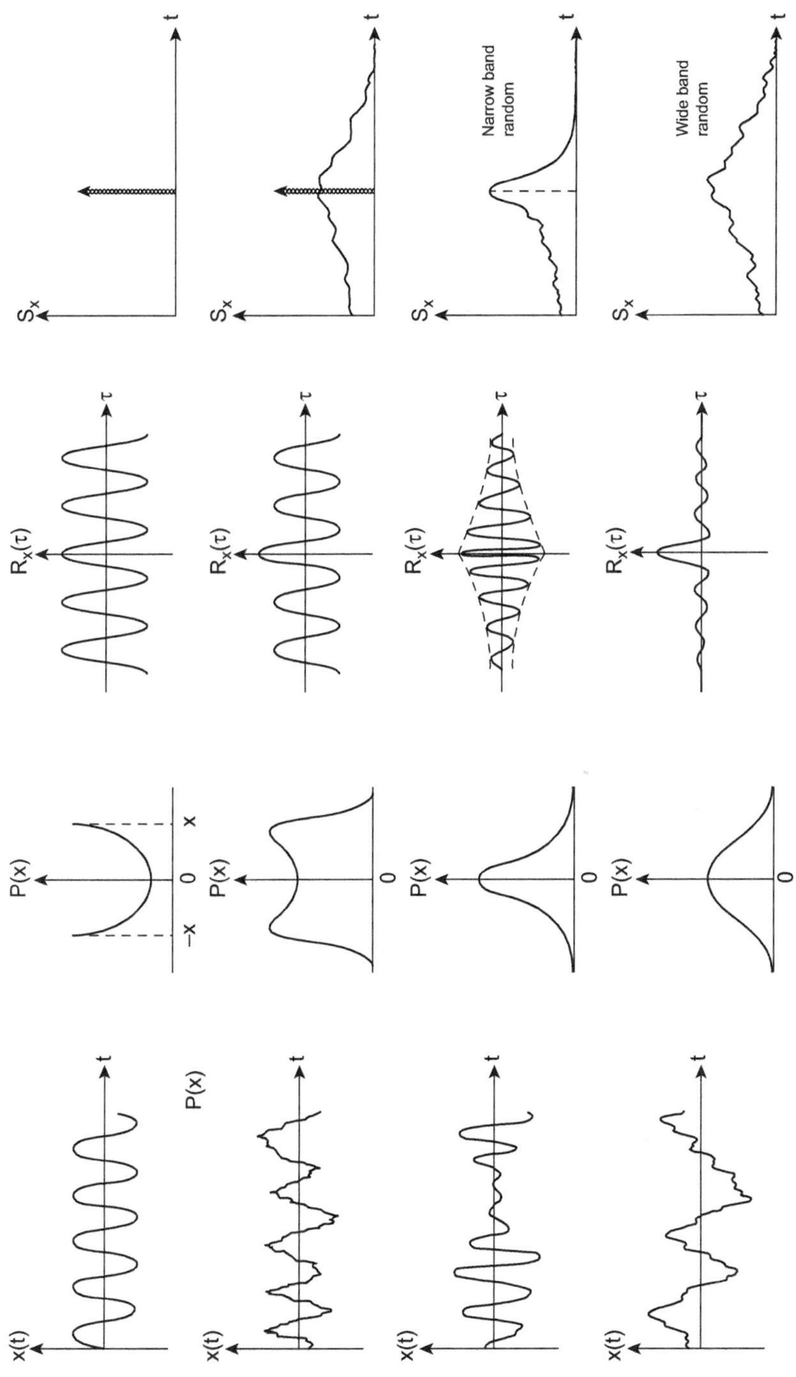

Figure 2.16 Probability density, autocorrelation, and PSD functions for four sample time histories

2.6.4 Joint Probability Density Function and Cross-Correlation Function

The joint probability density $p(x, y)$ of two random variables is the probability that both variables assume values within some defined pair of ranges at any instant of time. If we consider two random variables $x(t)$ and $y(t)$, the joint probability density has this property: the fraction of ensemble members for which $x(t)$ lies between x and $x+dx$ and $y(t)$ lies between y and $y + dy$ is $p(x, y)dxdy$. The joint probability densities are positive, and the probabilities of mutually exclusive events are additive. Also,

$$\int_{-\infty}^{\infty}\int_{-\infty}^{\infty} p(x, y)\,dxdy = 1. \qquad [2.178]$$

When two variables are statistically independent, the joint probability density is given by

$$p(x, y) = p(x)p(y). \qquad [2.179]$$

The cross-correlation function of two random variables indicates the general dependence of one variable on the other. The cross-correlation function of the time-history records $x(t)$ and $y(t)$ is shown in Figure 2.17 and given as

$$R_{xy}(\tau) = \lim_{T\to 0}\frac{1}{T}\int_{0}^{T} x(t)y(t+\tau)\,dt. \qquad [2.180]$$

The function $R_{xy}(\tau)$ is always a real-valued function and can be either negative or positive. Also, $R_{xy}(\tau)$ does not necessarily have a maximum at $\tau = 0$, nor is $R_{xy}(\tau)$ an even function. However, $R_{xy}(\tau)$ is symmetric; i.e.,

$$R_{xy}(-\tau) = R_{xy}(\tau). \qquad [2.181]$$

When $R_{xy}(\tau) = 0$, then $x(t)$ and $y(t)$ are uncorrelated.

We have previously defined the complex frequency response function $H(\omega)$, whose magnitude is equal to the amplitude ratio and whose ratio of imaginary to real parts is equal to the tangent of the phase angle φ. The Fourier transforms of the response and excitation [i.e., $X(\omega)$ and $F(\omega)$] are related through the frequency response function as

$$X(\omega) = H(\omega)F(\omega). \qquad [2.182]$$

This relation is valid for any arbitrary excitation $f(t)$. If the excitation is a stationary random process, then the response will also be a stationary

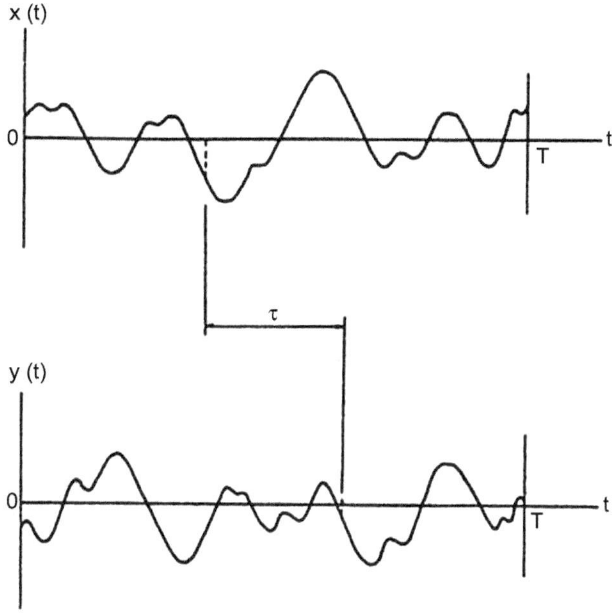

Figure 2.17 Cross-section measurements

random process. By using mathematical manipulations, it can be shown that for a linear system, the response mean-square spectral density $S_x(\omega)$ and the mean-square spectral density of the excitation $S_f(\omega)$ are related as follows:

$$S_x(\omega) = |H(\omega)|^2 S_f(\omega). \quad [2.183]$$

The mean-square value of the response can be obtained as follows:

$$R_x(0) = E[x^2(t)] = \frac{1}{2\pi}\int_{-\infty}^{\infty} |H(\omega)|^2 S_f(\omega)d\omega. \quad [2.184]$$

From Eqs. [2.183] and [2.184], it is evident that for a linear system, the response mean-square spectral density and mean-square value can be calculated from the mean-square spectral density of the excitation and the complex frequency response function $H(\omega)$, respectively.

If the excitation has a Gaussian probability distribution and the system is linear, then the response will also be Gaussian. This implies that for the stationary process the probability distribution of the response is completely defined by the mean and mean-square values of the response.

2.6.5 Response of an SDOF System to a Random Input

As an example, we can consider an SDOF system as shown in Figure 2.18, which is a simple model of a quarter vehicle traveling on a rough surface. The equation of motion is

$$m\ddot{y} + c(\dot{y} - \dot{z}) + k(y - z) = 0, \qquad [2.185]$$

which can be converted to be the following form:

$$m\ddot{x} + c\dot{x} + kx = -m\ddot{z}$$

where $x = y - z$. The PSD of the response is given by

$$S_y(\omega) = |H(\omega)|^2 S_z(\omega), \qquad [2.186]$$

where

$$|H(\omega)|^2 = (\omega/\omega_n)^4\left\{\left[1 - (\omega/\omega_n)^2\right]^2 + [2\xi(\omega/\omega_n)]^2\right\}^{-1}. \qquad [2.187]$$

If $S_z(\omega) = S_0$ (a constant), which represents a track input corresponding to white noise, then from Eqs. [2.186] and [2.187], we obtain

$$S_y(\omega) = \frac{S_0(\omega/\omega_n)^4}{\left[1 - (\omega/\omega_n)^2\right]^2 + [2\xi(\omega/\omega_n)]^2}. \qquad [2.188]$$

The mean-square value of the response will be

$$E[y^2(t)] = \frac{S_0}{2\pi}\int_{-\infty}^{\infty}\frac{(\omega/\omega_n)^4 d\omega}{\left[1 - (\omega/\omega_n)^2\right]^2 + [2\xi(\omega/\omega_n)]^2}. \qquad [2.189]$$

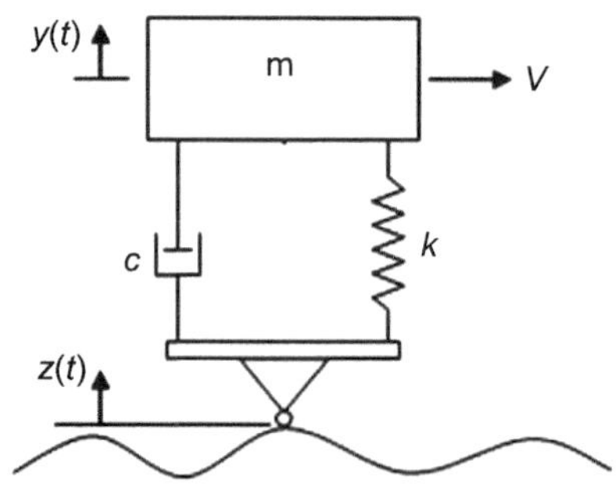

Figure 2.18 Schematic of a model of a quarter of a vehicle traveling over a rough road

An integration of Eq. [2.189] can be performed by using the residue theorem of complex variables, which gives

$$E[y^2(t)] = \frac{\pi\omega^4 S_0}{2\xi\omega_n^3}. \qquad [2.190]$$

Because the random process is Gaussian with zero mean value, the mean-square value equation described by Eq. [2.190] is sufficient to determine the shape of the response probability density function; this makes it possible to evaluate the probability that the response $y(t)$ might exceed a given displacement.

As an example, we consider a microsystem affected by the force of Brownian motion due to mechanical and electrical thermal noise. The following Langevin equation can be used to describe the effects due to Brownian motion, $m\ddot{y} + c\dot{y} + ky = F_N(t)$, in which random force $F_N(t)$ is due to Brownian motion and has the following properties:

$$\int_{-\infty}^{\infty} F_N(t)dt = 0 \qquad \int_{-\infty}^{\infty} F_N(t)F_N(t+\tau)d\tau = 2k_B Tc\delta(\tau),$$

where T is temperature, c is a damping coefficient, $\delta(\tau)$ is a Dirac delta function, and k_B is a Boltzmann constant, a physical constant relating energy at the individual particle level to temperature. We can get the power spectrum density of random force $F_N(t)$ as $S_N(\omega) = 4k_B Tc$. Then the response spectrum is

$$S_y(\omega) = S_N(\omega)/\left\{\left[1 - (\omega/\omega_n)^2\right]^2 + [2\xi(\omega/\omega_n)]^2\right\}$$

Next, we discuss the response of MDOF systems to random inputs.

We have already seen from Eqs. [2.71] and [2.72] that the uncoupled equation of motion of the rth mode of a dynamic system that has proportional damping is given by

$$\ddot{y}_r + 2\xi_r\omega_r\dot{y}_r + \omega_r^2 y_r = \overline{f}_r(t) = \left\{\overline{\varphi}^{(r)}\right\}^T\{F(t)\}. \qquad [2.191]$$

where $\left\{\overline{\varphi}^{(r)}\right\}$ represents the weighted rth modal vector of the undamped system. We introduce the following Fourier transforms of $y_r(t)$ and $\overline{f}_r(t)$, respectively, in the form

$$Y_r(\omega) = \int_{-\infty}^{\infty} y_r(t)e^{-i\omega t}dt$$

$$F_r(\omega) = \int_{-\infty}^{\infty} \overline{f}_r(t)e^{-i\omega t}dt = \sum_{j=1}^{n} \overline{\varphi}_j^{(r)} \int_{-\infty}^{\infty} F_j(t)e^{-i\omega t}dt. \qquad [2.192]$$

Then we obtain Fourier transforms on both sides of Eq. [2.191] as

$$Y_r(\omega)\left[-\omega^2 + i2\xi_r\omega + \omega_r^2\right] = \omega_r^2 F_r(\omega) \quad r = 1, 2, \ldots, n. \qquad [2.193]$$

Eq. [2.193] can be solved for $Y_r(\omega)$ as

$$Y_r(\omega) = H_r(\omega)F_r(\omega) \quad r = 1, 2, \ldots, n, \qquad [2.194]$$

where

$$H_r(\omega) = \left[1 - (\omega/\omega_r)^2 + i2\xi_r(\omega/\omega_r)\right]^{-1} \; r = 1, 2, \ldots, n. \qquad [2.195]$$

The response correlation matrix $[R_x(\tau)]$ is given as

$$[R_x(\tau)] = \lim_{T\to\infty} \frac{1}{T} \int_{-T/2}^{T/2} \{y(t)\}\,\{y(t+\tau)\}^T dt. \qquad [2.196]$$

Because the vector $\{x(t)\} = [\overline{\varphi}]\{y(t)\}$, we can write Eq. [2.196] as

$$\begin{aligned}[R_x(\tau)] &= \lim_{T\to\infty} \frac{1}{T} \int_{-T/2}^{T/2} [\overline{\varphi}]\{x(t)\}\,\{x(t+\tau)\}^T[\overline{\varphi}]^T dt \\ &= [\overline{\varphi}]\,[R_y(\tau)]\,[\overline{\varphi}]^T,\end{aligned} \qquad [2.197]$$

where

$$[R_y(\tau)] = \lim_{T\to\infty} \frac{1}{T} \int_{-T/2}^{T/2} \{y(t)\}\,\{y(t+\tau)\}^T dt \qquad [2.198]$$

is the response correlation matrix associated with generalized coordinates $y_r(t)$ ($r = 1, 2, \ldots, n$).

If $[H(\omega)]$ is the diagonal matrix of the frequency response function and $[H^*(\omega)]$ is its conjugate, then the correlation matrix is

$$[R_y(\tau)] = \frac{1}{2\pi} \int_{-\infty}^{\infty} [H^*(\omega)][S_f(\omega)][H(\omega)]e^{i\omega\tau} d\omega \qquad [2.199]$$

where $[S_f(\omega)]$ is an $n \times n$ excitation matrix associated with the generalized forces $\overline{f}_r(t)$. Now $[S_f(\omega)]$ can be expressed in terms of the Fourier transform of the excitation correlation matrix $[R_f(\tau)]$,

$$[S_f(\omega)] = \int_{-\infty}^{\infty} [R_f(\omega)]\; e^{-i\omega\tau} d\tau \qquad [2.200]$$

and $[R_f(\tau)]$ has the form

$$[R_f(\tau)] = \lim_{T\to\infty} \frac{1}{T} \int_{-T/2}^{T/2} \left\{\overline{f}(t)\right\}\left\{\overline{f}(t+\tau)\right\}^T dt, \qquad [2.201]$$

where $\left\{\overline{f}(t)\right\}$ is the vector of generalized forces $\overline{f}_r(t)$

Therefore,

$$\left\{\overline{f}(t)\right\} = [\overline{\varphi}]\{F(t)\} \qquad [2.202]$$

and

$$\left\{\overline{f}(t+\tau)\right\}^{\mathrm{T}} = \{F(t+\tau)\}^{\mathrm{T}}[\overline{\varphi}]^{\mathrm{T}}.$$

By substituting Eq. [2.202] into Eq. [2.201], we obtain

$$[R_f(\tau)] = [\overline{\varphi}]\,[R_F(\tau)]\,[\overline{\varphi}]^{\mathrm{T}}, \qquad [2.203]$$

where

$$[R_F(\tau)] = \lim_{T\to\infty} \frac{1}{T} \int_{-T/2}^{T/2} \{F(t)\}\{F(t+\tau)\}^{\mathrm{T}} dt. \qquad [2.204]$$

Introducing Eq. [2.203] into Eq. [2.200], we obtain

$$[S_f(\omega)] = [\overline{\varphi}] \int_{-\infty}^{\infty} [R_F(\tau)]e^{-i\omega\tau}d\tau[\overline{\varphi}]^{\mathrm{T}} = [\overline{\varphi}][S_F(\omega)][\overline{\varphi}]^{\mathrm{T}} \qquad [2.205]$$

where

$$[S_F(\omega)] = \int_{-\infty}^{\infty} [R_F(\tau)]e^{-i\omega\tau}d\tau \qquad [2.206]$$

is the excitation spectral matrix associated with the forces $F_i(t)$ $(i = 1, 2, \ldots, n)$.

The response correlation matrix is obtained by substituting Eqs. [2.199] and [2.205] into Eq. [2.197]:

$$[R_x(\tau)] = \frac{1}{2\pi}[\overline{\varphi}] \int_{-\infty}^{\infty} [H^*(\omega)][\overline{\varphi}][S_F(\omega)][\overline{\varphi}]^{\mathrm{T}}[H(\omega)]e^{i\omega\tau}d\omega[\overline{\varphi}]^{\mathrm{T}} \qquad [2.207]$$

and the autocorrelation function associated with the random response process $x_i(t)$ is

$$[R_{x_i}(\tau)] = \frac{1}{2\pi}[\overline{\varphi}_i] \int_{-\infty}^{\infty} [H^*(\omega)][\overline{\varphi}][S_F(\omega)][\overline{\varphi}]^T[H(\omega)]e^{i\omega\tau}d\omega[\overline{\varphi}_i]^T, \quad [2.208]$$

where $[\overline{\varphi}_i]$ is the ith row matrix [i.e., $[\overline{\varphi}_i] = \left[\overline{\varphi}_i^{(1)}\overline{\varphi}_i^{(2)}, \ldots, \overline{\varphi}_i^{(n)}\right]$], which for $\tau = 0$ yields the mean-square value

$$[R_{x_i}(0)] = \frac{1}{2\pi}[\overline{\varphi}_i] \int_{-\infty}^{\infty} [H^*(\omega)][\overline{\varphi}][S_F(\omega)][\overline{\varphi}]^T[H(\omega)]d\omega[\overline{\varphi}_i]^T. \quad [2.209]$$

2.7 NONLINEAR VIBRATION SYSTEMS

In vibration systems, nonlinear relationships generally exist, particularly when large deformation exists. In addition to the nonlinear behavior of inertia force, damping force, and stiffness force, nonlinearity could be due to geometry. A system with friction is usually nonlinear. One of the main reasons for modeling a system as nonlinear is that some phenomena occurring in nonlinear systems are not predictable by linear theory. These could consist of one or more of the following aspects: (i) the frequency of vibration may be dependent on the amplitude of vibration; (ii) when the frequency of excitation is increased, the amplitude of vibration may have a significant increase; (iii) when a harmonic force excites a system, the response will not only have the basic harmonic component, but also consist of superharmonics or subharmonics, as well as chaos; (iv) the system can become self-excited, and amplitudes of vibration may grow even without any external disturbance; and (v) the system can become unstable under certain conditions.

In this section, we will introduce some basic analytical methods for nonlinear systems, including the perturbation method and the method of variation of parameters. We present these methods by using several types of typical nonlinear equations, including the Duffing equation, van der Pol equation, and Mathieu equation, with a focus on illustrating the basic characteristics associated with these nonlinear systems.

2.7.1 Perturbation Method—Duffing Equation

Consider an undamped SDOF system under free conditions with a restoring force defined by a cubic function. The equation of motion can be written as

$$m\frac{d^2x}{dt^2} + \alpha x + \beta x^3 = 0. \quad [2.210]$$

Let

$$p = \sqrt{\alpha/m},\ T = pt,\ dt = dT/p,\ \varepsilon = \beta/mp^2. \quad [2.211]$$

Then Eq. [2.210] becomes

$$\ddot{x} + x + \varepsilon x^3 = 0. \quad [2.212]$$

Eq. [2.212] is known as *Duffing's equation*, and it is nonlinear with respect to the restoring force defined by the cubic function, which is shown in Figure 2.19. Next, we describe how to use the perturbation method to handle it. It is impossible to obtain an exact solution to Eq. [2.212], as in the case of a linear system.

Hence, we attempt an approximation solution. Assuming the initial conditions for the system are given by

$$x(0) = A;\ \dot{x}(0) = 0, \quad [2.213]$$

let $x_0(t)$ be the linear solution ($\beta = 0$). We can then perturb the solution about the linear solution in a series form as

$$x(T) = x_0(T) + \varepsilon x_1(T) + \varepsilon^2 x_2(T) + \cdots \quad [2.214]$$

The coefficient ε in Eq. [2.214] is a nonlinear parameter, and the solution assumed in Eq. [2.214] converges fast if $\varepsilon < 1$. Now substituting Eq. [2.214] into Eq. [2.212], we get

$$\begin{aligned}&(\ddot{x}_0 + \varepsilon\ddot{x}_1 + \varepsilon^2\ddot{x}_2 + \cdots) + (x_0 + \varepsilon x_1 + \varepsilon^2 x_2 + \cdots) + \\ &\varepsilon(x_0 + \varepsilon x_1 + \varepsilon^2 x_2 + \cdots)^3 = 0.\end{aligned} \quad [2.215]$$

Since ε is a parameter, sorting out the terms of ε^0, ε^1, ε^2, ..., we obtain

$$\begin{aligned}\ddot{x}_0 + x_0 &= 0 \\ \ddot{x}_1 + x_1 &= -x_0^3 \ldots.\end{aligned} \quad [2.216]$$

Since the first part of Eq. [2.216] is the linear part, and in view of the initial conditions [Eq. [2.213]], its solution takes the form

$$\begin{aligned} x_0 &= a_1 \cos T + b_1 \sin T \\ &= A \cos T. \end{aligned} \quad [2.217]$$

Substituting Eq. [2.217] into the second part of Eq. [2.216], we get

$$\begin{aligned} \ddot{x}_1 + x_1 &= -A^3 \cos^3 T \\ &= -\frac{3}{4}A^3 \cos T - \frac{1}{4}A^3 \cos 3T. \end{aligned} \quad [2.218]$$

From Eq. [2.217], we observe that there is an exciting term with a third harmonic of the natural frequency. Thus, the solution [Eq. [2.214]] contains higher harmonic terms. Eq. [2.218] is separated into two parts, one belonging to the harmonic solution and the other belonging to the higher harmonic. Thus,

$$\begin{aligned} \ddot{x}_{11} + x_{11} &= -\frac{3}{4}A^3 \cos T \\ \ddot{x}_{12} + x_{12} &= -\frac{1}{4}A^3 \cos 3T \\ x_1 &= x_{11} + x_{12}. \end{aligned} \quad [2.219]$$

It may be noted here that Eq. [2.212] is not a classical forced vibration problem of a system, and the terms on the right side of Eq. [2.219] are due to the perturbation of the nonlinear free vibration problem. Hence, we must consider both the transient and particular the integral parts of the equation in Eq. [2.219] for the final solution. Noting that the first equation

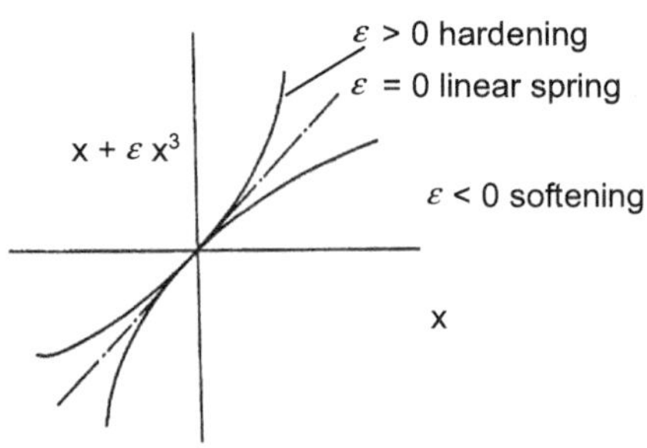

Figure 2.19 Hardening and softening spring characteristics

in Eq. [2.219] is excited by forcing the term at a resonant frequency, we write the following:

$$x_{11} = T(a_2 \cos T + b_2 \sin T).$$

Hence,

$$\ddot{x}_{11} = -2a_2 \sin T + 2b_2 \cos T - x_{11}.$$

Substituting this latter expression into *the first* equation of Eq. [2.219], we get

$$x_{11} = -\frac{3}{8}A^3 \; T \sin T. \qquad [2.220]$$

Similarly, for the higher harmonic, we have

$$x_{12} = a_3 \cos T + b_3 \sin T + \frac{A^3}{32}\cos 3T.$$

Since

$$x_{12} = 0; \dot{x}_{12} = 0 \quad \text{at} \quad T = 0,$$

we have

$$x_{12} = -\frac{1}{32}A^3(\cos T - \cos 3T). \qquad [2.221]$$

Hence, the complete solution for x_1 is given by

$$x_1 = -\frac{3}{8}A^3 \; T \sin T - \frac{1}{32}A^3 \; (\cos T - \cos 3T) \qquad [2.222]$$

The first term on the right side of Eq. [2.222] grows with time. Such a term is called *secular*. It is not physically possible to have a growing solution for the system in Figure 2.19, as the amplitude of vibration should be limited. This difficulty is overcome by introducing frequency-amplitude interaction by the Lindstedt-Poincaré method as follows.

For the linear system with $\varepsilon = 0$, in Eq. [2.212], the solution becomes periodic, with period 2π. For $\varepsilon \neq 0$, we introduce an unspecified function ω as follows:

$$\tau = \omega T \qquad [2.223]$$

We choose ω such that the secular term in Eq. [2.222] can be avoided. With the help of Eq. [2.223], we can now write Eq. [2.212] as

$$\omega^2 \ddot{x} + x + \varepsilon x^3 = 0. \quad [2.224]$$

As before, we write the solution to Eq. [2.224] as

$$x(\tau) = x_0(\tau) + \varepsilon\, x_1(\tau) + \varepsilon^2\, x_2(\tau) + \ldots \quad [2.225]$$

and also choose the function for ω in series as

$$\omega = \omega_0 + \varepsilon\, \omega_1 + \varepsilon^2\, \omega_2 + \ldots \quad [2.226]$$

Therefore,

$$\begin{aligned}&\left(\omega_0 + \varepsilon\omega_1 + \varepsilon^2\omega_2 + \cdots\right)^2 \left(\ddot{x}_0 + \varepsilon\, \ddot{x}_1 + \varepsilon^2\, \ddot{x}_2 + \cdots\right)\\ &+ \left(x_0 + \varepsilon\, x_1 + \varepsilon^2\, x_2 + \cdots\right) + \varepsilon\left(x_0 + \varepsilon\, x_1 + \varepsilon^2\, x_2 + \cdots\right)^3 = 0.\end{aligned} \quad [2.227]$$

As before, we now separate terms of ε^0, ε^1, ε^2, ... to obtain

$$\begin{aligned}&\omega_0^2\, \ddot{x}_0 + x_0 = 0\\ &\omega_0^2\, \ddot{x}_1 + x_1 = -2\omega_0\omega_1\ddot{x}_0 - x_0^3\\ &\omega_0^2\, \ddot{x}_2 + x_2 = -\left(2\omega_0\omega_2 + \omega_1^2\right)\ddot{x}_0 - 2\omega_0\omega_1\ddot{x}_1 - 3x_0^2x_1 \ldots.\end{aligned} \quad [2.228]$$

In addition to the initial condition [Eq. [2.213]], we now prescribe

$$x_i(\tau + 2\pi) = x_i(\tau), \quad [2.229]$$

since the solution should be periodic in τ with time period 2π.

The solution of the first part of Eq. [2.228] gives

$$x_0 = a_1 \cos\frac{\tau}{\omega_0} + b_1 \sin\frac{\tau}{\omega_0}.$$

In view of the initial conditions [Eq. [2.213]] and the periodic condition [Eq. [2.229]], this gives

$$x_0 = A \cos\tau \quad [2.230]$$

$$\omega_0 = 1. \quad [2.231]$$

With the help of these two equations, the second equation in Eq. [2.228] becomes

$$\begin{aligned}\ddot{x}_1 + x_1 &= -2\omega_1 A \cos\tau - A^3 \cos^3\tau\\ &= -\left(2\omega_1 A - \frac{3}{4}A^3\right)\cos\tau - \frac{1}{4}A^3 \cos 3\tau.\end{aligned} \quad [2.232]$$

The bracketed term in Eq. [2.232] gives rise to a secular term in the solution. Hence, in order to avoid this, we select

$$\omega_1 = \frac{3}{8}A^2 \qquad [2.233]$$

so that there is no term in the solution, which grows with time.

Now the solution of Eq. [2.232] is governed by the third harmonic excitation term, and it can be written as

$$\begin{aligned} x_1 &= a_2 \cos \tau + b_2 \sin \tau + \frac{A^3}{32} \cos 3\tau \\ &= \frac{A^3}{32}(-\cos \tau + \cos 3\tau). \end{aligned} \qquad [2.234]$$

Substituting Eq. [2.234] into the third part of Eq. [2.228], we determine ω_2 and x_2. Considering these three terms only, the solution of Eq. [2.224] is given by

$$\begin{aligned} x(\tau) = &\left(A - \frac{1}{32}\varepsilon A^3 + \frac{23}{1024}\varepsilon^2 A^5\right)\cos \tau + \left(\frac{1}{32}\varepsilon A^3 - \frac{3}{128}\varepsilon^2 A^5\right)\cos 3\tau \\ &+ \frac{1}{1024}\varepsilon^2 A^5 \cos 5\tau \end{aligned} \qquad [2.235]$$

$$\omega = 1 + \frac{3}{8}\varepsilon A^2 - \frac{21}{256}\varepsilon^2 A^4. \qquad [2.236]$$

Eq. [2.236] shows the dependency of frequency ω on amplitude A given initially. If $\varepsilon > 0$, we have a hardening spring; if $\varepsilon < 0$, then the spring is the softening type, as shown in Figure 2.20. Eq. [2.236] is plotted in Figure 2.20 for both cases.

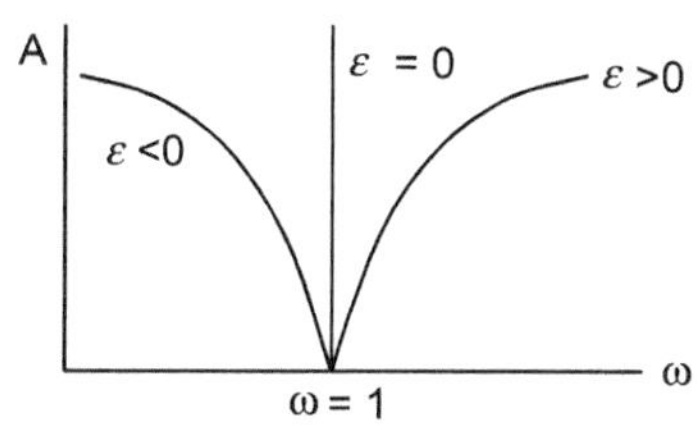

Figure 2.20 Amplitude-frequency relations.

The free vibration response of Eq. [2.210] is, therefore, given by

$$x(t) = \left(A - \frac{1}{32}\varepsilon A^3 + \frac{23}{1024}\varepsilon^2 A^5\right)\cos\sqrt{\frac{\alpha}{m}}\,t + \left(\frac{1}{32}\varepsilon A^3 - \frac{3}{128}\varepsilon^2 A^5\right)\cos 3\sqrt{\frac{\alpha}{m}}t + \frac{1}{1024}\varepsilon^2 A^5 \cos 5\sqrt{\frac{\alpha}{m}}\,t. \quad [2.237]$$

2.7.2 Amplitude–Frequency Dependent and Jump Phenomenon

Next, we consider the damped Duffing's equation under external harmonic excitation:

$$\ddot{x} + c\dot{x} + \alpha x \pm \beta x^3 = F\cos(\omega t + \varphi) = A_1 \sin\omega t + A_2\cos\omega t. \quad [2.238]$$

Assuming the first approximation to be

$$x_1 = A\ cos\ \omega t, \quad [2.239]$$

where A is assumed to be fixed and ω to be found, it can be derived that

$$\left[(\alpha - \omega^2)A \pm \frac{3}{4}\beta A^3\right]^2 + (c\omega A)^2 = A_1^2 + A_2^2 = F^2. \quad [2.240]$$

The response curves from Eq. [2.240] are depicted in Figure 2.21. It shows that the nonlinear system exhibits surprising phenomena that occur in linear systems: the amplitude of vibration of the system described by Eq. [2.238] could increase or decrease suddenly as the excitation frequency ω is increased or decreased gradually. This behavior is known as the *jump phenomenon*. It is clear that there could be two amplitudes of vibration for a given forcing frequency, as depicted in the bend regions of the curves of Figure 2.21 (cases $\beta \neq 0$).

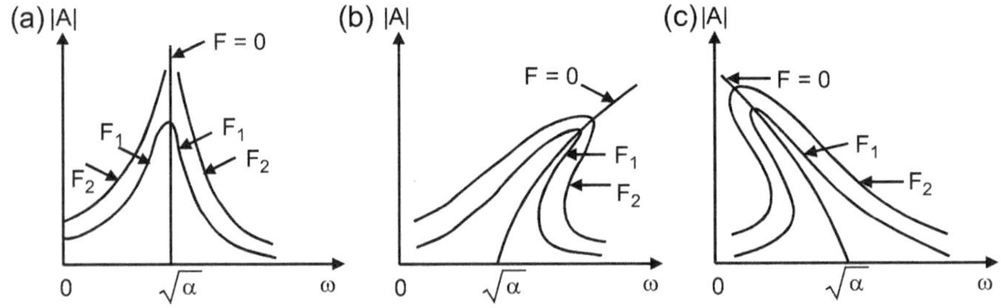

Figure 2.21 Response curves

2.7.3 van der Pol's Equation

As another application example of the perturbation method, we consider a self-excited system described by van der Pol's equation:

$$\ddot{x} - \varepsilon(1 - x^2)\dot{x} + \omega_0^2 x = 0. \qquad [2.241]$$

The initial conditions are $x(0) = 0$, $\dot{x}(0) = A\omega_0$.

Applying the Lindstedt-Poincaré method gives

$$\begin{aligned} x(\tau) &= x_0(\tau) + \varepsilon\, x_1(\tau) + \varepsilon^2 x_2(\tau) + \cdots \\ \omega^2 &= \omega_0^2 + \varepsilon\,\omega_1^2 + \varepsilon^2\,\omega_2^2 + \cdots \end{aligned} \qquad [2.242]$$

By inserting this series into Eq. [2.241], and equating coefficients of like terms, the result to the order ε^2 is

$$\begin{aligned} x(t) =& \left(2 - \frac{29\varepsilon^2}{96\omega_0^2}\right)\sin\,\omega_0 t + \frac{\varepsilon}{4\omega_0}\cos\,\omega_0 t \\ &+ \frac{\varepsilon}{4\omega_0}\left(\frac{3\varepsilon}{4\omega_0}\sin\,3\omega_0 t - \cos\,3\omega_0 t\right) - \frac{5\varepsilon^2}{124\omega_0^2}\sin\,5\omega_0 t. \end{aligned} \qquad [2.243]$$

In the next sections, we will further elaborate on the features of van der Pol's equation.

2.7.4 Method of Variation of Parameter

The perturbation method gives a steady-state solution, while the variation of parameter method described below allows small changes in amplitude and phase angle in the time solution. Consider the equation of a nonlinear system given by

$$\ddot{x} + \omega_0^2\,x + \varepsilon\,\varphi(x, \dot{x}, t) = 0. \qquad [2.244]$$

where $\varphi\,(x, \dot{x}, t)$ is the nonlinearity and ε is a constant. If we disregard $\varphi\,(x, \dot{x}, t)$, the solution corresponding to the linear equation $\ddot{x} + \omega_0^2\,x = 0$ is

$$x = A\,\cos(\omega_0 t + \theta) \qquad [2.245]$$

$$\dot{x} = -\omega_0\,A\,\sin(\omega_0 t + \theta). \qquad [2.246]$$

Assuming that A and θ are time-dependent functions rather than constants, differentiating Eq. [2.245] with respect to time t gives

$$\dot{x} = \dot{A}\,\cos\,\Psi - (A\,\sin\,\Psi)\left(\omega_0 + \dot{\theta}\right), \qquad [2.247]$$

where $\Psi = \omega_0\,t + \theta$ and $\dot{\Psi} = \omega_0 + \dot{\theta}$.

Subtracting Eq. [2.246] from Eq. [2.247], we obtain

$$\dot{A}\cos\Psi - \dot{\theta}A\sin\Psi = 0. \qquad [2.248]$$

Differentiating Eq. [2.246] with respect to t gives

$$\ddot{x} = -\omega_0\dot{A}\sin\Psi - (\omega_0 A\cos\Psi)\left(\omega_0 + \dot{\theta}\right). \qquad [2.249]$$

Substituting Eqs. [2.245] and [2.246] into Eq. [2.244] and subtracting the resultant equation from Eq. [2.249], we get

$$-\dot{A}\omega_0\sin\Psi - \dot{\theta}\omega_0 A\cos\Psi = -\varepsilon\varphi(A\cos\Psi, -\omega_0 A\sin\Psi, t). \qquad [2.250]$$

Solving Eqs. [2.248] and [2.250] for $\dot{A}$ and $\dot{\theta}$, we obtain

$$\dot{A} = \frac{\varepsilon}{\omega_0}\sin\Psi\,\varphi(A\cos\Psi, -\omega_0 A\sin\Psi, t)$$

and

$$\dot{\theta} = \frac{\varepsilon}{\omega_0 A}\cos\Psi\,\varphi(A\cos\Psi, -\omega_0 A\sin\Psi, t). \qquad [2.251]$$

Thus, the second-order differential equation is transferred into a first-order differential equation.

An approximate solution of Eq. [2.251] is obtained by assuming that ε is small and that A and θ do not change rapidly. The expression of $\dot{A}, \dot{\theta}$ can be expanded into a Fourier series. Since the motion is over a single cycle and the terms under the summation signs in the series are of the same period and consequently vanish, and if the changes in A and θ are small over one cycle, then the approximate average values of $\dot{A}$ and $\dot{\theta}$ equal the instantaneous values. Hence,

$$\dot{A} \approx \frac{\varepsilon}{2\pi\omega_0}\int_0^{2\pi}\sin\Psi\,\varphi(A\cos\Psi, -\omega_0 A\sin\Psi, t)\,d\Psi \qquad [2.252a]$$

$$\dot{\theta} \approx \frac{\varepsilon}{2\pi\omega_0 A}\int_0^{2\pi}\cos\Psi\,\varphi(A\cos\Psi, -\omega_0 A\sin\Psi, t)\,d\Psi, \qquad [2.252b]$$

where A is assumed to be constant in the integrand.

Next, we apply the method of variation of parameters to solve Rayleigh's equation:

$$\ddot{x} - (\alpha - \beta\dot{x}^2)\dot{x} + \omega^2 x = 0. \qquad [2.253]$$

By applying these procedures, we have the following relationships:

$$\dot{A} = \frac{-1}{2\pi\omega_0}\int_0^{2\pi} \cos^2\Psi(-\alpha + \beta A^2\omega^2\cos^2\Psi)A\,\omega\,d\Psi$$
$$= A(\alpha - 3\beta A^2\omega^2/4)/2. \qquad [2.254]$$

This equation can be integrated directly:

$$t = 2\int_{A_0}^{A} \frac{dA}{A(\alpha - \gamma A^2)} = \frac{1}{\alpha}\ln\frac{A^2}{\alpha - \gamma A^2}\Bigg|_{A_0}^{A}, \qquad [2.255]$$

solving for A as follows:

$$A = \frac{\alpha}{\gamma}\left[\frac{1}{1 + \left(\frac{\alpha}{\gamma A_0^2} - 1\right)e^{-\alpha t}}\right]^{1/2}, \qquad [2.256]$$

where

$$\gamma = 3\beta^2\omega/4. \qquad [2.257]$$

Actually, the Rayleigh equation can be transferred to the van der Pol equation through proper transformation. We will further elaborate the features of Rayleigh equation next.

2.7.5 Phase Plot, Limit Cycles, Self-Excited Oscillations and Chaos

In addition to the closed-form approximate solutions derived in the previous discussion, a graphical method is very useful in determining the qualitative information about the behavior of the nonlinear dynamic systems. An SDOF system requires two parameters to describe the state of motion completely: namely, the displacement and velocity of the system. When these two parameters are used as coordinate axes, then the resulting graphical representation of the motion is called the *phase plane representation*. Here, each point in the phase plane represents a possible state of the system. As time changes, the state of the system will change. A typical representative point in the phase plane moves and traces a curve known as the *trajectory*. The trajectory shows how the solution of the system varies with time. Figure 2.22 is the phase plot of the solutions of Rayleigh's equation for two given initial conditions.

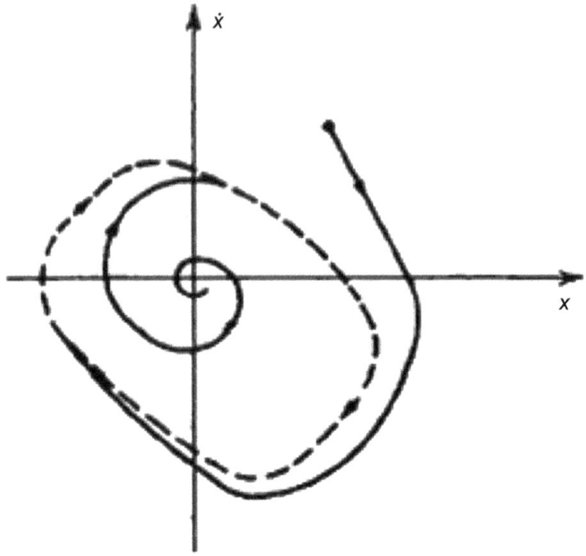

Figure 2.22 Limit cycle for Rayleigh's equation; the broken line shows a limit cycle.

The limit cycles are the trajectories which, due to different initial conditions, form a family of concentric closed curves in a phase plane. The broken line in Figure 2.22 shows a stable limit cycle. For some of the initial points inside or outside the closed curves, they could approach and then stay on the closed curve. In contrast with the forced vibration, the nonlinear system is autonomous. A limit cycle is a nonconservative and nonlinear phenomenon. As shown in Figure 2.22 there is only one closed trajectory, which is independent of the initial conditions. The initial conditions could be inside or outside the closed trajectory. Once initiated, the system will eventually "lock" itself into a limit cycle with a constant amplitude and frequency. A limit cycle occurs if, over one cycle, the net energy input from the excitation is equal to the energy dissipation within the system. Self-excitation is a function of the motion itself, such as displacement and/or velocity. If the excitation force in a function of velocity is $F\dot{x}$, we get

$$m\,\ddot{x} + c\,\dot{x} + kx = F\dot{x} \qquad [2.258]$$

or

$$m\,\ddot{x} + (c - F)\dot{x} + kx = 0. \qquad [2.259]$$

The equation is autonomous. If $(c-F)>0$, the system is stable. If $(c-F)<0$, we have negative damping and the amplitude increases with each oscillation. If $(c-F)=0$, sustained oscillation is possible. The oscillation occurs near the natural frequency of the system.

Chaos refers to behavior of a nonlinear system that is inherently unpredictable. Generally, the term means the dynamic behavior of a nonlinear system whose response, although described by a deterministic equation, becomes unpredictable because the nonlinearities in the equation enormously amplify the errors in the initial conditions of the system. The typical characteristic of chaotic vibration is that, if the numerical solution (and, presumably, the physical system that it represents) is started twice at nearly identical initial conditions, the two solutions will diverge exponentially with time.

The attractor of a nonlinear dynamic system is defined as a point toward which all nearby solutions move as time progresses. The attractor of a nonlinear dynamic system could be equilibrium or rest points.

In phase plots, the trajectory of a nonlinear system could exhibit as a periodic circle, two or more periodic circles, or even strange abstractors that represent geometric points to allow chaotic trajectories to be attracted. A nonlinear system could exhibit a phenomenon known as *bifurcation,* which means that the system's period could become two, and then three, and then even more with the continuous change in system parameters and excitation parameters. The system could reach a chaotic state through a series of bifurcations.

The Poincaré map represents points in phase space that are sampled stroboscopically or observed at equal intervals of time, which denote the fundamental period of the characteristic function such as a forcing periodic function. Specially, if the system is periodic, the Poincaré map will exhibit one dot. If the system is two periodic, the Poincaré map will exhibit two dots. If the system is chaos, the Poincaré map generally exhibits certain special striking patterns, rather than a random scatter of points.

Next, we present the chaotic behavior of forced Duffing's equation. Consider the following equation:

$$\ddot{x}(t) + 0.2\dot{x}(t) - x(t) + x^3(t) = 0.3 \cos 1.29t. \qquad [2.260]$$

Figure 2.23 represents a Poincaré map, which indicates points that occur at equal intervals of time $\tau = 2\pi/(1.29)$.

Usually, chaos can be perceived from a Poincaré map. The other evidence of the existence of chaos includes the existence of a positive Lyapunov exponent, distinct peaks and broadband spectrum. Some of chaos

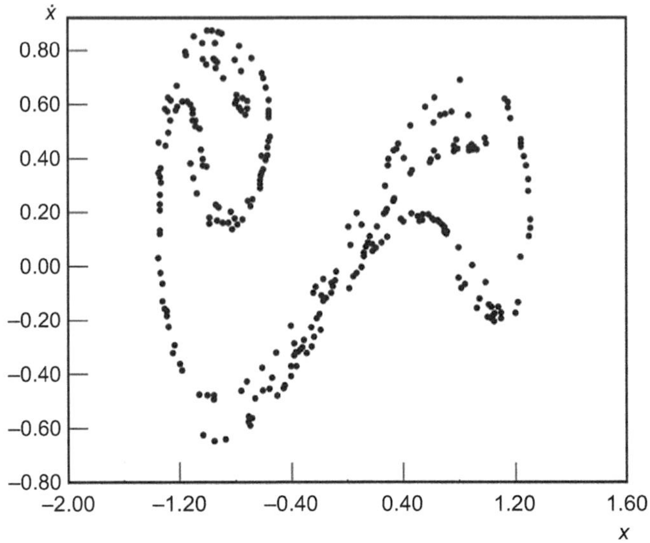

Figure 2.23 The Poincaré map of Eq. [2.60]

phenomena associated with friction system will be further illustrated in Chapters 4, 5 and 6.

2.7.6 Stability of Equilibrium

A nonlinear system can have more than one equilibrium position. Therefore, it is necessary to define the equilibrium positions and determine the stability of the system for each of its equilibriums, the condition $\{\dot{x}, \ddot{x}\} = \{\dot{x}_1, \dot{x}_2\} = \{0, 0\}$. Hence, an equilibrium position is called a *singular point*. The other points in the phase plane are called regular points.

To determine the equilibrium positions, let us consider a single-degree-of-freedom system described by two first-order differential equations:

$$\dot{x}_1 = X_1(x_1, x_2) \quad \text{and} \quad \dot{x}_2 = X_2(x_1, x_2), \qquad [2.261]$$

where X_1 and X_2 are functions of (x_1, x_2). Let (s_1, s_2) be an equilibrium position. It follows that

$$\dot{x}_1 = X_1(s_1, s_2) = 0 \quad \text{and} \quad \dot{x}_2 = X_2(s_1, s_2) = 0. \qquad [2.262]$$

The values of s_1 and s_2 can be found from Eq. [2.262].

If the origin of the equations is transferred to (s_1, s_2) by a coordinate translation

$$x_1 = y_1 + s_1 \quad \text{and} \quad x_2 = y_2 + s_2, \qquad [2.263]$$

then Eq. [2.261] becomes

$$\dot{x}_1 = \dot{y}_1 = X_1(s_1 + y_1,\ s_2 + y_2)$$
$$\dot{x}_2 = \dot{y}_2 = X_2(s_1 + y_1,\ s_2 + y_2).$$

Expanding X_1 and X_2 about (s_1, s_2) by Taylor's series, we obtain

$$\dot{y}_1 = a_{11}\, y_1 + a_{12}\, y_2 + \varphi_1(y_1, y_2)$$
$$\dot{y}_2 = a_{21}\, y_1 + a_{22}\, y_2 + \varphi_2(y_1, y_2)$$

or

$$\begin{bmatrix} \dot{y}_1 \\ \dot{y}_2 \end{bmatrix} = \begin{bmatrix} a_{11} & a_{12} \\ a_{21} & a_{22} \end{bmatrix} \begin{bmatrix} y_1 \\ y_2 \end{bmatrix} + \begin{bmatrix} \varphi_1 \\ \varphi_2 \end{bmatrix}. \qquad [2.264]$$

This can be written as

$$\{\dot{y}\} = A\{y\} + \{\varphi\}, \qquad [2.265]$$

where (y_1, y_2) are the state variables referred to the origin at (s_1, s_2), the elements in the matrix A are constants, and the element in $\{\varphi\}$ consists of the nonlinear terms. Hence, the linearized equations of the system about the equilibrium at (s_1, s_2) are

$$\{\dot{y}\} = A\{y\}. \qquad [2.266]$$

The stability of the system about (s_1, s_2) can be examined by the trajectory of Eq. [2.266] in a phase plane. Applying linear theory, the solution of Eq. [2.266] is

$$\{y\} = \{y_0\}\, e^{\lambda t}, \qquad [2.267]$$

where λ is an eigenvalue and $\{y_0\}$ a vector of constants. Substituting Eq. [2.267] into [2.266], and simplifying it, we get

$$[\lambda I - A]\{y_0\} = \{0\}, \qquad [2.268]$$

where I is a 2×2 -unit matrix. The values of λ are obtained from the characteristic equation

$$\Delta(\lambda) = |\lambda I - A| = 0. \qquad [2.269]$$

There are two eigenvalues, λ_1 and λ_2. The system is stable about the equilibrium (s_1, s_2) only if the real part of roots λ_1 and λ_2 from Eq. [2.269] is either zero or a negative number.

Assume a similarity transformation

$$\{y\} = B\{z\}, \text{ or } \begin{bmatrix} y_1 \\ y_2 \end{bmatrix} = \begin{bmatrix} b_{11} & b_{12} \\ b_{21} & b_{22} \end{bmatrix} \begin{bmatrix} z_1 \\ z_2 \end{bmatrix} \qquad [2.270]$$

where B is a nonsingular matrix and the equations in the variables (z_1, z_2) are uncoupled. Substituting Eq. [2.270] into [2.266] and premultiplying by B^{-1}, the resulting uncoupled equations are

$$\{\dot{z}\} = B^{-1}AB\{z\}, \qquad [2.271]$$

$$\begin{bmatrix} \dot{z}_1 \\ \dot{z}_2 \end{bmatrix} = \begin{bmatrix} \lambda_1 & 0 \\ 0 & \lambda_2 \end{bmatrix} \begin{bmatrix} z_1 \\ z_2 \end{bmatrix}, \qquad [2.272]$$

or

$$\dot{z}_1 = \lambda_1 z_1 \quad \text{and} \quad \dot{z}_2 = \lambda_2 z_2. \qquad [2.273]$$

The stability of the system about an equilibrium (s_1, s_2) can be examined in the (y_1, y_2) or (z_1, z_2) phase plane. The system stability can be examined in either plane. The eigenvalues of Eqs. [2.266] and [2.271] are identical because the matrices A and $B^{-1}AB$ possess the same determinant.

Recalling $|B^{-1}| = |B|^{-1}$, we have $|B^{-1}AB| = |B^{-1}|\,|A|\,|B| = |B|^{-1}\,|A|\,|B| = |A|$.

Here, the trajectories are mapped from one plane to the other. The next sections discuss the stability of the four cases.

Case 1. Node: $\lambda_{1,2}$ Real and Same Sign

A node occurs if λ_1 and λ_2 are real and have the same sign. From Eq. [2.273], we have

$$\frac{dz_2}{dz_1} = \frac{\lambda_2 z_2}{\lambda_1 z_1} \text{ or } \frac{dz_2}{z_2} = \frac{\lambda_2}{\lambda_1}\frac{dz_1}{z_1},$$

which can be integrated directly to yield

$$z_2 = C\, z_1^{\lambda_2/\lambda_1}, \qquad [2.274]$$

where C is a constant. The exact plot depends on the values of C and the ratio λ_2/λ_1. For instance, if $\lambda_2/\lambda_1 = 2$ and $C = 1$, we get $z_2 = z_1^2$. If $\lambda_2 = \lambda_1$, the trajectories are simply radial lines from the origin.

The directions of the trajectories can be deduced from Eq. [2.273]. If both λ_1 and λ_2 are negative, the system is stable and the trajectories converge toward the equilibrium. Conversely, if $\lambda_{1,2}$ is positive, the system is unstable and the trajectories point away from the equilibrium.

Noting that the trajectories in the (z_1, z_2) plane are symmetrical, but those in the (y_1, y_2) plane are governed by the characteristics discussed earlier.

Noting that since (y_1, y_2) are obtained from (x_1, x_2) by a coordinate translation, the x's and y's have the same physical interpretation. The characteristics of phase trajectories, as discussed in the previous section, do not apply to the plots in the (z_1, z_2) phase plane.

Case 2. Saddle Point: $\lambda_{1,2}$ Real and Opposite Sign

A saddle point occurs if $\lambda_{1,2}$ are real and have opposite signs. Since one of the λs is positive, the system in the neighborhood of a saddle point is always unstable. Since the ratio λ_2/λ_1 is negative, Eq. [2.274] can be expressed as

$$z_2 = C\, z_1^{-|\lambda_2/\lambda_1|}$$

or

$$z_2\, z_1^{|\lambda_2/\lambda_1|} = C. \qquad [2.275]$$

This clearly shows that the trajectory is a hyperbola.

Case 3. Vortex: $\lambda_{1,2}$ Imaginary

A *vortex*, or *center*, occurs if $\lambda_{1,2} = \pm j\,\beta$ are imaginary, where $j = \sqrt{-1}$ and β is a constant. From Eq. [2.273], we get

$$\dot{z}_1 = j\beta z_1 \quad \text{and} \quad \dot{z}_2 = -j\beta z_2$$

or

$$z_1 = z_{10}\, e^{j\beta t} \quad \text{and} \quad z_2 = z_{20}\, e^{-j\beta t},$$

where (z_{10}, z_{20}) are constants. The factor $e^{\pm j\beta t}$ represents a harmonic motion of unit magnitude with circular frequency β. Therefore, the resulting motion in the (z_1, z_2) plane is a combination of two harmonic motions, which is an ellipse. The motion in the neighborhood of equilibrium in the (z_1, z_2) or (y_1, y_2) plane forms a closed curve. Hence by definition, the system is stable.

Case 4. Focus: $\lambda_{1,2}$ Complex Conjugates

A focus occurs if $\lambda_{1,2} = \alpha \pm j\,\beta$ are complex conjugates, where $j = \sqrt{-1}$ and α and β are constants. From Eq. [2.273], we get

$$\dot{z}_1 = (\alpha + j\,\beta)\, z_1 \quad \text{and} \quad \dot{z}_2 = (\alpha - j\,\beta)\, z_2 \qquad [2.276]$$

or

$$z_1 = (z_{10}\, e^{\alpha t})\, e^{j\beta t} \quad \text{and} \quad z_2 = (z_{20}\, e^{\alpha t}) e^{-j\beta t},$$

where (z_{10}, z_{20}) are constants. The factor $e^{\pm j\beta t}$ represents a harmonic motion of unit magnitude with circular frequency β as before. If $\alpha > 0$, $e^{\alpha t}$ increases exponentially with time t and the trajectory in the (z_1, z_2) plane is a divergent logarithmic spiral. Hence for $\alpha > 0$, the equilibrium is unstable. If $\alpha < 0$, the trajectory is a convergent logarithmic spiral, and the system is therefore asymptotically stable.

In conclusion, the stability about an equilibrium at (s_1, s_2) can be examined from the roots of the characteristic equation. Substituting $A = [a_{ij}]$ from Eq. [2.264] into Eq. [2.269], the characteristic equation and the roots $\lambda_{1,2}$ are

$$\lambda^2 - (a_{11} + a_{22})\lambda + (a_{11}a_{22} - a_{12}a_{21}) = 0 \qquad [2.277]$$

$$\lambda_{1,2} = \frac{1}{2}\left\{(a_{11} + a_{22}) \pm \left[(a_{11} + a_{22})^2 - 4(a_{11}a_{22} - a_{12}a_{21})\right]^{1/2}\right\}. \qquad [2.278]$$

Introducing the parameters u and v as

$$u = (a_{11} + a_{22}) \quad \text{and} \quad v = (a_{11}a_{22} - a_{12}a_{21}), \qquad [2.279]$$

Eqs. [2.277] and [2.278] become

$$\lambda^2 - u\,\lambda + v = 0 \qquad [2.280]$$

$$\lambda_{1,2} = \frac{1}{2}\left\{u \pm \sqrt{u^2 - 4v}\right\}. \qquad [2.281]$$

These four cases can be summarized as follows:

- ***Case 1*** Node: $\lambda_{1,2}$ real and same sign. This requires $u^2 > 4v$ and $v > 0$, with $u^2 = 4v$ as a limiting case. The node is stable if $u < 0$ and unstable if $u > 0$.
- ***Case 2*** Saddle point: $\lambda_{1,2}$ real and opposite sign. This requires $u^2 > 4v$ and $v < 0$. This is always unstable.
- ***Case 3*** Center: $\lambda_{1,2}$ imaginary. This requires $u^2 < 4v$ and $v = 0$. By definition, the system is stable.
- ***Case 4*** Focus: $\lambda_{1,2}$ complex conjugates. This requires $u^2 < 4v$ and $v \neq 0$. The system is stable if $u < 0$ and unstable if $u > 0$.

The stability of the cases can be mapped in the (u, v) plane. The demarcation between the nodes and foci is given by

$$u^2 = 4v, \qquad [2.282]$$

whereby the center is mapped in the positive real axis and the stable regions are given in the fourth quadrant of the map plot.

2.7.7 Parametrically Excited Systems and the Mathieu Equation

Next, we consider parametrically excited systems. The simplest case of a parametrically excited system can be described by the following equation:

$$\ddot{x} + \omega^2(\lambda + \gamma \cos \omega t)x = 0, \qquad [2.283]$$

where the stiffness of the system is a harmonic function. Eq. [2.283] is known as the *Mathieu equation*.

For convenience, let $\omega = 2$. Then Eq. [2.283] reduces to the standard form of the Mathieu equation:

$$\ddot{x} + (\delta + 2\varepsilon \cos 2t)x = 0, \varepsilon << 1. \qquad [2.284]$$

The dynamic stability properties of the equation can be investigated by using parameter group (δ,ε).

Next, we seek the solution by the Lindstedt-Poincaré method, assuming the solution takes the form

$$\mathrm{x(t) = x_0(t) + \varepsilon\, x_1(t) + \varepsilon^2\, x_2(t) + \ldots,} \qquad [2.285]$$

and in addition, choose the function for δ in series as

$$\delta = \delta_0 + \varepsilon\, \delta_1 + \varepsilon^2 \delta_2 + \ldots, \sqrt{\delta_0} = 0, 1, 2, \ldots \qquad [2.286]$$

Inserting Eqs. [2.285] and [2.286] into the Mathieu equation, and equating the coefficients of like power of ε to zero, we obtain the following sets of equations:

$$\ddot{\mathrm{x}}_0 + \delta_0\, \mathrm{x}_0 = 0 \qquad [2.287a]$$

$$\ddot{\mathrm{x}}_1 + \delta_0\, \mathrm{x}_1 = -(\delta_1 + 2\cos 2\mathrm{t})\mathrm{x}_0 \qquad [2.287b]$$

$$\ddot{\mathrm{x}}_2 + \delta_0 \mathrm{x}_2 = -(\delta_1 + 2\cos 2\mathrm{t})\mathrm{x}_1 - \delta_2 \mathrm{x}_0, \ \sqrt{\delta_0} = 0, 1, 2, \ldots \qquad [2.287c]$$

This should be solved sequentially for the various values of $\sqrt{\delta_0}\ (= 0, 1, 2, \ldots)$. The zero-order approximation is given by

$$x_0 = \begin{cases} \cos\sqrt{\delta_0}t \\ \sin\sqrt{\delta_0}t \end{cases}, \sqrt{\delta_0} = 0, 1, 2, \ldots \qquad [2.288]$$

The stability boundary curve are obtained by introducing the solution $x_0 = \cos\sqrt{\delta_0}t$ and $x_0 = \sin\sqrt{\delta_0}t\left(\sqrt{\delta_0} = 0, 1, 2, \ldots\right)$ into Eq. [2.287a] and allowing the $x_i(t)$, $(\mathrm{i} = 1, 2, \ldots)$ to be periodic. It yields an infinite number of solution pairs, with one pair for every value of $\sqrt{\delta_0}$ except for $\sqrt{\delta_0} = 0$. Considering the case $\sqrt{\delta_0} = 0$, in which case $x_0 = 1$, Eq. [2.287b] yields

$$\ddot{x}_1 = -\delta_1 - 2\cos 2t. \qquad [2.289]$$

Since x_1 should be periodic, we must have $\delta_1 = 0$. Then the solution is

$$x = (1/2)\cos 2t.$$

Thus Eq. [2.287b] becomes

$$\ddot{x}_2 = (-2\cos 2t)((1/2)\cos 2t) - \delta_2 = -(1/2 + \delta_2) - (1/2)\cos 4t. \quad [2.290]$$

For x_2 being periodic, the constant term in Eq. [2.290] should be set equal to zero, which gives $\delta_1 = -1/2$. Thus, there is following equation for $\sqrt{\delta_0} = 0$:

$$\delta = -(1/2)\varepsilon^2 + \ldots, \qquad [2.291]$$

which is a parabola passing through the origin of the parameter plane.

Next, we consider $\sqrt{\delta_0} = 1$, using a similar approach, we obtain two transition curves:

$$\delta = 1 - \varepsilon - (1/8)\varepsilon^2 + \ldots \text{ corresponding to } x_0 = \cos t \qquad [2.292]$$

$\delta = 1 + \varepsilon - (1/8)\varepsilon^2 + \ldots$ corresponding to $x_0 = \sin t$

Furthermore, we obtain two transition curves:

$$\delta = 4 + (5/12)\varepsilon^2 + \ldots \text{ corresponding to } x_0 = \cos 2t \qquad [2.293]$$

$\delta = 4 - (1/12)\varepsilon^2 + \ldots$ corresponding to $x_0 = \sin 2t$

Following the same pattern, transition curves can be obtained for $x_0 = \cos\sqrt{\delta_0}t$ and $x_0 = \sin\sqrt{\delta_0}t\left(\sqrt{\delta_0} = 0, 1, 2, \ldots\right)$. The curves are plotted in Figure 2.24. For pairs of parameters that are inside shaded areas, the motion is unstable.

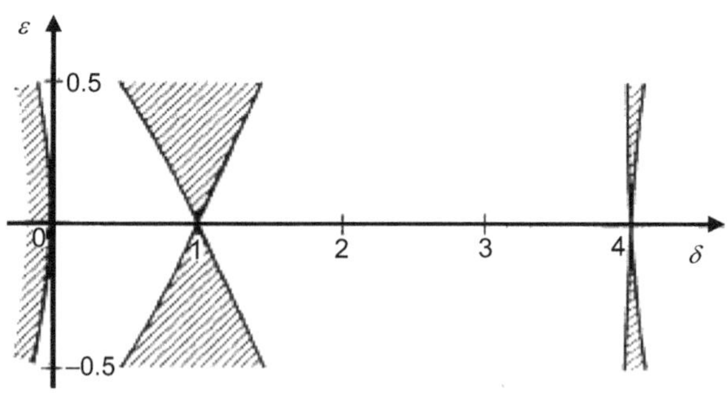

Figure 2.24 Stability chart for Mathieu's equation

2.7.8 Transient and Nonstationary Vibrations

Up to now, we have discussed nonlinear vibration systems with constant frequency and amplitude. In friction-involved systems, transient or nonstationary phenomena could occur due to the instantly dynamic transition of coupling of two components under external operational conditions or under system interactions, in addition to the conventional time-varying properties such as stiffness or friction. For simplification, we only illustrate the case with an excitation that has time-dependent frequencies and amplitudes. The motion equation of the most simplified case can be represented as

$$\ddot{x} + \omega_0^2 x = \varepsilon f(x, \dot{x}) + 2\varepsilon a(\varepsilon t)\cos[\theta(t, \varepsilon t)]. \qquad [2.294]$$

This system leads to nonstationary or nonperiodic resonance. This could be a representative of the cases where a modification of the response near the resonance condition or a passage through resonance takes place. For this system, steady-state solution does not exist, and the solution is aperiodic.

2.7.9 MDOF Systems

The general form of the equations of motion for a nonlinear MDOF system is governed by

$$[m]\{\ddot{x}\} + [c(\dot{x}, x)]\{\dot{x}\} + [k(\dot{x}, x)]\{x\} = \{F(t)\}, \qquad [2.295]$$

where the internal set of forces opposing the displacements and velocities are assumed to be nonlinear functions of $\{x\}$ and $\{\dot{x}\}$. This type of equations is usually solved by numerical approaches. In order to find the displacement and velocity vector $\{x\}$, $\{\dot{x}\}$ that satisfy the nonlinear equilibrium in Eq. [2.295], it is necessary to perform an equilibrium iteration sequence in each time step.

There are several numerical methods available for the solution of the response of linear or nonlinear MDOF systems. Many explicit and implicit integration schemes are available. Typical explicit schemes include the central difference methods, two-cycle iteration with trapezoidal rule, and the fourth-order Runge-Kutta method. The implicit schemes include the Wilson-θ method, the Newmark-β method, and high-order methods. The Newmark-β method will be detailed in next chapter of this book with applications illustrated for friction-involved nonlinear dynamics systems.

The MDOF nonlinear system, as shown in Eq. [2.295], may exhibit many similar nonlinear phenomena such as that in SDOF systems (for example, subharmonics or superharmonics) presented previously. In particular, the MDOF nonlinear systems may have internal resonance. For these cases, the corresponding modes are strongly coupled. In free vibrations, the energy keeps on changing among the corresponding modes. For quadratic nonlinearity, the condition for internal resonance is

$$\omega_m \approx 2\omega_k, or\ \omega_q \approx \omega_p \pm \omega_m. \quad [2.296]$$

For cubic nonlinearity, the condition for internal resonance is

$$\omega_m \approx 3\omega_k, or\ \omega_q \approx 2\omega_p \pm \omega_m, or\ \omega_q \approx \omega_p \pm \omega_m \pm \omega_k. \quad [2.297]$$

For harmonic excited nonlinear system with exciting frequency Ω, besides primary resonance ($\Omega \approx \omega_m$) and secondary resonance ($p\Omega = q\omega_m, p, q$ are integers), there are combination resonances:

$$\begin{aligned} p\Omega &= q_1\omega_1 + q_2\omega_2 + \cdots, \\ &\quad (p, q_i \text{ are integers subject to certain conditions}). \end{aligned} \quad [2.298]$$

When a system involves both internal resonance and combination resonance, some fractal harmonics may exist. For example, in a system with quadratic nonlinearity, there are components of $\Omega/3, 2\Omega/3$. For a system with cubic nonlinearity, there are components of $\Omega/5$, $2\Omega/5$ or $\Omega/2$, $3\Omega/2$.

2.8 ADVANCED DYNAMICS

2.8.1 Rigid Body Kinematics

In this section, we present the fundamentals of the kinematics of rigid body by using both vector and matrix methods. The position vector, velocity, and acceleration of a rigid body can be described using different reference coordinates, which in turn can be converted into each other using coordinate transformation through matrix operation. Two sets of Euler angles are introduced to describe the motion of a rigid body, and finally, the transformation matrix approach is discussed briefly [6–12].

2.8.1.1 Position Vector, Velocity, and Acceleration

Consider a particle moving in a three-dimensional (3D) space, as shown in Figure 2.25, in which O is the origin of the coordinate system. If we choose

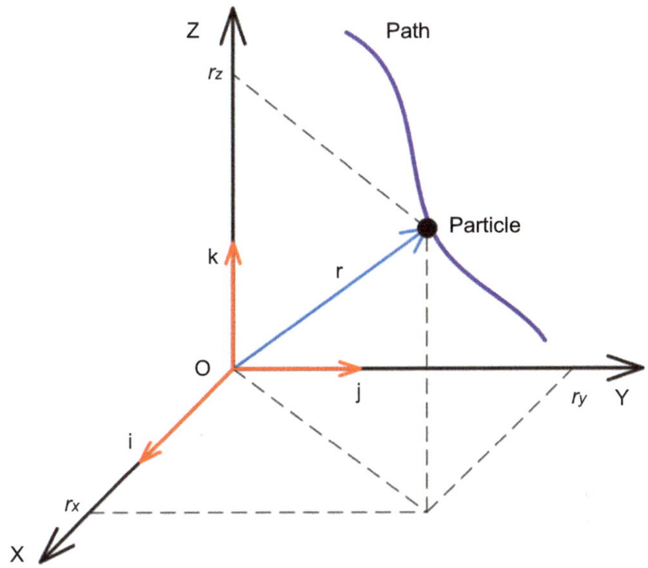

Figure 2.25 Schematic of particle moving in a three dimension space.

Cartesian coordinate system xyz in Figure 2.25 as an inertial or fixed reference frame, the absolute motion of the particle in Figure 2.25 can be described by a position vector **r** as follows:

$$\mathbf{r} = r_x i + r_y j + r_z k, \qquad [2.299]$$

where i, j, and k are the unit vectors of x, y, and z, respectively; and r_x, r_y, and r_z are scalar components of r in x-, y-, and z-coordinates. The position vector r can be alternatively presented in matrix form:

$$\mathbf{r} = \left[r_x \; r_y \; r_z\right]^T. \qquad [2.300]$$

The particle motion can be described using a cylindrical coordinate $\rho\theta z$, as shown in Figure 2.26, in which unit vectors are $\mathbf{e}_\rho$, $\mathbf{e}_\theta$, and $\mathbf{e}_z$. Then the position vector **r** in Eq. [2.299] can be expressed in terms of $\mathbf{e}_\rho$ $\mathbf{e}_\theta$ $\mathbf{e}_z$ as

$$\mathbf{r} = \rho\mathbf{e}_\rho + z\mathbf{e}_z. \qquad [2.301]$$

The velocity of the particle can be obtained as a time derivative of the position vector **r** as follows:

$$\mathbf{v} = \frac{d\mathbf{r}}{dt} = \dot{r}_x \mathrm{i} + \dot{r}_y \mathrm{j} + \dot{r}_z \mathrm{k} = \dot{\rho}\mathbf{e}_\rho + \rho\dot{\theta}\mathbf{e}_\theta + \dot{z}\mathbf{e}_z. \qquad [2.302]$$

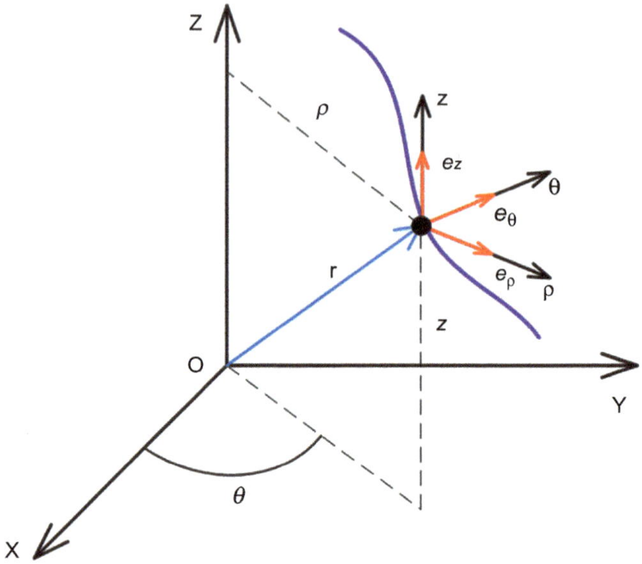

Figure 2.26 Schematic of a particle moving in a 3D space described by cylindrical coordinates

The acceleration of the particle a can be obtained as the time derivative of the velocity **v**, given by

$$\mathbf{a} = \frac{d\mathbf{v}}{dt} = \ddot{r}_x\mathrm{i} + \ddot{r}_y\mathrm{j} + \ddot{r}_z\mathrm{k} = \left(\ddot{\rho} - \rho\dot{\theta}^2\right)\mathrm{e}_\rho + \left(\rho\ddot{\theta} + 2\dot{\rho}\dot{\theta}\right)\mathrm{e}_\theta + \ddot{z}\mathrm{e}_z. \quad [2.303]$$

2.8.1.2 Angular Velocity of a Rigid Body

Compared with the 3D translational motion of particle as mentioned previously, a rigid body can perform 3D translational motion and rotational motion. The translational velocities of a rigid body can be described using the similar formulas [Eqs. [2.299]–[2.302]. The angular velocity vector of a rigid body ω can be expressed as

$$\omega = \frac{d\theta}{dt}\mathbf{n} = \omega_1\mathbf{e}_1 + \omega_2\mathbf{e}_2 + \omega_3\mathbf{e}_3, \quad [2.304]$$

where ω_1, ω_2, and ω_3 are components of the angular velocity in an arbitrary rectangular coordinate system with unit vectors $\mathbf{e}_1$, $\mathbf{e}_2$, and $\mathbf{e}_3$. The angular velocity can be expressed in matrix form:

$$\widetilde{\omega} = \begin{bmatrix} 0 & -\omega_3 & \omega_2 \\ \omega_3 & 0 & -\omega_1 \\ -\omega_2 & \omega_1 & 0 \end{bmatrix} \quad [2.305]$$

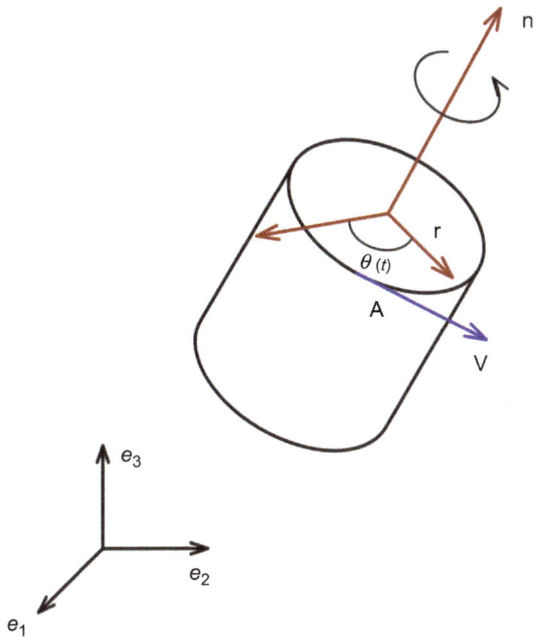

Figure 2.27 Angular velocity of a rigid body

Figure 2.27 shows a rigid body moving in an arbitrary rectangular coordinate system with unit vectors $\mathbf{e}_1$, $\mathbf{e}_2$, and $\mathbf{e}_3$. For a point A on a rigid body, its velocity can be expressed as

$$\mathbf{v} = \boldsymbol{\omega} \times \mathbf{r} = \tilde{\boldsymbol{\omega}}\mathrm{r} \quad [2.306]$$

Eq. [2.306] can be expressed in matrix form:

$$\mathbf{v} = \begin{pmatrix} v_1 \\ v_2 \\ v_3 \end{pmatrix} = \begin{bmatrix} 0 & -\omega_3 & \omega_2 \\ \omega_3 & 0 & -\omega_1 \\ -\omega_2 & \omega_1 & 0 \end{bmatrix} \begin{pmatrix} r_1 \\ r_2 \\ r_3 \end{pmatrix}. \quad [2.307]$$

2.8.1.3 Moving Coordinate Systems

A moving coordinate system xyz used to describe the motion of a rigid body can be expressed as one of the following three types, depending on its motion with respect to the fixed or inertial frame XYZ with origin [10–11]:

Translating coordinate systems (Figure 2.28a)

Rotating coordinate systems (Figure 2.28b)

Translating and rotating coordinate systems (Figure 2.29)

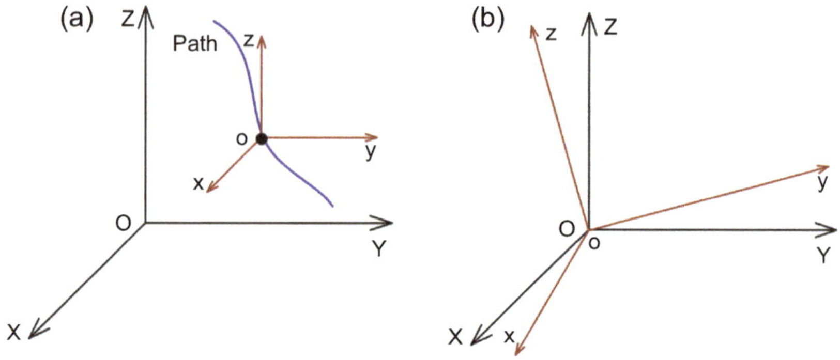

Figure 2.28 (a) Translating coordinate systems; (b) rotating coordinate systems

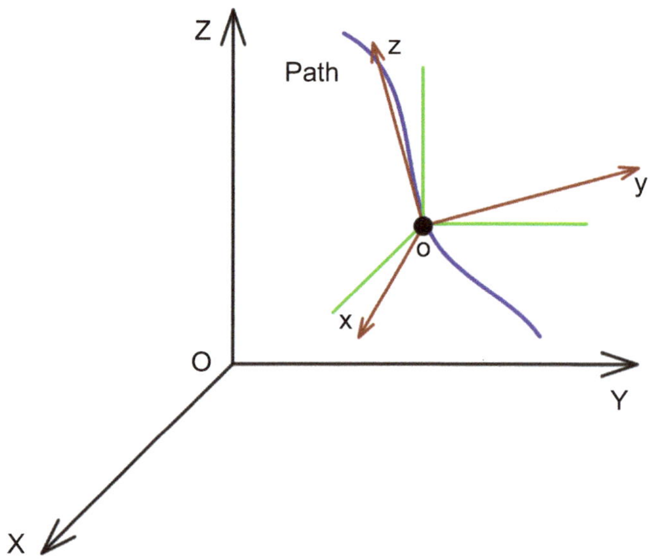

Figure 2.29 Translating and rotating a coordinate system

A moving coordinate system *xyz* with origin o attached to a moving body is normally used as a reference frame to describe the kinematics of the rigid body.

Normally, origin o is chosen to be the body's gravity center.

Figure 2.30 shows a particle moving in 3D space. *XYZ* is a fixed frame with unit vectors **ijk**; and *xyz* is the moving frame with unit vectors $\mathbf{e}_1$, $\mathbf{e}_2$, $\mathbf{e}_3$. If the angular velocity of *xyz* is $\boldsymbol{\omega}$ and the position vector **r** is

$$\mathbf{r} = \mathbf{R} + \boldsymbol{\rho}, \tag{2.308}$$

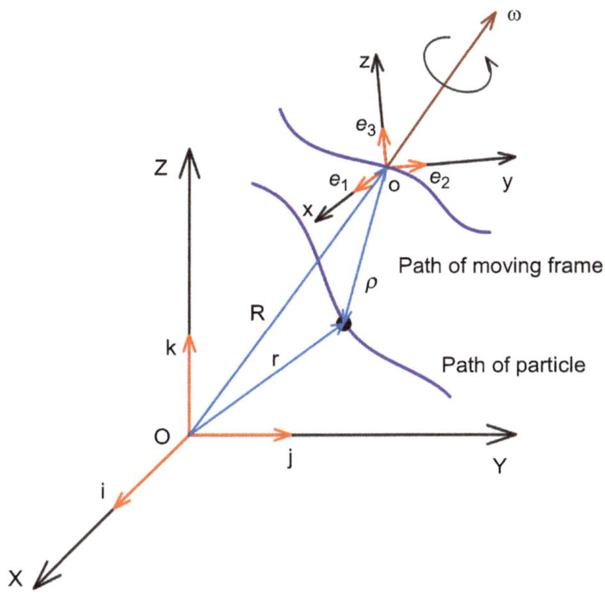

Figure 2.30 Moving coordinate systems

the velocity **v** of the particle is

$$\begin{aligned}\mathbf{v} &= \frac{d\mathbf{r}}{dt} \\ &= \frac{d\mathbf{R}}{dt} + \frac{d\boldsymbol{\rho}}{dt} \\ &= \dot{\mathbf{R}} + \mathbf{v}_r + \boldsymbol{\omega} \times \boldsymbol{\rho},\end{aligned} \qquad [2.309]$$

where $\dot{\mathbf{R}}$ is the velocity of the origin *o* of the reference frame *xyz;* $\boldsymbol{\omega}$ is its angular velocity; $\mathbf{v}_r$ is the relative velocity of the particle with respect to *xyz* or relative velocity seen by the observer moving (both translating and rotating) with *xyz;* whereas

$d\boldsymbol{\rho}/dt$ is the relative velocity seen by the observer, who translates but does not rotate with *xyz*; and $\boldsymbol{\omega}\times\boldsymbol{\rho}$ is the difference between these two relative velocities. If $\boldsymbol{\rho}$ is

$$\rho = \rho_1\mathbf{e}_1 + \rho_2\mathbf{e}_2 + \rho_3\mathbf{e}_3, \qquad [2.310]$$

then

$$\mathrm{v}_r \equiv \left(\frac{d\boldsymbol{\rho}}{dt}\right)_{rel} = \dot{\rho}_1\mathbf{e}_1 + \dot{\rho}_2\mathbf{e}_2 + \dot{\rho}_3\mathbf{e}_3 \qquad [2.311]$$

$$\begin{aligned} \mathrm{a} &= \frac{d\mathrm{v}}{dt} \\ &= \ddot{\mathbf{R}} + \mathbf{a}_r + \boldsymbol{\omega} \times \mathbf{v}_r + \boldsymbol{\omega} \times \frac{d\boldsymbol{\rho}}{dt} + \frac{d\boldsymbol{\omega}}{dt} \times \boldsymbol{\rho} \end{aligned} \quad [2.312]$$

$$\frac{d\rho}{dt} = \mathrm{v}_r + \omega \times \rho. \quad [2.313]$$

Finally, we have

$$\mathrm{a} = \ddot{\mathbf{R}} + \mathbf{a}_r + \boldsymbol{\omega} \times \boldsymbol{\omega} \times \boldsymbol{\rho} + \dot{\boldsymbol{\omega}} \times \boldsymbol{\rho} + 2\boldsymbol{\omega} \times \mathbf{v}_r, \quad [2.314]$$

where $\mathbf{a}_r = \ddot{\rho}_1\mathbf{e}_1 + \ddot{\rho}_2\mathbf{e}_2 + \ddot{\rho}_3\mathbf{e}_3$.

We can make the following observations about the physical meaning of each term in Eq. [2.314]:

$\ddot{\mathbf{R}}$ is the acceleration of the origin o of the moving reference frame xyz.

$\mathbf{a}_r$ is the relative acceleration of the particle as observed in the moving reference frame xyz.

$\boldsymbol{\omega} \times \boldsymbol{\omega} \times \boldsymbol{\rho}$ is a centripetal acceleration, or the correction term for the local position vector $\boldsymbol{\rho}$, considering that the observer rotates with the moving reference frame.

$\dot{\boldsymbol{\omega}} \times \boldsymbol{\rho}$ is another correction term for the angular acceleration vector $\dot{\omega}$ of the moving reference frame.

$2\boldsymbol{\omega} \times \mathbf{v}_r$ is the Coriolis acceleration, which is the other correction term from two sources, both of which measure the rotation of the basis (unit) vectors of the moving reference frame and associate with an interaction of motion along more than one coordinate curve.

2.8.1.4 Coordinate Transformation

In this section, we consider a general 3D coordinate transformation. Specifically, we want to establish a transformation matrix **C** that transforms components of a vector in one system of coordinates to another system of coordinates. Assume fixed or inertial frame XYZ with unit vectors IJK, and xyz is a rotating coordinate system with unit vectors ijk, as shown in Figure 2.31.

An arbitrary vector **r** in Figure 2.31 can be expressed in the two frames as

$$\mathrm{r} = r_X\mathrm{I} + r_Y\mathrm{J} + r_Z\mathrm{K} = r_x\mathrm{i} + r_y\mathrm{j} + r_z\mathrm{k} \quad [2.315]$$

It can be expressed in a matrix form as

$$\begin{pmatrix} r_X \\ r_Y \\ r_Z \end{pmatrix} = \mathbf{C}\begin{pmatrix} r_x \\ r_y \\ r_z \end{pmatrix}, \quad [2.316]$$

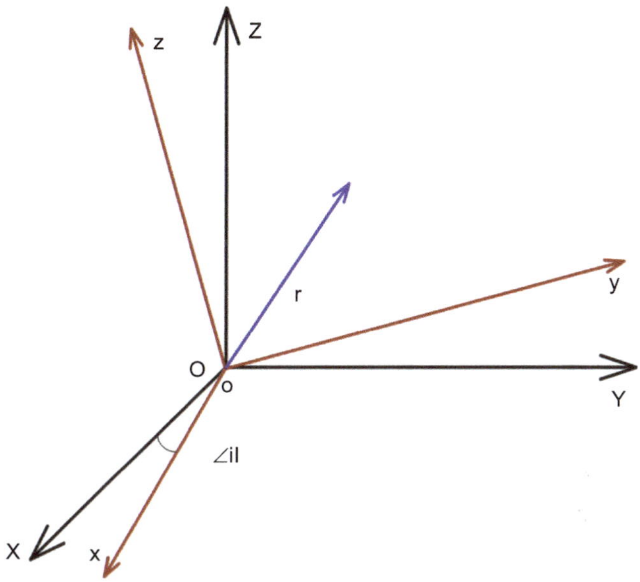

Figure 2.31 A vector **r** in two sets of coordinate systems *XYZ* and *xyz*

where **C** is a matrix of directional cosines called a *coordinate transformation matrix,* given by

$$C = \begin{bmatrix} \cos\angle \mathrm{iI} & \cos\angle \mathrm{jI} & \cos\angle \mathrm{kI} \\ \cos\angle \mathrm{iJ} & \cos\angle \mathrm{jJ} & \cos\angle \mathrm{kJ} \\ \cos\angle \mathrm{iK} & \cos\angle \mathrm{jK} & \cos\angle \mathrm{kK} \end{bmatrix}. \quad [2.317]$$

Given that **C** is the orthogonal matrix where $\mathbf{C}^{-1} = \mathbf{C}^T$, this yields

$$\begin{pmatrix} r_x \\ r_y \\ r_z \end{pmatrix} = \mathrm{C}^T \begin{pmatrix} r_X \\ r_Y \\ r_Z \end{pmatrix}. \quad [2.318]$$

2.8.1.5 First Set of Euler Angles—Precession-Nutation-Spin (φθψ)

This set of Euler angles is normally used to describe the microgyroscopic systems. The sequence of rotations shown in Figure 2.32 is as follows:

Precession: Rotation about the z-axis by $\varphi(t)$ to get $x^I y^I z^I$ or $x^I y^I Z$

Nutation: Rotation about the x^I axis by $\theta(t)$ to get $x^{II} y^{II} z^{II}$ or $x^I y^{II} z^{II}$

Spin: Rotation about the z^{II} axis by $\psi(t)$ to get xyz or xyz^{II}

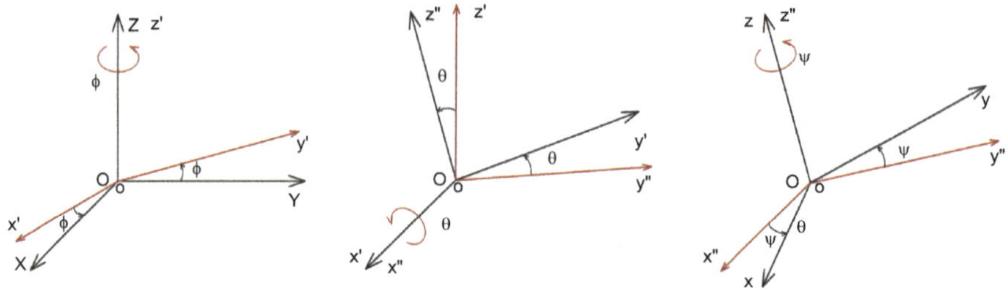

Figure 2.32 First set of Euler's angles and sequence of rotation

The coordinate transformations are, then,

$$\begin{pmatrix} r_X \\ r_Y \\ r_Z \end{pmatrix} = \begin{bmatrix} cos\varphi & -sin\varphi & 0 \\ sin\varphi & cos\varphi & 0 \\ 0 & 0 & 1 \end{bmatrix} \begin{pmatrix} r_{x'} \\ r_{y'} \\ r_{z'} \end{pmatrix} = C_1 \begin{pmatrix} r_{x'} \\ r_{y'} \\ r_{z'} \end{pmatrix} \quad [2.319a]$$

$$\begin{pmatrix} r_{x'} \\ r_{y'} \\ r_{z'} \end{pmatrix} = \begin{bmatrix} 1 & 0 & 0 \\ 0 & cos\theta & -sin\theta \\ 0 & sin\theta & cos\theta \end{bmatrix} \begin{pmatrix} r_{x''} \\ r_{y''} \\ r_{z''} \end{pmatrix} = C_2 \begin{pmatrix} r_{x''} \\ r_{y''} \\ r_{z''} \end{pmatrix} \quad [2.319b]$$

$$\begin{pmatrix} r_{x''} \\ r_{y''} \\ r_{z''} \end{pmatrix} = \begin{bmatrix} cos\psi & -sin\psi & 0 \\ sin\psi & cos\psi & 0 \\ 0 & 0 & 1 \end{bmatrix} \begin{pmatrix} r_x \\ r_y \\ r_z \end{pmatrix} = C_3 \begin{pmatrix} r_x \\ r_y \\ r_z \end{pmatrix}. \quad [2.319c]$$

Combining these three equations, we have

$$\begin{pmatrix} r_X \\ r_Y \\ r_Z \end{pmatrix} = C_1C_2C_3 \begin{pmatrix} r_x \\ r_y \\ r_z \end{pmatrix}. \quad [2.320]$$

2.8.1.6 Second Set of Euler Angles—Yaw-Pitch-Row ($\psi\theta\varphi$)

This set of Euler angles is normally used to describe the dynamics of microvehicles. The sequence of rotations as shown in Figure 2.33 is as follows:

Yaw: rotation about the z-axis by $\psi(t)$ to get $x^l y^l z^l$ or $x^l y^l Z$
Pitch: rotation about the y^l axis by $\theta(t)$ to get $x^{ll} y^{ll} z^{ll}$ or $x^{ll} y^l z^{ll}$
Roll: rotation about the x^{ll} axis by $\varphi(t)$ to get xyz or $x^{ll} yz$

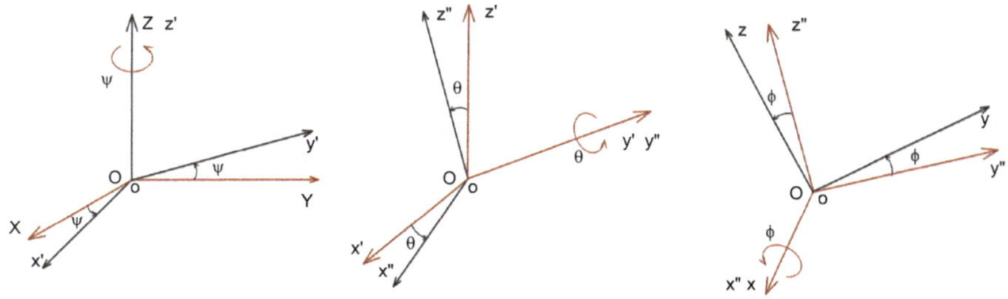

Figure 2.33 Yaw, pitch, and roll axes of microvehicle dynamics

The coordinate transformations are, then,

$$\begin{pmatrix} r_X \\ r_Y \\ r_Z \end{pmatrix} = \begin{bmatrix} cos\psi & -sin\psi & 0 \\ sin\psi & cos\psi & 0 \\ 0 & 0 & 1 \end{bmatrix} \begin{pmatrix} r_{x'} \\ r_{y'} \\ r_{z'} \end{pmatrix} = [R_\psi] \begin{pmatrix} r_{x'} \\ r_{y'} \\ r_{z'} \end{pmatrix}$$

$$\begin{pmatrix} r_{x'} \\ r_{y'} \\ r_{z'} \end{pmatrix} = \begin{bmatrix} cos\theta & 0 & sin\theta \\ 0 & 1 & 0 \\ -sin\theta & 0 & cos\theta \end{bmatrix} \begin{pmatrix} r_{x''} \\ r_{y''} \\ r_{z''} \end{pmatrix} = [R_\theta] \begin{pmatrix} r_{x''} \\ r_{y''} \\ r_{z''} \end{pmatrix} \qquad [2.321]$$

$$\begin{pmatrix} r_{x''} \\ r_{y''} \\ r_{z''} \end{pmatrix} = \begin{bmatrix} 1 & 0 & 0 \\ 0 & cos\varphi & -sin\varphi \\ 0 & sin\varphi & cos\varphi \end{bmatrix} \begin{pmatrix} r_x \\ r_y \\ r_z \end{pmatrix} = [R_\varphi] \begin{pmatrix} r_x \\ r_y \\ r_z \end{pmatrix}.$$

We have, then,

$$\begin{pmatrix} r_X \\ r_Y \\ r_Z \end{pmatrix} = [R_\psi][R_\theta][R_\varphi] \begin{pmatrix} r_x \\ r_y \\ r_z \end{pmatrix}. \qquad [2.322]$$

2.8.1.7 Angular Velocity Related to Euler Angles

For the first set of Euler angles, the absolute angular velocity $\boldsymbol{\omega}$ of xyz-coordinates is given by

$$\begin{aligned} \boldsymbol{\omega} &= \dot{\varphi}\mathrm{k} + \dot{\theta}\mathrm{e}_{x'} + \dot{\psi}\mathrm{e}_z \\ &\equiv \omega_x \mathrm{e}_x + \omega_y \mathrm{e}_y + \omega_z \mathrm{e}_z. \end{aligned} \qquad [2.323]$$

Rewriting $e_{x'}$ and k in terms of e_x, e_y and e_z, we get

$$\begin{pmatrix} \omega_x \\ \omega_y \\ \omega_z \end{pmatrix} = \begin{bmatrix} sin\theta sin\psi & cos\psi & 0 \\ sin\theta cos\psi & -sin\varphi & 0 \\ cos\theta & 0 & 1 \end{bmatrix} \begin{pmatrix} \dot{\varphi} \\ \dot{\theta} \\ \dot{\psi} \end{pmatrix}. \qquad [2.324]$$

Similarly, for the second set of Euler angles, the absolute angular velocity ω of xyz-coordinates is given by

$$\begin{aligned} \boldsymbol{\omega} &= \dot{\psi}\mathrm{k} + \dot{\theta}\mathrm{e}_{y'} + \dot{\varphi}\mathrm{e}_x \\ &= \omega_x \mathrm{e}_x + \omega_y \mathrm{e}_y + \omega_z \mathrm{e}_z. \end{aligned} \qquad [2.325]$$

Similarly, rewriting $e_{y'}$ and **k** in terms of $\mathbf{e}_x$, $\mathbf{e}_y$, and $\mathbf{e}_z$ yields

$$\begin{pmatrix} \omega_x \\ \omega_y \\ \omega_z \end{pmatrix} = \begin{bmatrix} 1 & 0 & -sin\theta \\ 0 & cos\varphi & cos\theta sin\varphi \\ 0 & -sin\varphi & cos\theta cos\varphi \end{bmatrix} \begin{pmatrix} \dot{\varphi} \\ \dot{\theta} \\ \dot{\psi} \end{pmatrix}. \qquad [2.326]$$

Eqs. [2.324] and [2.326] relate the Euler angles, the rotation measured in real applications, with the components of the angular velocity, ω_x, ω_y, and ω_z, in the reference coordinate system.

2.8.1.8 A Finite Motion

A general motion of any rigid body can be resolved into the translation **u** of an arbitrary point on the body and a finite rotation φ about this point, as shown in Figure 2.34a. In the next section, we introduce the transformation matrix expressed for a finite rotation, including both translation and rotation [10–11].

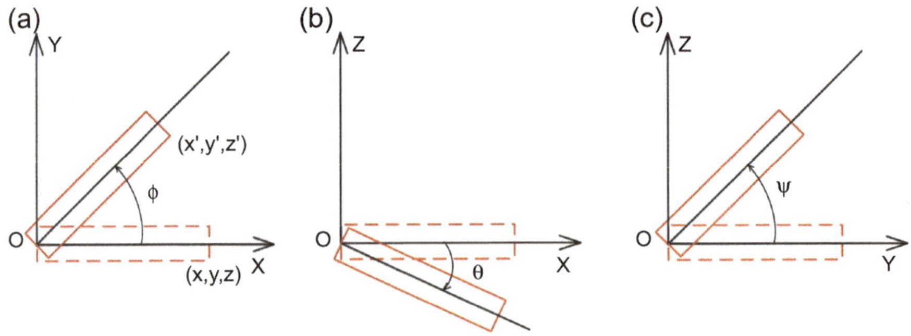

Figure 2.34 Schematic of finite rotation of a rigid body (a) about the *z*-axis; (b) about the *y*-axis; (c) about the *x*-axis

Considering a position vector of an arbitrary point P on a rigid body before and after the rotation as $\mathbf{r}_p$ and $\mathbf{r}_p{}'$, respectively, a transformation matrix $\mathbf{T}$ relating $\mathbf{r}_p$ and $\mathbf{r}'$is given by

$$\mathbf{r}_p{}' = T\boldsymbol{r}_p$$

The transformation matrix $\mathbf{T}$ has the following properties:

Because of no deformation of a rigid body, $\mathbf{T}$ is the same for any point$p$ in the body. Hence, we have

$$\mathbf{r}' = \mathbf{Tr}. \qquad [2.327]$$

The rotation should be invertible:

$$\mathbf{r} = \mathbf{T}^{-1}\mathbf{r}' \qquad [2.328]$$

The length of $\mathbf{r}$ is unchanged; hence,

$$\mathbf{r}\cdot\mathbf{r} = \mathbf{r}^T\mathbf{r} = \mathbf{r}'\cdot\mathbf{r}' = (\mathbf{r}')^T\mathbf{r}' \qquad [2.329a]$$

$$\mathbf{T}^T\mathbf{T} = \mathbf{I}. \qquad [2.329b]$$

This indicates that $\mathbf{T}^{-1} = \mathbf{T}^T$ or $\mathbf{T}$ is the orthogonal matrix.

The transformation $\mathbf{T}$ for a rotation about the Z-, Y-, and X-axes can be determined subsequently as follows.

Rotation About the *Z*-axis With φ

If the previous coordinates of any point p is $\mathbf{r} = [x\ y\ z]^T$ and the new coordinates of this point is $\mathbf{r}' = [x'\ y'\ z']^T$ (as seen in Figure 2.34a), then

$$\begin{pmatrix} x' \\ y' \\ z' \end{pmatrix} = \begin{bmatrix} \cos\varphi & -\sin\varphi & 0 \\ \sin\varphi & \cos\varphi & 0 \\ 0 & 0 & 1 \end{bmatrix} \begin{pmatrix} x \\ y \\ z \end{pmatrix} = \mathrm{T}_1 \begin{pmatrix} x \\ y \\ z \end{pmatrix} \qquad [2.330]$$

$$\mathbf{T}_1 = \begin{bmatrix} \cos\varphi & -\sin\varphi & 0 \\ \sin\varphi & \cos\varphi & 0 \\ 0 & 0 & 1 \end{bmatrix}. \qquad [2.331]$$

Rotation About the *Y*-axis With *θ*

From Figure 2.34b, we obtain the transformation matrix $\mathbf{T}_2$ for a rotation about the Y-axis as follows:

$$\mathbf{T}_2 = \begin{bmatrix} \cos\theta & 0 & \sin\theta \\ 0 & 1 & 0 \\ -\sin\theta & 0 & \cos\theta \end{bmatrix}. \qquad [2.332]$$

Rotation About the X-axis With ψ

From Figure 2.34c, the transformation matrix $\mathbf{T}_3$ for a rotation about the X-axis is as follows:

$$\mathbf{T}_3 = \begin{bmatrix} 1 & 0 & 0 \\ 0 & cos\psi & -sin\psi \\ 0 & sin\psi & cos\psi \end{bmatrix}.$$

In general, the transformation matrices do not commute i.e., $T_1T_2 \neq T_2T_1$. However, for the infinitesimal angular displacement, $cos\varphi \approx 1$, $sin\varphi \approx \varphi$, and so on. Also, $\varphi \approx \omega_z \Delta t$, $\theta \approx \omega_y \Delta t$, and $\psi \approx \omega_x \Delta t$. In this case, $\mathbf{T}_1$, $\mathbf{T}_2$, and $\mathbf{T}_3$ do commute. If the rigid body has an infinitesimal rotation about an arbitrary axis during time Δt, the new position vector $\mathbf{r}^l$ related to the previous position vector $\mathbf{r}$ is

$$\mathbf{r}^l = \mathbf{r}(t + \Delta t) = \mathbf{T}_1\mathbf{T}_2\mathbf{T}_3\mathbf{r}(t). \qquad [2.333]$$

Substitution of $\mathbf{T}_1$, $\mathbf{T}_2$, and $\mathbf{T}_3$ into Eq. [2.333] yields

$$\mathbf{r}(t + \Delta t) = \begin{bmatrix} 1 & -\omega_z\Delta t & \omega_y\Delta t \\ \omega_z\Delta t & 1 & -\omega_x\Delta t \\ -\omega_y\Delta t & \omega_x\Delta t & 1 \end{bmatrix} \begin{pmatrix} x(t) \\ y(t) \\ z(t) \end{pmatrix}. \qquad [2.334]$$

Therefore, the velocity $\mathbf{v}$ is given by

$$\mathbf{v} = \lim_{\Delta t \to 0} \frac{\mathbf{r}(t + \Delta t) - \mathbf{r}(t)}{\Delta t} = \begin{bmatrix} 0 & -\omega_z & \omega_y \\ \omega_z & 0 & -\omega_x \\ -\omega_y & \omega_x & 0 \end{bmatrix} \begin{pmatrix} x(t) \\ y(t) \\ z(t) \end{pmatrix} = \tilde{\boldsymbol{\omega}}\mathbf{r}. \qquad [2.335]$$

This is consistent with Eqs. [2.304]–[2.306].

A general motion of a rigid body (as shown in Figure 2.35) can be divided into two parts: a translation $\mathbf{u}$ and a finite rotation θ.

The position vector $\mathbf{r}$ describing the finite rotation of the rigid body is, then,

$$\mathbf{r} = \mathbf{u} + \boldsymbol{\rho}^l = \mathbf{u} + \mathbf{A}\boldsymbol{\rho}, \qquad [2.336]$$

where $\mathbf{A}$ is the transformation matrix of the rotation relating $\boldsymbol{\rho}$ and $\boldsymbol{\rho}'$. In addition, $\mathbf{r} = [x\ y\ z]^T$ and $\mathbf{u} = [u_x\ u_y\ u_z]^T$, expressed in the inertial reference

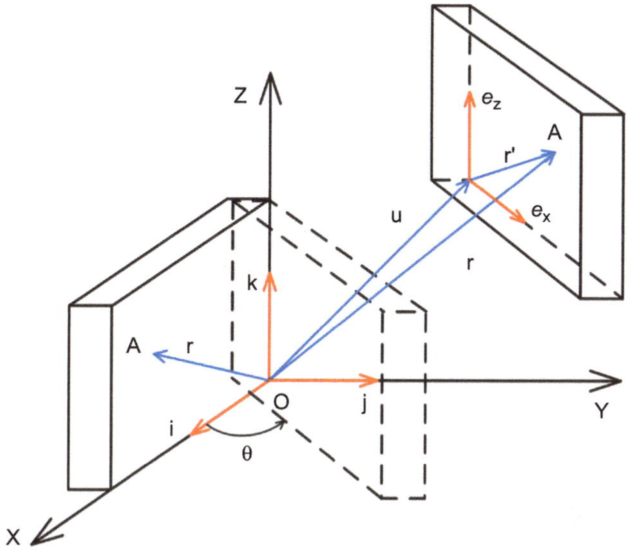

Figure 2.35 Schematic of finite motion of a rigid body

coordinates xyz, and $\boldsymbol{\rho} = [\rho_x \; \rho_y \; \rho_z]^T$, expressed in the local coordinate system $\mathbf{e}_x\mathbf{e}_y \; \mathbf{e}_z$. Substituting the component vectors into Eq. [2.336], the finite motion in Figure 2.35 is governed by

$$\begin{pmatrix} x \\ y \\ z \end{pmatrix} = \begin{pmatrix} u_x \\ u_y \\ u_z \end{pmatrix} + \begin{bmatrix} cos\theta & -sin\theta & 0 \\ sin\theta & cos\theta & 0 \\ 0 & 0 & 1 \end{bmatrix} \begin{pmatrix} \rho_x \\ \rho_y \\ \rho_z \end{pmatrix} \qquad [2.337]$$

If $\mathbf{r}$ and $\boldsymbol{\rho}$ are expanded as $\mathbf{r} = [x \; y \; z \; 1]^T$ and $\boldsymbol{\rho} = [\rho_x \; \rho_y \; \rho_z \; 1]^T$, Eq. [2.337] can be rewritten as

$$\mathbf{r}_{4\times1} = \mathbf{T}_{4\times4}\, \rho_{4\times1}, \qquad [2.338]$$

where $\mathbf{T}$ is the transformation matrix for a general finite motion given by

$$\mathbf{T} = \left[\begin{array}{ccc|c} & & & u_x \\ & A_{3x3} & & u_y \\ & & & u_z \\ \hline 0 & 0 & 0 & 1 \end{array}\right].$$

For a finite translation, the transformation matrix is simply

$$\mathbf{T}_1 = \left[\begin{array}{ccc|c} 1 & 0 & 0 & u_x \\ 0 & 1 & 0 & u_y \\ 0 & 0 & 1 & u_z \\ \hline 0 & 0 & 0 & 1 \end{array}\right].$$

For a finite rotation, the transformation matrix is simply

$$\mathbf{T}_2 = \left[\begin{array}{c|c} A_{3x3} & \begin{array}{c} 0 \\ 0 \\ 0 \end{array} \\ \hline \begin{array}{ccc} 0 & 0 & 0 \end{array} & 1 \end{array}\right].$$

It should be noted that $\mathbf{T} = \mathbf{T}_1\mathbf{T}_2$.

2.8.2 Linear and Angular Momentums

In the next section, we discuss the angular momentum of a rigid body and mass moment of inertia. These topics are used as fundamentals for a study of the dynamics of a rigid body and a multibody mechanical system.

Figure 2.36 shows a rigid body moving in 3D coordinates. Let the angular velocity of the body be $\boldsymbol{\omega}$.

Considering a rigid body, its angular momentum is

$$\mathbf{H}_o = \int \mathbf{r} \times \mathbf{v}dm, \qquad [2.339]$$

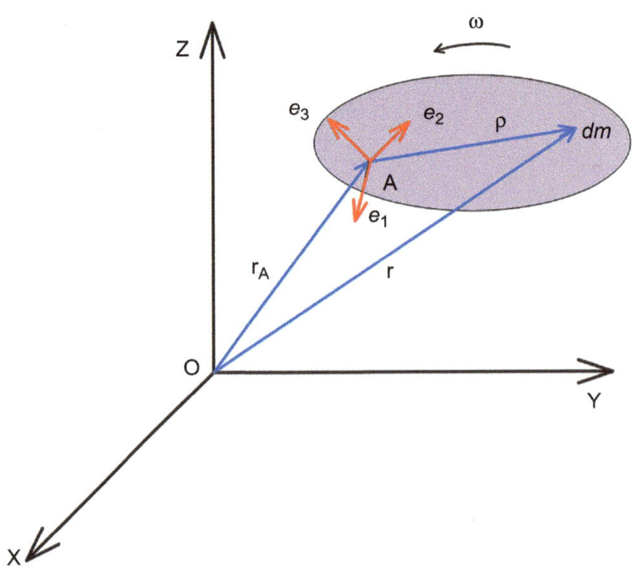

Figure 2.36 A rigid body moving in 3D coordinates

where $\mathbf{r} = \mathbf{r}_A + \boldsymbol{\rho}$ is a position vector from fixed origin o to differential mass dm of the body. Let $\mathbf{e}_1\mathbf{e}_2\mathbf{e}_3$ in Figure 2.36 be the rotating reference frame, with its origin located at an arbitrary point A on the body. If the reference coordinate system and the body have the same angular velocity (i.e., $\boldsymbol{\omega}$), then

$$\mathbf{v} = \dot{\mathbf{r}} = \mathbf{v}_A + \boldsymbol{\omega} \times \boldsymbol{\rho}. \qquad [2.340]$$

It should be noted that the velocity of the differential mass dm relative to $\mathbf{e}_1\mathbf{e}_2\mathbf{e}_3$ is zero. This is because: (i) the rigid body has no deformation and (ii) $\mathbf{e}_1\mathbf{e}_2\mathbf{e}_3$ rotates synchronously with the body. Substitution of Eq. [2.340] into Eq. [2.339] yields

$$\begin{aligned}
\mathbf{H}_o &= \int [(\mathbf{r}_A + \boldsymbol{\rho}) \times (\mathbf{v}_A + \boldsymbol{\omega} \times \boldsymbol{\rho})]dm \\
&= \mathbf{r}_A \times \mathbf{v}_A \int dm + \left[\int \boldsymbol{\rho} dm\right] \times \mathbf{v}_A \\
&\quad + \mathbf{r}_A \times \left[\boldsymbol{\omega} \times \int \boldsymbol{\rho} dm\right] + \int \boldsymbol{\rho} \times (\boldsymbol{\omega} \times \boldsymbol{\rho}) dm \\
&= \mathbf{r}_A \times m\mathbf{v}_A + m\boldsymbol{\rho}_c \times \mathbf{v}_A + \mathbf{r}_A \times [\boldsymbol{\omega} \times m\boldsymbol{\rho}_c] + \int \boldsymbol{\rho} \times (\boldsymbol{\omega} \times \boldsymbol{\rho}) dm,
\end{aligned} \qquad [2.341]$$

where $m = \int dm$ is the mass of the rigid body and $\boldsymbol{\rho}_c$ is the position of the body's $C.G.$ measured with respect to A and given by $\boldsymbol{\rho}_c = \frac{1}{m}\int \boldsymbol{\rho} dm$.

Furthermore, the rotation of a rigid body can be considered as two different cases: pure rotation and general motion with both rotation and translation. These are discussed next.

2.8.2.1 Pure Rotation About Fixed Point O

In this case, if we choose point A in Figure 2.36 fixed at o, $\mathbf{r}_A = \mathbf{v}_A = \mathbf{0}$, and Eq. [2.341] becomes

$$\mathbf{H}_o = \int \boldsymbol{\rho} \times (\boldsymbol{\omega} \times \boldsymbol{\rho}) dm. \qquad [2.342]$$

2.8.2.2 General Motion

In this case, if point A in Figure 2.36 is fixed at the body's $C.G$ (i.e., point c), $\mathbf{r}_A = \mathbf{r}_c$ and $\boldsymbol{\rho}_c = \mathbf{0}$. Eq. [2.341] then becomes

$$\mathbf{H}_o = \mathbf{r}_c \times m\mathbf{v}_c + \int \boldsymbol{\rho} \times (\boldsymbol{\omega} \times \boldsymbol{\rho}) dm. \qquad [2.343]$$

For a rigid body, the angular momentum about c is defined as

$$\mathbf{H}_c \equiv \int \boldsymbol{\rho} \times (\boldsymbol{\omega} \times \boldsymbol{\rho}) dm. \quad [2.344]$$

Hence,

$$\mathbf{H}_o = \mathbf{r}_c \times m\mathbf{v}_c + \mathbf{H}_c. \quad [2.345]$$

Due to a constant geometric property of the rigid body, the angular momentum can be simplified further as follows. First, note that the angular momentum about o [Eq. [2.342]], in the case of pure rotation, and the angular momentum about c [Eq. [2.344]], in the case of general motion, take the same form. Therefore, we will remove the subscripts in Eqs. [2.342] and [2.344] for convenience and generally rewrite both equations as

$$\mathbf{H} = \int \boldsymbol{\rho} \times (\boldsymbol{\omega} \times \boldsymbol{\rho}) dm = \int (\boldsymbol{\rho} \cdot \boldsymbol{\rho})\boldsymbol{\omega} - (\boldsymbol{\rho} \cdot \boldsymbol{\omega})\boldsymbol{\rho} dm. \quad [2.346]$$

Here, $\boldsymbol{\rho}$ and $\boldsymbol{\omega}$ can be expressed in terms of components with respect to the **e**1**e**2**e**3 coordinate system as follows:

$$\boldsymbol{\rho} = \begin{bmatrix} x & y & z \end{bmatrix}^T, \boldsymbol{\omega} = \begin{bmatrix} \omega_x & \omega_y & \omega_z \end{bmatrix}^T$$

$$\mathbf{H} = \left(\int \begin{bmatrix} y^2 + z^2 & -xy & -xz \\ -xy & x^2 + z^2 & -yz \\ -xz & -yz & x^2 + y^2 \end{bmatrix} dm \right) \boldsymbol{\omega} = \begin{bmatrix} I_{11} & I_{12} & I_{13} \\ I_{21} & I_{22} & I_{23} \\ I_{31} & I_{32} & I_{33} \end{bmatrix} \boldsymbol{\omega} = \mathbf{I}\boldsymbol{\omega}, \quad [2.347]$$

where **I** is the matrix of moments of inertia.

Any three orthogonal axes ($\mathbf{e}'_1\mathbf{e}'_2\mathbf{e}'_3$) that yield all zero cross-products of inertia (i.e., $I_{12} = I_{13} = I_{23} = 0$) are called *principal axes*. In this case, $I_{11} = I_1$, $I_{22} = I_2$, and $I_{33} = I_3$ are called the *principal inertias*. I_1, I_2, and I_3 can be determined from the eigenvalues of matrix **I**. With the principal inertias, the angular momentum of a rigid body can be simplified as

$$\mathbf{H} = I_1\omega_1\mathbf{e}'_1 + I_2\omega_2\mathbf{e}'_2 + I_3\omega_3\mathbf{e}'_3. \quad [2.348]$$

2.8.3 Dynamics of a Rigid Body and Euler Equations

In this section, the dynamics of a rigid body are treated by using the Newton-Euler equations. According to the laws of angular momentum, we can formulate the dynamics equations governing the motion of a rigid body.

For a rigid body having pure rotation about o with angular velocity ω, the governing equation is

$$\sum \mathbf{M}_o = \dot{\mathbf{H}}_o. \qquad [2.349]$$

Eq. [2.349] is the law of angular momentum for a rigid body. In this case, we choose the reference coordinate system $\{\mathbf{e}_1\mathbf{e}_2\mathbf{e}_3\}$, with its origin fixed at o, which rotates with the body with the same angular velocity ù. Hence, the angular momentum about o can be simplified as

$$\mathbf{H}_o = \mathbf{I}_o\boldsymbol{\omega}, \qquad [2.350]$$

where $\mathbf{I}_o$ is the constant matrix of moments of inertia about o whose components are along the $\{\mathbf{e}_1\mathbf{e}_2\mathbf{e}_3\}$ axes.

For a general motion of a rigid body, the equations governing both translation and rotation are

$$\sum \mathbf{F} = m\dot{\mathbf{v}}_c \qquad [2.351]$$

$$\sum \mathbf{M}_c = \dot{\mathbf{H}}_c \qquad [2.352]$$

Eq. [2.351] is the law of linear momentum for a rigid body, and Eq. [2.352] is the law of angular momentum for a rigid body. For general motion, we normally choose the reference coordinate system such that its origin is fixed at the $C.G.$ of the body and its coordinates rotate with the body. If the rigid body has angular velocity ω, then

$$\mathbf{H}_c = \mathbf{I}_c\boldsymbol{\omega}, \qquad [2.353]$$

where $\mathbf{I}_c$ is the constant matrix of moments of inertia about c whose components are along the reference coordinates.

For both cases of motion, if the reference coordinates (i.e., $\mathbf{e}_1\mathbf{e}_2\mathbf{e}_3$) are in directions such that they are the principal axes, then $\mathbf{H}_o$ and $\mathbf{H}_c$ in Eqs. [2.350] and [2.353] are simply

$$\mathbf{H}_o = \mathbf{I}_o\boldsymbol{\omega} = I_{1o}\omega_1\mathbf{e}_1 + I_{2o}\omega_2\mathbf{e}_2 + I_{3o}\omega_3\mathbf{e}_3 \qquad [2.354]$$

and

$$\mathbf{H}_c = \mathbf{I}_c\boldsymbol{\omega} = I_{1c}\omega_1\mathbf{e}_1 + I_{2c}\omega_2\mathbf{e}_2 + I_{3c}\omega_3\mathbf{e}_3. \qquad [2.355]$$

Hence, Eqs. [2.349] and [2.352] can be simplified further as

$$\sum \mathrm{M}_o = \begin{pmatrix} M_{1o} \\ M_{2o} \\ M_{3o} \end{pmatrix} = \begin{pmatrix} I_{1o}\dot{\omega}_1 \\ I_{2o}\dot{\omega}_2 \\ I_{3o}\dot{\omega}_3 \end{pmatrix} + \begin{bmatrix} 0 & -\omega_3 & \omega_2 \\ \omega_3 & 0 & -\omega_1 \\ -\omega_2 & \omega_1 & 0 \end{bmatrix} \begin{pmatrix} I_{1o}\omega_1 \\ I_{2o}\omega_2 \\ I_{3o}\omega_3 \end{pmatrix} \qquad [2.356]$$

and

$$\sum \mathrm{M}_c = \begin{pmatrix} M_{1c} \\ M_{2c} \\ M_{3c} \end{pmatrix} = \begin{pmatrix} I_{1c}\dot{\omega}_1 \\ I_{2c}\dot{\omega}_2 \\ I_{3c}\dot{\omega}_3 \end{pmatrix} + \begin{bmatrix} 0 & -\omega_3 & \omega_2 \\ \omega_3 & 0 & -\omega_1 \\ -\omega_2 & \omega_1 & 0 \end{bmatrix} \begin{pmatrix} I_{1c}\omega_1 \\ I_{2c}\omega_2 \\ I_{3c}\omega_3 \end{pmatrix} \qquad [2.357]$$

Alternatively, they can be written in scalar form as

$$\begin{aligned} M_{1o} &= I_{1o}\dot{\omega}_1 + (I_{3o} - I_{2o})\omega_2\omega_3 \\ M_{2o} &= I_{2o}\dot{\omega}_2 + (I_{1o} - I_{3o})\omega_1\omega_3 \\ M_{3o} &= I_{3o}\dot{\omega}_3 + (I_{2o} - I_{1o})\omega_1\omega_2 \end{aligned} \qquad [2.358]$$

$$\begin{aligned} M_{1c} &= I_{1c}\dot{\omega}_1 + (I_{3c} - I_{2c})\omega_2\omega_3 \\ M_{2c} &= I_{2c}\dot{\omega}_2 + (I_{1c} - I_{3c})\omega_1\omega_3 \\ M_{3c} &= I_{3c}\dot{\omega}_3 + (I_{2c} - I_{1c})\omega_1\omega_2 \end{aligned} \qquad [2.359]$$

Eqs. [2.358] and [2.359] are called *Euler's equations.*

2.8.4 Lagrange Equations

In addition to Euler equations, the equations of motion of rigid-body dynamics system can often be derived in a simple manner by the use of Lagrange equations. Lagrange equations are usually formulated in terms of generalized coordinates and the energy of a system.

The equations of motion of a rigid-body dynamics system can be formulated in a number of different coordinate systems. A rigid body performing free motion without any constraints in 3D space needs six variables or coordinates to describe its motion; i.e., three for translations and another three for rotations. The independent coordinates that used to describe the motion are defined as degrees of freedom. A rigid body performing free motion without any constraints has 6 degrees of freedom. For a system of *N*-rigid bodies having 3D free motion, the number of degrees of freedom is $6 \times N$.

If any two rigid bodies are connected to each other, the mechanism connecting the bodies is called *constraint*. The constraint imposes additional relative motion of one body with respect to another. With constraints, the motion of each rigid body in all six coordinates is not independent; hence, the number of degrees of freedom for each body is reduced to less than 6.

The constraints are usually represented by constraints equations, which describe the relative motions of any two connected bodies.

Generally speaking, the number of degrees of freedom of a multibody system is

$$M = 6 \times N - \sum C,$$

where M is the number of degrees of freedom, N is the number of rigid bodies, and $\sum C$ is the number of all constraint equations.

N independent coordinates are necessary to describe the motion of a system having n degrees of freedom. Any set of N independent coordinates is called *generalized coordinates,* usually designated as $q_1, q_2, \ldots, q_M$ (where M is the number of degrees of freedom).

The generalized coordinates may be lengths, angles, or any other set of numbers that uniquely define the configuration of the system at any time. They are also independent of the conditions of constraint.

The kinetic energy of a rigid body, as shown in Figure 2.37, is given by

$$T = \frac{1}{2}\int \mathbf{v}\boldsymbol{\cdot}\mathbf{v}dm, \qquad [2.360]$$

where $\mathbf{v}$ is the velocity of element dm.

From Figure 2.37, the velocity $\mathbf{v}$ of mass dm is

$$\mathbf{v} = \mathbf{v}_c + \boldsymbol{\omega} \times \boldsymbol{\rho}. \qquad [2.361]$$

Then,

$$\begin{aligned} T &= \frac{1}{2}\int (\mathbf{v}_c + \boldsymbol{\omega} \times \boldsymbol{\rho})\boldsymbol{\cdot}(\mathbf{v}_c + \boldsymbol{\omega} \times \boldsymbol{\rho})dm \\ &= \frac{1}{2}\int \mathbf{v}_c\boldsymbol{\cdot}\mathbf{v}_c dm + \int \mathbf{v}_c\boldsymbol{\cdot}\boldsymbol{\omega} \times \boldsymbol{\rho} dm + \frac{1}{2}\int (\boldsymbol{\omega} \times \boldsymbol{\rho})\boldsymbol{\cdot}(\boldsymbol{\omega} \times \boldsymbol{\rho})dm \\ &= \frac{1}{2}\mathbf{v}_c\boldsymbol{\cdot}\mathbf{v}_c \int dm + (\mathbf{v}_c \times \boldsymbol{\omega})\boldsymbol{\cdot} \int \boldsymbol{\rho} dm + \frac{1}{2}\int \boldsymbol{\omega}\boldsymbol{\cdot}(\boldsymbol{\rho} \times \boldsymbol{\omega} \times \boldsymbol{\rho})dm. \end{aligned} \qquad [2.362]$$

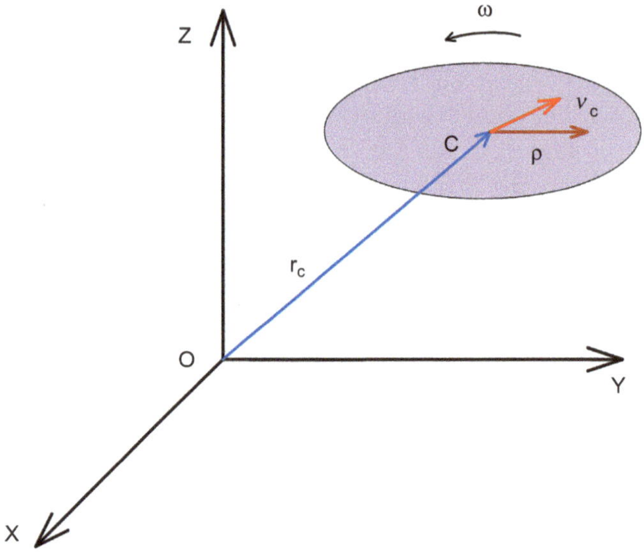

Figure 2.37 Schematic of a rigid body moving in a 3D coordinate system

The terms $\int \boldsymbol{\rho} dm = 0$ and $\int dm = m$

$$\frac{1}{2}\int \boldsymbol{\omega}\bullet(\boldsymbol{\rho}\times\boldsymbol{\omega}\times\boldsymbol{\rho})dm = \frac{1}{2}\boldsymbol{\omega}\bullet\int \boldsymbol{\rho}\times(\boldsymbol{\omega}\times\boldsymbol{\rho})dm,$$

where $\int \boldsymbol{\rho}\times(\boldsymbol{\omega}\times\boldsymbol{\rho})dm = \mathrm{H}_c$ is the angular momentum about the $C.G.$

Hence, the kinetic energy of a rigid body [Eq. [2.362]] is simply

$$T = \frac{1}{2}m\mathbf{v}_c\bullet\mathbf{v}_c + \frac{1}{2}\mathbf{H}_c\bullet\boldsymbol{\omega}. \qquad [2.363]$$

Eq. [2.363] can be put in the matrix form as

$$T = \frac{1}{2}m\mathbf{v}_c^T\mathbf{v}_c + \frac{1}{2}\boldsymbol{\omega}^T\mathbf{I}_c\boldsymbol{\omega}. \qquad [2.364]$$

Some active forces, such as gravitational forces and internal forces due to elastic deformation, can be represented by a gradient operator of a scalar potential energy function V as

$$\mathbf{F} = -\boldsymbol{\nabla}V(\mathrm{r}), \qquad [2.365]$$

where $\boldsymbol{\nabla}$ is the gradient operator given (in Cartesian coordinates) by

$$\boldsymbol{\nabla} = \frac{\partial}{\partial x}\mathbf{i} + \frac{\partial}{\partial y}\mathbf{j} + \frac{\partial}{\partial z}\mathbf{k}.$$

The forces in Eq. [2.365] are called *conservative forces,* and the function V is called *potential energy*. Consequently, work done by the conservative forces is independent to the path of motion and equal to the change of the potential energy ΔV. Therefore, the work done by the conservative force **F**, resulting in any path of motion from A to B, is

$$W = \int_A^B \mathbf{F} \cdot \delta \mathbf{r} \equiv V(\mathbf{r}_A) - V(\mathbf{r}_B). \qquad [2.366]$$

Examples of the conservative forces **F** and their corresponding potential energy V are as follows.

1. Gravity near the Earth's surface:

$$\mathbf{F} = -\mathrm{mg}\mathbf{e}_z, \mathrm{V} = \mathrm{mgz}. \qquad [2.367a]$$

2. Gravitational force:

$$\mathbf{F} = \frac{-\mathrm{GMm}}{\mathrm{r}^2}\mathrm{e}_\mathrm{r}, \mathrm{V} = \frac{-\mathrm{GMm}}{\mathrm{r}}. \qquad [2.367b]$$

3. Elastic spring force:

$$F = -kx, V = \frac{1}{2}kx^2. \qquad [2.367c]$$

4. Electric force between two charges:

$$F = \frac{Q_1 Q_2}{4\pi\varepsilon_0 r^2}, V = \frac{-Q_1 Q_2}{4\pi\varepsilon_0 r}. \qquad [2.367d]$$

where $\frac{1}{4\pi\varepsilon_0} = 8.99 \times 10^9$ in mks units.

When external forces act on the system, the configuration of the system changes. The new configuration of the system can be obtained by changing the generalized coordinates q_j by $\delta q_j j = 1, 2, \ldots, N$ where N denotes the number of generalized coordinates (or degrees of freedom) of the system. If U denotes the work done in changing the generalized coordinate q_j by the amount δq_j, the corresponding generalized force can be defined as

$$Q_j = \frac{U_j}{\delta q_j}, j = 1, 2, \ldots N, \qquad [2.368]$$

where Q_j is a force or moment when q_j is a linear or angular displacement.

The equations of motion of rigid-body dynamics system can be derived in a simple manner in terms of generalized coordinates by using Lagrange equations. If general coordinates are defined and the energy expressions are available, the equations of motion of the rigid-body dynamics can be derived by using Lagrange equations.

Lagrange equations can be stated, for an n degree of freedom system, as

$$\frac{d}{dt}\frac{\partial T}{\dot{\partial q_j}} - \frac{\partial T}{\partial q_j} + \frac{U_j}{\partial q_j} = Q_j^n \; j = 1, 2, \ldots N, \qquad [2.369]$$

where $\dot{q}_j$ is the generalized velocity and Q_j^n is the nonconservative generalized force corresponding to the generalized coordinate. The forces represented by may be dissipative (damping) forces or other external forces that are not derivable from a potential function.

Eq. [2.369] represents a system of n differential equations, one corresponding to each of the n generalized coordinates.

REFERENCES

[1] Rao, S.S., 1986. Mechanical Vibrations. Addison-Wesley, Reading, MA.
[2] Nayfeh, A.H., Mook, D.T., 1979. Nonlinear Oscillations. Wiley, New York.
[3] Shi, H.M., Sheng, G., Wu, Y., 1991. Mechanical Vibration System: Analysis, Measurement, Modeling, and Control. HUST Press, Wuhan, China.
[4] Shi, H.M., et al., 1990. Modeling of Dynamic Mechanical Systems. MEATU Press, Beijing.
[5] Soedel, W., 1993. Vibrations of Shells and Plates, second ed. Marcel Dekker, Inc., New York.
[6] Dukkipati, R.V., et al., 2008. Road Vehicle Dynamics. SAE International, Warrendale, PA.
[7] Greenwood, D.T., 1988. Principles of Dynamics. Prentice Hall.
[8] Ginsberg, J.H., 1998. Advanced Engineering Dynamics. Cambridge University Press, Cambridge, U.K.
[9] Moon, F.C., 1998. Applied Dynamics: With Applications to Multibody and Mechatronic Systems. John Wiley & Sons, New York.
[10] Jintanawan, T., 2005. Advanced Dynamics, 2103-617. Chulalongkorn University.
[11] Beer, F., et al., 2012. Vector Mechanics for Engineers: Dynamics, tenth ed. McGraw-Hill, New York.
[12] Jazar, R.N., 2011. Advanced Dynamics: Rigid Body, Multibody, and Aerospace Applications. Wiley, New York.

CHAPTER 3

Friction

3.1 INTRODUCTION

Friction is defined as tangential resistant force from a sliding interface in a dynamical system. To investigate friction, we need to characterize surface roughness, which has statistical properties. Friction has been demonstrated to be concerned with surface topography, and friction investigation has been conventionally attributed to the determination of the real area of contact and the understanding of the mechanism of mating contacts. For instance, surface physics explains friction in terms of the formation of adhesive junctions by interacting asperities and their breaking away by shearing, whereas continuum mechanics interpret friction by the interlocking and fracture of asperities.

To address the problem of real sliding, asperity contact is a very difficult task involving complex dynamics. The feasible approach is to assume the contact to be of a quasi-static nature to a certain extent. In many applications with relatively smooth surfaces, the deformation of contacting asperities can be assumed to be linear and elastic. For many problems, the contact has to be extended to nonelastic and nonlinear conditions and has to involve dynamics.

In this chapter, we present the fundamentals of contact and friction between two contact surfaces in the context of quasi-static states by assuming that normal motion is ignored. We focus on the mechanics of contact and friction by outlining the mechanical attributes of various friction processes in the context of problems of friction dynamics. For advanced topics related to tribological materials, chemistry, and physics, readers are recommended to refer to dedicated tribology references [1−30].

After the introduction, the second section briefs the mathematical description of surface roughness and introduces the fundamental contact mechanics including Hertz analytical solution of single asperity contact and treatment of multiple asperity contact between two solid surfaces. The third section presents the basic principles of friction of two contact surfaces. The basic physics of adhesion of two surfaces are presented. The friction of two dry contact surfaces is presented, and then wet friction or the friction of two

Friction Dynamics
ISBN 978-0-08-100285-8
http://dx.doi.org/10.1016/B978-0-08-100285-8.00003-1

liquid mediated surfaces is described. Finally, a variety of friction models are presented.

3.2 CONTACT BETWEEN TWO SOLID SURFACES

3.2.1 Description of Surfaces

In nature and in engineering practice, there are no topographically smooth surfaces. All solid surfaces, however smooth on a given scale, are comparatively rough on a smaller scale. In other words, solid surfaces are almost always comprised of random variations in surface height on some scale, as shown in Figure 3.1. This feature of the surface repeats down to the nanoscale. A surface that visually looks smooth is actually quite rough under microscale investigation. These surface height deviations are called *surface roughness*, which is characterized by asperities and valleys of various magnitudes. Surface roughness is associated with a length scale. The magnitude of the surface height deviations can be different at different length scales for the same surface. Usually, the surface fluctuations with a long wavelength are called the *waviness of a surface*, whereas those having a shorter wavelength are commonly associated with the roughness.

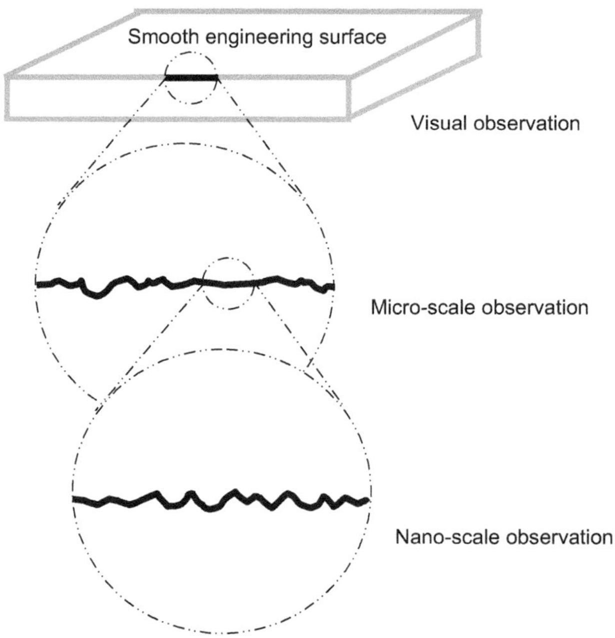

Figure 3.1 Roughness on different scales

Surface roughness is usually quantified using statistics. Statistic parameters like the distribution of the surface heights are used to describe the vertical deviations of surface height with respect to a reference plane. Most surface roughness is random, and the distribution of the surface heights usually follows a Gaussian distribution. The most common measures of surface roughness include the centerline-average R_a characterizing the average roughness and root-mean-square (rms) roughness R_q. We consider a surface height profile $z(x)$ over a length L, in which profile heights is measured from a mean line. The centerline-average is defined as

$$R_a = \frac{1}{L}\int_0^L |z - m|dx, \qquad [3.1]$$

in which the mean is given by

$$m = \frac{1}{L}\int_0^L z dx. \qquad [3.2]$$

The rms roughness is defined as

$$R_q = \sqrt{\frac{1}{L}\int_0^L z^2 dx}. \qquad [3.3]$$

The variation of the distribution is characterized by the standard deviation of the height of the surface from the centerline, σ, which is defined as

$$\sigma = \sqrt{\frac{1}{L}\int_0^L (z - m)^2 dx} = \sqrt{R_q^2 - m^2}. \qquad [3.4]$$

Skewness (dimensionless) of roughness is

$$\gamma = \frac{1}{L\sigma^3}\int_0^L (z - m)^3 dx. \qquad [3.5]$$

Kurtosis (dimensionless) of roughness is

$$\kappa = \frac{1}{L\sigma^4}\int_0^L (z - m)^4 dx. \qquad [3.6]$$

A more detailed statistical analysis is given next. If we denote probability density function by $p(z)$, the probability that the height of a particular point in the surface will lie between z and $z + dz$, then the probability that the height of a point on the surface is lower than z_0 is given by the cumulative probability function:

$$P(z \leq z_0) = \int_{-\infty}^{z_0} p(z)dz = P(z_0). \qquad [3.7]$$

It has been found that many real surfaces exhibit a height distribution that is close to the following normal or Gaussian probability function:

$$p(z) = \frac{1}{\sigma\sqrt{2\pi}} e^{\left[-\frac{(z-m)^2}{2\sigma^2}\right]}. \qquad [3.8]$$

3.2.2 Contact Mechanics of Two Solid Surfaces

When two elastic spheres are brought together in contact and then loaded, local deformation at the contact regime will occur. The contact point will enlarge into a surface of contact. Its area is a circle when the contacting bodies are spheres. The contact area size and the pressure between the spheres can be determined from their geometry and elastic properties. The solutions for deformation, area of contact, pressure distribution, and stresses at the contact area are quantified by conventional Hertz theory. The related assumption is made that the deformation is elastic, and the dimensions of the contact area are small, relative to the radii of curvature and to the overall dimensions of the bodies. Thus, the radii, though varying, may be taken as constant over the very small areas surrounding the contact area. The deflection integral is going to be derived for a plane surface.

Assume two contact spheres of different materials and radii R_1 and R_2. Figure 3.2 shows the spheres after loading, with the radius a of the contact area greatly exaggerated for illustration purposes.

The effective modulus involving Young's moduli and Poisson ratios, E_1, E_2, ν_1 and ν_2 is defined as follows:

$$E^* = \left[\frac{(1-\nu_1^2)}{E_1} + \frac{(1-\nu_2^2)}{E_2}\right]^{-1}. \qquad [3.9]$$

This expression describes the compliance of the system at a fixed contact area. Furthermore, Hertz theory gives the relationship between the load

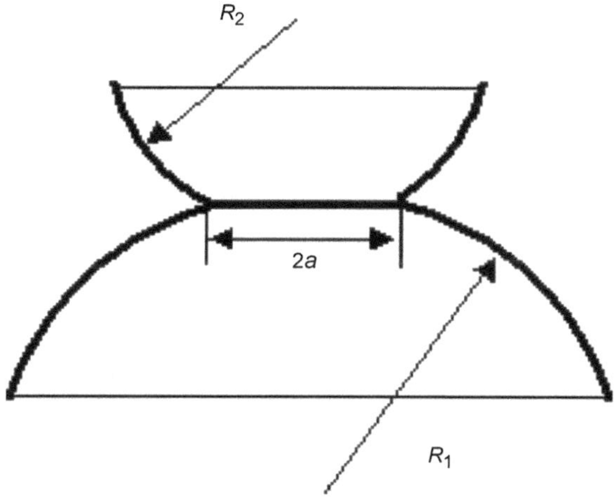

Figure 3.2 Contact of two spheres

and the deformation of the two contact spheres as the two surfaces are brought into contact under loading. It had been derived that the contact radius a is given by

$$a = \left(\frac{3RP}{4E^*}\right)^{1/3}, \quad [3.10]$$

in which P is the load, while R is a function of R_1 and R_2, called the *radii of curvature:*

$$R = \left[\frac{1}{R_1} + \frac{1}{R_2}\right]^{-1}.$$

Defining the contact displacement as

$$\delta = R_1 + R_2 - d, \quad [3.11]$$

where d is the distance between the centers of the two spheres. The compressive displacement or the normal approach is given by

$$\delta = \frac{a^2}{R}. \quad [3.12]$$

The contact area is given by

$$A = \pi a^2 = \pi R\delta. \quad [3.13]$$

Eq. [3.13] can be used for many other cases, such as one surface being concave. When one surface is concave, the contact area is on the inside of a surface. For this situation, the numerical value of its radius is to be taken as negative in all equations.

In this theory, the assumption is that the two surfaces in contact are frictionless, so shear stresses are assumed to be undeveloped at the interface. This assumption is often appropriate for very stiff materials where adhesive forces are relatively unimportant, but that is often not the case for softer materials, where adhesive forces play a very important role. In these cases, a full-friction boundary condition, where sliding of the two surfaces is not allowed, is often more appropriate. Hertz model is applicable to the nonadhesive cases of load and displacement. Actual values of load and displacement will generally differ from these values when the contacting surfaces adhere to one another. There are numerous studies [26–80] dedicated to extending the application of Hertz theory, some of which will be presented in following sections in this chapter.

All engineering surfaces are not smooth to a certain extent. The asperities on the surface of very compliant solids could be squashed flat by the contact pressure, and perfect contact likely can be obtained through the contact area. It is unusual to flatten rough surfaces by elastic or plastic deformation of the asperities, as shown in Figure 3.3.

In general, contact between solid surfaces is discontinuous and the real area of contact is a small fraction of the nominal contact area. The real contact surfaces are actually the asperity contact summation and are usually

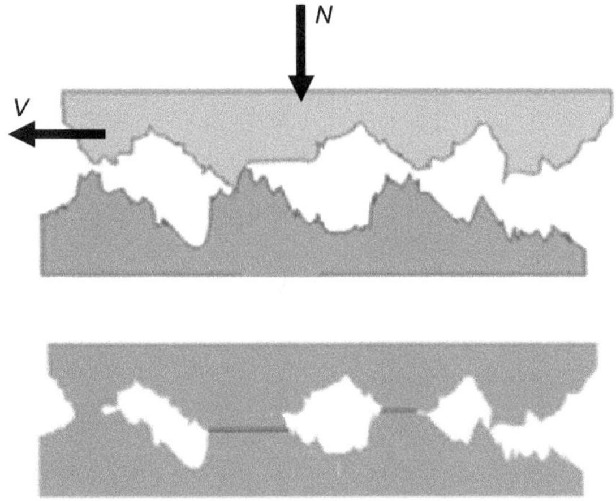

Figure 3.3 Contact surfaces (top: unflatten; bottom: flatten)

much smaller than the apparent surface. Next, we discuss the contact of identical asperities and random asperities.

Although all surfaces generally have roughness, certain simplifications can be achieved if the contact between a single rough surface and a perfectly smooth surface is considered. Moreover, the problem can be simplified further by introducing a theoretical model for the rough surface in which the asperities are considered as sphere arrays so that their elastic deformation characteristics may be defined by Hertz theory. It is further assumed that there is no interaction between separate asperities—that is, the displacement due to a load on one asperity does not affect the heights of the neighboring asperities.

Figure 3.4 shows a surface of unit nominal area consisting of an array of identical spherical asperities, all of the same height z with respect to a reference plane xx'. The number of asperities is given as N. As the smooth surface approaches the rough surface, due to the application of a load, the normal approach is given by $(z - d)$, where d is the current separation between the smooth surface and the reference plane.

Clearly, each asperity is deformed equally and carries the same load P_i, so that for N asperities per unit area, the total load P will be equal to NP_i. For each asperity, the load P_i and the area of contact A are known from applying Hertz theory. Thus, if R is the asperity radius, then

$$P_i = \frac{4E^*}{3} R^{1/2} \delta^{3/2} \qquad [3.14]$$

and $A_i = \pi R \delta$. The total load is

$$P = \frac{4E^* N A_i^{3/2}}{3\pi^{3/2} R}. \qquad [3.15]$$

The load can be related to the total contact area $A = NA_i$ by

$$P = \frac{4E^* A^{3/2}}{3\pi^{3/2} N^{1/2} R}. \qquad [3.16]$$

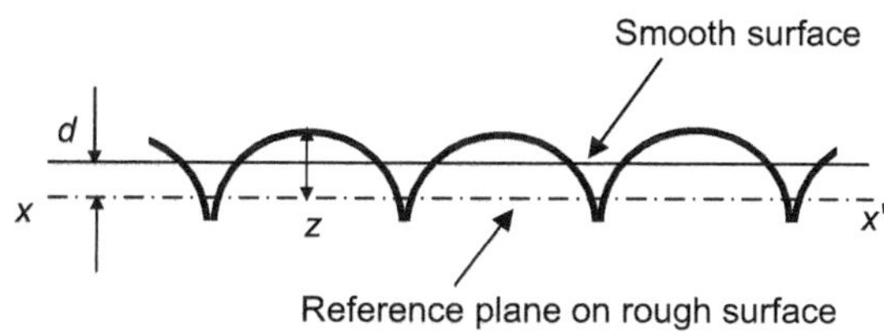

Figure 3.4 Contact of a sphere array with a smooth surface

This result indicates that the real area of contact is related to the two third power of the load, when the deformation is elastic.

It has been discussed that the asperities of real surfaces have different heights characterized by a probability distribution of their peak heights. Therefore, the simple surface models should be modified accordingly, and the analysis of its contact must now include probability estimation as to the number of the asperities in contact. If the separation between the smooth surface and the reference plane xx' is d, then there will be contact at any asperity whose height was originally greater than d, as shown in Figure 3.5.

The conventional statistical model dealing with contact surfaces with asperities is the Greenwood and Williamson model (GW model), which assumes that the summit of asperities on the surface is spherical, the asperity summits on two surfaces have a constant radius (R_{p1} and R_{p2}, respectively), and a composite radius (as defined here) is used to characterize the interface:

$$R_p = \left[\frac{1}{R_{p1}} + \frac{1}{R_{p2}}\right]^{-1}. \qquad [3.17]$$

The height is assumed to be random and with Gaussian distribution, with standard deviations σ_{p1}, σ_{p2} and an equivalent standard deviation is defined as follows:

$$\sigma_p = \left(\sigma_{p1}^2 + \sigma_{p2}^2\right)^{1/2}. \qquad [3.18]$$

For elastic contact in static conditions or in dynamic conditions without shear stress, the GW model quantifies the relationship among apparent pressure p_a, apparent contact area A_a, mean real pressure p_r, real contact area A_r, and number of contact number n as a function of separation d. Then the problem reduces to a plate in contact with asperity. If $p(z)$ is the probability density of the asperity peak height distribution, then the

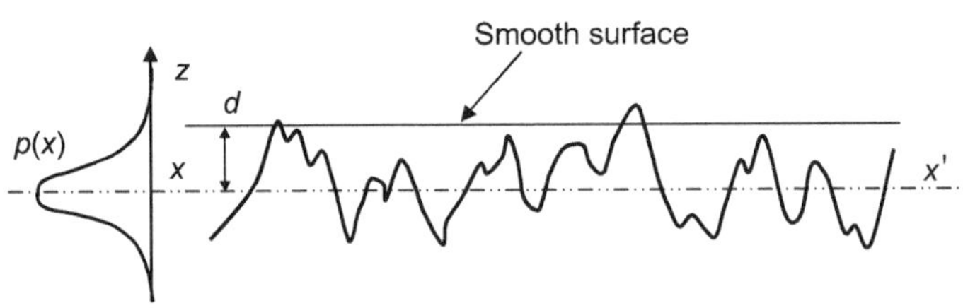

Figure 3.5 Rough surface

probability that a particular asperity has a height between z and $z + dz$ above the reference plane will be $P(z)$. Thus, the probability of contact for any asperity of height z is

$$P(z > d) = \int_d^\infty p(z)dz. \qquad [3.19]$$

If we consider a unit nominal area of the surface containing asperities and assume that the total asperity number is N, then the number of contacts n will be given by

$$n = N\int_d^\infty p(z)dz. \qquad [3.20]$$

Since the normal approach is $(z - d)$ for any asperity, the total area of contact and the expected load will be given by

$$A_r = \pi NR_p \int_d^\infty (z-d)p(z)dz \qquad [3.21]$$

$$P = A_r p_r = A_a p_a = \frac{4}{3}NE^* R_p^{1/2} \int_d^\infty (z-d)^{3/2} p(z)dz. \qquad [3.22]$$

It is convenient to use nondimensional variables. According to [3.22], the derived relationship can be written as

$$\frac{p_a}{(\eta R_p \sigma_p)E^*(\sigma_p/R_p)^{1/2}} = \frac{4}{3}F_{3/2}(D) \qquad [3.23]$$

$$\frac{p_r}{E^*(\sigma_p/R_p)^{1/2}} = \frac{4}{3\pi}F_{3/2}(D)/F_1(D) \qquad [3.24]$$

$$\frac{A_r E^*(\sigma_p/R_p)^{1/2}}{p_a A_a} = \frac{3\pi}{4}F_1(D)/F_{3/2}(D) \qquad [3.25]$$

$$(A_r/n)R_p\sigma_p = \pi F_1(D)/F_0(D) \qquad [3.26]$$

$$\frac{nR_p\sigma_p E^*(\sigma_p/R_p)^{1/2}}{p_a A_a} = 3F_0(D)/4F_{3/2}(D) \qquad [3.27]$$

Using these equations, one may evaluate the total real area, load, and number of contact spots for any given height distribution, in which $D = d/\sigma_p$ is the dimensionless separation; $\eta = N/A_a$ is the density of asperity summits per unit area on a surface with smaller density; and $F_m(D)$ is a function defined as

$$F_m(D) = \int_D^{\infty} (s - D)^m p^*(s) ds, \qquad [3.28]$$

where $p^*(s)$ is the standardized peak-height probability density function in which the height distribution has been scaled to make its standard deviation unit. A usually used approximation is given by

$$A_r \approx \frac{3.2 p_a A_a}{E^* (\sigma_p / R_p)^{1/2}}, \qquad [3.29]$$

from which the real area of contact is proportional to the load. The relationship between the real area of contact and the load will depend on both the mode of deformation and the distribution of the surface profile. More sophisticated analysis is based on fractal roughness description and continuum plasticity theory. When the asperities deform elastically, depending on the deformation mode within the contact, its real area can be estimated from

$$A_r = c\left(\frac{P}{E^*}\right)^n, \qquad [3.30]$$

in which $2/3 < n < 1$, the linearity between the load and the real area of contact occurs only where the distribution approaches an exponential form, and this is very often true for many practical engineering surfaces.

When two surfaces of high elastic modulus materials are pressed together, they come into contact only at the tips of the asperities. The total contact area is A_r and $A_r << A_a$, the nominal area of contact. For this reason, the stress on the asperities is generally large. When the stress exceeds yield pressure, the plastic deformation occurs.

When analyzing the real contact between two engineering surfaces, it is usually assumed that they are covered with asperities having random height distributions and deforming elastically or plastically under normal loads. The sum of all microcontacts created by individual asperities constitutes the real area of contact, which is usually only a tiny fraction of the apparent geometrical area of contact. There are two groups of properties (namely, deformation properties of the materials in contact and surface topography characteristics), which define the magnitude of the real contact area under a given normal load P. Generally, the contact behavior of solid in contact is determined by a value defined as plasticity index, ψ:

$$\psi = \left(\frac{E^*}{H}\right)\left(\frac{\sigma_p}{R_p}\right)^{1/2}, \quad [3.31]$$

in which E^* is the composite or effective elastic modulus, H is hardness, and σ_p and R_p are the composite standard deviations and composite radii of the summits (surface heights).

If the plasticity index $\psi < 0.6$, then the deformation is largely elastic. When $\psi > 1.0$, the predominant deformation mode within the contact zone is plastic deformation. The index depends on both the mechanical properties and the surface roughness of the contact surfaces.

The mechanical property ratio E^*/H and the surface roughness determine the extent of plasticity in the contact region. For most metals and ceramics, $E^*/H > 100$. For polymers, it is on the order of 10. Thus, the plasticity index for polymer is on the order of one-tenth of that of metal and ceramics, and therefore, the contact is primarily elastic except for very rough surfaces. On the other hand, for surfaces with very little surface roughness, like the diamondlike carbon-coated surfaces of magnetic disks, the contact is usually elastic.

For plastic contact, we have the following approximation:

$$A_r = c\frac{P}{H}, \quad [3.32]$$

where c is the proportionality constant. When the asperities deform plastically, the load is linearly related to the real area of contact for any distribution of asperity heights.

The introduction of an additional tangential load produces a phenomenon called *junction growth*, which is responsible for a significant increase in the asperity contact areas. The magnitude of the junction growth of metallic contact can be estimated from the expression

$$A = A_{p0}\left[1 + \alpha\left(\frac{F_f}{P}\right)^2\right]^{1/2}, \quad [3.33]$$

in which A_{p0} is the real area of contact without any shear stress, α is a constant, F_f is friction force, and P is normal load.

3.3 FRICTION BETWEEN TWO SOLID SURFACES

3.3.1 Adhesion

3.3.1.1 Solid–Solid Adhesion

When bringing two solid surfaces into contact, adhesion or bonding across the interface develops. This leads to the adhesive force that is perpendicular to the surface, in addition to the applied normal force.

Adhesion could occur in both solid–solid contact and liquid-involved solid–solid contact. Two clean solid surfaces tend to create a strong bond, whereas contaminated or boundary film-covered surfaces tend to yield weak bonds.

The adhesive junctions of asperities on solid–solid contact are caused by interatomic and intermolecular force attractions. Generally, the adhesion could be chemical or physical. The former includes covalent bonds, ionic or electrostatic bonds, metallic bonds, and hydrogen bonds, the physical interaction involves van der Waals bonds.

The adhesion is generally proportional to the normal force. This is because the applied normal force increases the real contact area that promotes the bonds. On the other hand, the real contact area also increases as a result of interatomic attraction. Moreover, a shear force in addition to the normal force usually increases the adhesive force, as the shear effect in addition to the compressive effect tends to increase the real contact area.

There are many references on the solid–solid adhesion friction [25–115], and some recent developments incorporate the results of numerical analysis and experimental investigation for the elastic–plastic contact, adhesion, and sliding asperity in modern statistical representation of roughness. In the following section, we describe some of the basic theories that cover interfacial forces such as electrostatic forces and van der Waals forces.

Electrostatic bonding occurs when positively charged cations and negatively charged anions interact. Electrostatic attractive forces across the interface can arise from a difference in work functions or from electrostatic charging of opposed surfaces. Differences in the work function lead to the formation of an electrical double layer by a net transfer of electrons from one surface to the other.

With small separations, the electrostatic pressure between flat surfaces is generally less than the van der Waals pressure. It can be assumed that all objects are free of charge before contact begins. However, after contacting the objects, contact electrification and tribo-electrification occur, and forces due to these charges could occur.

The free energy for electrostatic interaction between two charged atoms or ions with a distance of x is given by

$$E_s = \frac{z_1 z_2 e^2}{4\pi\varepsilon_0 \varepsilon x} \quad [3.34]$$

The electrostatic force is obtained by differentiating the energy with respect to distance x as

$$F_s = -\frac{z_1 z_2 e^2}{4\pi\varepsilon_0 \varepsilon x^2} \quad [3.35]$$

in which e is the charge of a single electron, $e = 1.6 \times 10^{-19}$ C, z_1 and z_2 are the ionic valences, ε_0 is the permittivity of a vacuum 8.854×10^{-12} C^2/Nm^2, and ε is the dielectric constant of the medium, 1.00059 in air at 1 atm.

Considering electrostatic force between two parallel plates with surface charge density of σ on one of the plates, the electric field E is given by

$$E = \frac{\sigma}{\varepsilon_0 \varepsilon}. \quad [3.36]$$

Then the electrostatic force per unit area is given by

$$F_s = -\frac{1}{2}\varepsilon\varepsilon_0 E^2. \quad [3.37]$$

If the potential difference between the plates is V, then the electric field is $E = V/x$.

van der Waals force always occurs between molecules and is much less than the binding force between atoms. The van der Waals force is effective from comparatively large separations (up to 10 nm) down to interatomic spacing (about 0.2 nm). The van der Waals dispersion forces between two bodies are caused by mutual electric interaction of the induced dipoles in the two bodies. Dispersion forces generally dominate over orientation and induction forces except for strongly polar molecules. Here, van der Waals forces exist for every material in every environmental condition and depend on the object geometry, material type, and separation distance.

When two atoms are brought sufficiently close together, they start to experience the intermolecular forces interacting with each other. At the beginning, it is an attraction force, and its strength increases with decreasing distance until a maximum point is reached, and then it decreases with decreasing distance. When the distance is reduced further, the force becomes repulsive and increasingly stronger. This is reflected in Figure 3.6, which shows the potential energy between two atoms as a function of their distance. One usual expression used to describe this potential is the Lennard-Jones potential, in which the attractive van der Waals potential is modeled as an inverse sixth-power term and the repulsive potential is modeled as an inverse twelfth-power term,

$$W = -\frac{C}{r^6} + \frac{D}{r^{12}}, \qquad [3.38]$$

in which $C = 10^{-77}$ Jm^6 and $D = 10^{-134}$ Jm^{12} are two constants for atoms in a vacuum.

For interfaces with spacing of a few nanometers, the repulsive potential term can be ignored, and we obtain the pure van der Waals potential:

$$W = -\frac{C}{r^6}. \qquad [3.39]$$

Eq. [3.38] expresses the potential between two atoms, and it can be integrated over an infinitely long and infinitely deep half-space to obtain the potential between an atom and an infinite plate:

$$W = -\frac{\pi C \rho_1}{6h^6} + \frac{\pi D \rho_1}{45h^9}, \qquad [3.40]$$

where ρ_1 is the number density of atoms in the infinite plate and h is the distance between the atom and the plate. Eq. [3.40] can be further

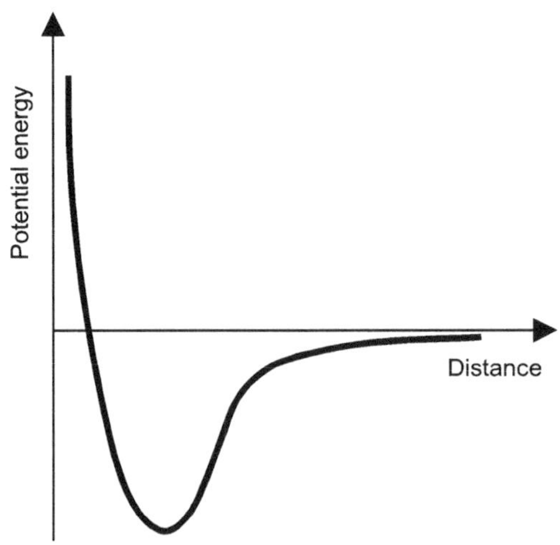

Figure 3.6 Potential energy of molecules

integrated over a volume of material to get the potential between an amount of material and an infinite plate.

To obtain the intermolecular force between an amount of material with surface S and the plate, we need to differentiate the integrated potential in a direction perpendicular to the plate. After that, the intermolecular force between each S of material and the plate can be written as

$$F_v = \frac{dW_v}{dz} = \frac{A}{6\pi}\iint_S \frac{dxdy}{h^3} - \frac{B}{45\pi}\iint_S \frac{dxdy}{h^9}, \qquad [3.41]$$

where A is the Hamaker constant and B is the constant related to the repulsive term. The first term on the right side of Eq. [3.41] is the attractive van der Waals force, and the second term is the repulsive intermolecular force. The attractive and repulsive portions of the force have different acting ranges. The attractive van der Waals force has a much longer acting range than the repulsive portion.

Based on Eq. [3.41], the intermolecular force per unit area can be simplified as

$$p_v = \frac{A}{6\pi h^3} - \frac{B}{45\pi h^9}, \qquad [3.42]$$

in which A is the Hamaker constant and B is the repulsive term-related constant.

The two solid surfaces first experience the attraction force when the distance between them is less than about 10 nm. The strength of the attractive force increases with the reduction of the spacing until $h_a = (2B/5A)^{1/6}$, reaching a maximum attractive value. When the spacing is further reduced, the short-range repulsive force becomes effective, and at the end, it makes the intermolecular force zero at $h_0 = (2B/15A)^{1/6}$. Below this threshold value, the repulsive force will be dominant, and then other interactions like the rearrangement of the surface could happen due to the rapid increase of the repulsive force.

As an approximation to solids, assuming $A = 10^{-19}$ J, and $B = 10^{-76}$ Jm^6, we get $h_a \approx 0.3$ nm. An approximation of the interaction energy due to van der Waals forces per unit area between two parallel plates in the nonretarded regime ($h < 20$ nm) can be given only by attractive term:

$$E_v = -\frac{A}{12\pi h^2}$$

and the intermolecular force per unit area between the plates can be approximated by

$$p_v = -\frac{A}{6\pi h^3}, \quad [3.43]$$

in which A is the Hamaker constant. For most solids and liquids, the Hamaker constant lies in the range of $0.4–4 \times 10^{-19}$ J.

An approximation to the precise calculation of the van der Waals force is to apply the concept of free surface energy [3,4,7]. Assuming that two materials have free surface energies per unit area of γ_1, γ_2, the energy of adhesion per unit area is defined as

$$W_a = \Delta\gamma = \gamma_1 + \gamma_2 - \gamma_{12}, \quad [3.44]$$

in which $\Delta\gamma$ equals a reduction in the surface energy of the interface per unit area. It is negative and represents the energy that needs to be applied to separate a unit-bonded interface. Furthermore, it can be approximated as $\Delta\gamma = C(\gamma_1 + \gamma_2)$, in which C is the compatibility parameter for the two materials and always falls between 1 and 0.

In most cases, a kind of assumption can be made for microasperities, but an exact solution would be more proper for the nanoasperities. The intermolecular attractive forces depend on atomic spacing and the corresponding surface energies of materials. For extra smooth surfaces, the adhesion can be very large, leading to the virtual welding of one surface to another. For rough or particle-involved interfaces, its effects can be disregarded.

For sufficiently small contacts, the normal adhesion forces between the surfaces affect the contact conditions. Various adhesion models, between an elastic sphere and a flat surface, have been introduced to extend the Hertz model. The model by Johnson, Kendall, and Roberts (JKR model) [27] assumes that the attractive intermolecular surface forces cause elastic deformation beyond that predicted by Hertz theory and produce a subsequent increase of the contact area. This model also assumes that the attractive forces are confined to the contact area and are zero outside the contact area.

When the surface adhesion force is considered, the stress distribution between two surfaces is tensile at the edge of the contact area and remains compressive in the center. The JKR model is based on these considerations.

Considering the energy release rate, w, which has the units of a surface energy (energy/area), and describes the amount of energy that is needed to decrease the contact area, A, by a unit amount:

$$w = \frac{\partial}{\partial A}(U_E + U_M), \tag{3.45}$$

where U_E is the elastic energy of the system and U_M is the mechanical potential energy associated with the applied load. In terms of the current values of P, R, and a, the contact radius was found to be

$$a = \left[\frac{3R}{4E^*}\left(P + 3\pi Rw + \sqrt{6\pi RPw + (3\pi Rw)^2}\right)\right]^{\frac{1}{3}}, \tag{3.46}$$

where w is the energy per unit contact area, equal to the thermodynamic work of adhesion, and E^* is the equivalent elastic modulus of the two spheres. As a result of the surface forces, the contact size is larger than the value in the Hertz model and will be finite for no external force.

The contact force P_1 between the surfaces is bigger than normal load P:

$$P_1 = P + 3\pi Rw + \sqrt{6\pi RwP + (3\pi Rw)^2}. \tag{3.47}$$

The corresponding contact displacement is

$$\delta = \frac{(3P_1)^{\frac{2}{3}}}{(4E^*)^{\frac{2}{3}}R^{\frac{1}{3}}} - \frac{1}{2E^*}\left(\frac{4E^*}{3RP_1}\right)^{\frac{1}{3}}(P_1 - P), \tag{3.48}$$

which depends nonlinearly on the external force P. The JKR model is based on the analysis at equilibrium and has been verified by static and quasi-static experiments.

Another model of this type assumes that the contact area does not change due to the attractive surface forces and remains the same as in Hertz theory; thereby, the attractive forces are assumed to act only outside the contact area. More models were developed to fill the gap between the above two models.

3.3.1.2 Liquid-Mediated Adhesion

A liquid-involved interface has extra adhesion in addition to the previously discussed adhesions acting on dry solids [116–162]. Liquid-mediated adhesion generally consists of meniscus force and viscous force.

Viscous force is rate-dependent and is only significant for high-viscous liquid. Meniscus force is due to surface tension. The presence of the capillary condensation and liquid film could significantly increase the adhesion between surfaces.

In particular, even for a dry interface, moisture (or other liquid vapor) could condense onto the interface as a bulky liquid, and exhibit annular-shaped capillary condensation around the contact asperities. The schematic of liquid condensation in the interface is shown in Figure 3.7. This phenomenon is very common in engineering interfaces, especially for exposure to ambient environments where water moisture is unavoidable.

A thin liquid layer between two solid plates can work as an adhesive. If the contact angle between liquid and solid is less than 90 degrees, as shown in Figure 3.8, the pressure inside the liquid will be lower than outside and a net attractive force between the plates will appear.

Assume that θ is the contact angle between the liquid and the solid in air, h is liquid thickness, A is the wet area, and F is the external force. The

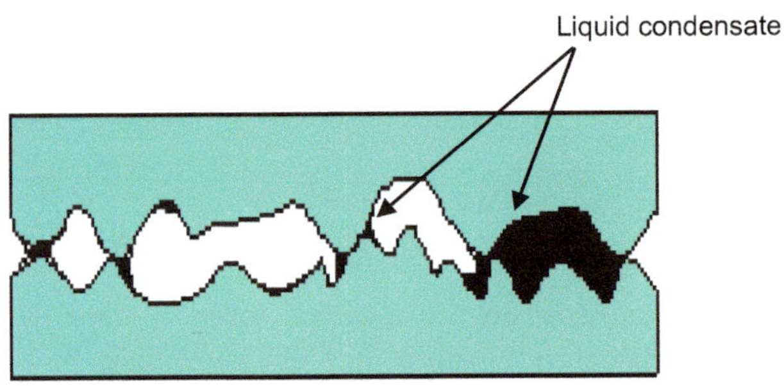

Figure 3.7 Liquid condensation in the interface

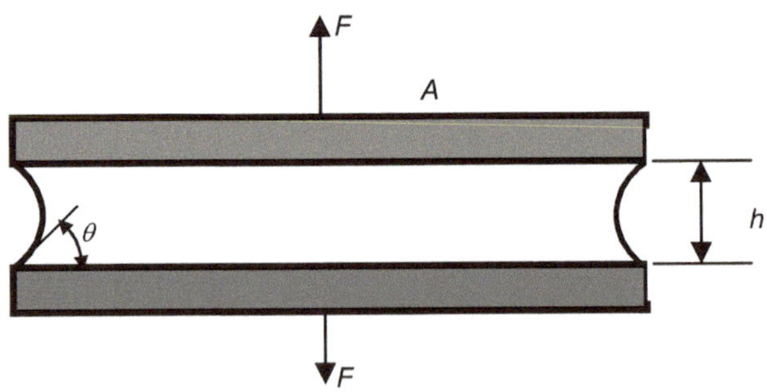

Figure 3.8 A thin-layer liquid working as an adhesive between two plates

pressure difference Δp_{la} at the liquid–air interface is given by the Laplace equation:

$$\Delta p_{la} = \frac{\gamma_{la}}{r}, \quad [3.49]$$

where γ_{la} is the surface tension of the liquid–air interface, and r is the radius of curvature of the meniscus (negative if concave).

Considering that the liquid is between the plates and the liquid contacts the solid at the fixed contact angle, from simple geometry, it follows that

$$r = -\frac{h}{2\cos\theta}. \quad [3.50]$$

In equilibrium, an external force F separating the plates must be applied to counterbalance the capillary pressure forces:

$$F = -\Delta p_{la}A = \frac{2A\gamma_{la}\cos\theta}{h}, \quad [3.51]$$

where A is the wet area. It should be noted that a positive force F corresponds to a negative Laplace pressure. The pressure inside the liquid is lower than outside and the plates are pushed together by pressure forces.

Eq. [3.51] can also be used to evaluate the meniscus between a solid sphere contacting a plate with liquid meniscus.

For stiction calculations, it is convenient to calculate the surface energy stored at the interface that is bridged by a drop of liquid. Considering a drop of liquid placed on a solid surrounded by air in equilibrium, the contact angle between the liquid and the solid is determined by the balance between the surface tensions of the three interfaces. This balance is expressed by Young's equation:

$$\gamma_{sa} = \gamma_{sl} + \gamma_{la}\cos\theta, 0 < \theta < \pi, \quad [3.52]$$

where γ_{sa} is the surface tension of the solid–air interface and γ_{sl} is the surface tension of the solid–liquid interface. Young's equation is also valid for configurations other than the typical one. The contact angle is the same on a curved or irregularly shaped surface.

Consider a general case of solid sphere proximity to a surface with a separation of D forming a meniscus. The surface has continuous liquid film h, as shown in Figure 3.9, the meniscus is given by

$$F = -2\pi R\gamma(1 + \cos\theta). \quad [3.53]$$

When surface has no liquid film, or $h = 0$, the meniscus is given by

$$F = -4\pi R\gamma\cos\theta/(1 + D/(S - D)). \quad [3.54]$$

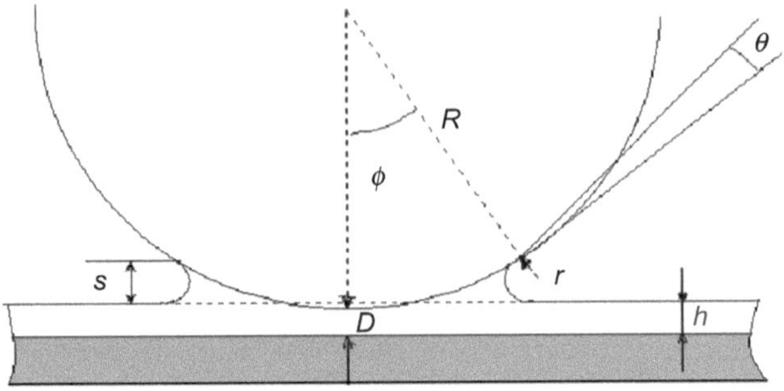

Figure 3.9 Schematic of an isolated meniscus in the presence of a liquid film

In a real interface, stiction can show great dependence on the relative humidity of air. The friction measurements of silicon and ceramics show that the static friction coefficient is strongly dependent on relative humidity. This is caused by the meniscus force due to capillary condensation. Both the normal and horizontal components of the meniscus force could contribute to the friction. Liquids that are wet or have a small contact angle on surfaces will spontaneously condense into cracks, pores, and small gaps surrounding the points of contact between the surfaces. At equilibrium, the meniscus curvature is characterized by the Kelvin radius:

$$r_K = \frac{V\gamma_{la}}{RT\ln(p/p_s)}, \qquad [3.55]$$

where V is the molar volume ($1.804 \times 10^{-5}\text{m}^3/\text{mol}$ at 20°C), p is the vapor pressure, and p_s is the saturation vapor pressure. At room temperature, $V\gamma_{la}/RT = 0.54$ nm for water. The meniscus curvature strongly depends on the relative vapor pressure p/p_s. At 100% relative humidity, a water film can form all over the surface. The amount of condensed liquid in thermodynamic equilibrium is determined by both the Kelvin radius and the contact angle. The meniscus curvatures are equal to the Kelvin radius, and the contact angles satisfy Young's equation.

Models for the adhesion force due to capillary condensed liquid have been fully developed in hard disk drive tribology.

The kinetic meniscus force can be estimated as follows [4]. When a surface approaches another one with the liquid mediation, there is a kinetic process leading to the final equilibrium. During this process, the flow of liquid could increase the meniscus. The driving pressure can be represented as

$$p = -\frac{\gamma_{la}}{r} - \frac{A}{6\pi h^3}. \qquad [3.56]$$

The first term is the result of the Laplace equation, the second term originates from the van der Waals effect.

Assuming the volume of liquid to be v, its flow rate is

$$\frac{dv}{dt} = -\frac{2\pi rh^3}{3\eta_l}\frac{\partial}{\partial r}\left[\frac{\gamma_{la}}{r} + \frac{A}{6\pi h^3}\right]. \qquad [3.57]$$

Then the radius of the projected region of meniscus, the neck radius $x(t)$, can be derived as a function of time with respect to its equilibrium value $(x)_{eq}$. Finally, the meniscus force can be given by

$$F(t) = 2\pi R\gamma(1 + \cos\theta)\left[\frac{x(t)}{(x)_{eq}}\right]^2. \qquad [3.58]$$

A drop of lubricant filling the gap between the two surfaces will form a toe-dipping regime and thus initiate the stiction. A typical example is the meniscus developed in the slider-disk interface of the hard disk drive. With the increase of the contact time, in addition to the solid—solid asperity microdisplacement, the microflow and the diffusion of the lubricant will redistribute the interface lubricant to form an equilibrium state, and therefore form a pillbox regime accompanying the micro-descent process. This is schematically shown in Figure 3.10. Many efforts to quantify the meniscus force have been reported.

Another meniscus model is given in the following. As shown in Figure 3.9, the hemispherical surface of an asperity with radius R on the disk is nearly in contact with (distance D) the flat surface of the slider. The lubricant buildup is with a transient thickness h on the slider flat surface.

The meniscus force can be derived as

$$F_m = 2\pi R^2\gamma(1 + \cos\theta)\Phi\frac{d\Phi}{dD}, \qquad [3.59]$$

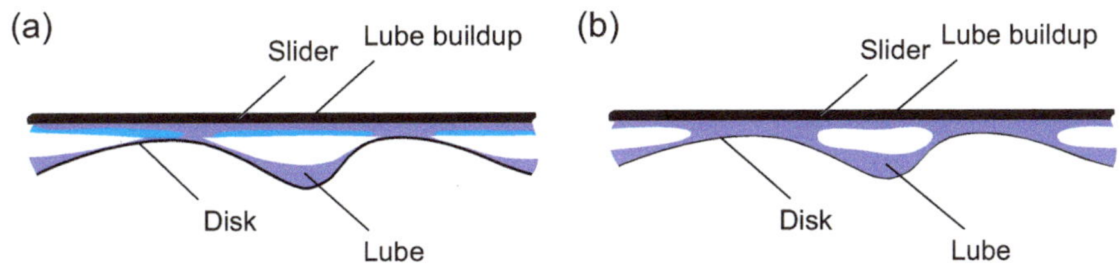

Figure 3.10 Schematic of rough asperities on a disk in contact with a slider surface; (a) short-term stiction: toe-dipping regime; b) long-term stiction: pillbox regime

where γ is the surface tension of the lubricant, θ is the contact angle formed by the lubricant, and D is the distance between the flat plane and the hemispherical surface of the asperity.

In addition to the meniscus force, the viscous force contributes to the adhesive force by the following relationship:

$$F_v = \frac{\beta \eta_l}{t_s}, \quad [3.60]$$

in which β is a proportional constant, η_l is the dynamic viscosity of the liquid, and t_s is the time needed to separate the two surfaces. Obviously, t_s is inversely proportional to the breakaway acceleration and velocity of interface.

3.3.2 Dry Friction

3.3.2.1 Friction Mechanisms

When there is contact sliding between two bodies under a normal load P, a resistant force always exists; this force is called *frictional force* and is represented by F. Frictional force always acts on a component in a direction opposite to the motion of the component. Two basic facts about friction have been experimentally established: (i) the frictional force is a function of the normal load on the contact, $F = \mu P$, where μ is the coefficient of friction and P is the normal load; and (ii) the frictional force is independent of a nominal area of contact.

These two statements constitute what are known as the laws of sliding friction under dry conditions. Sliding friction has been studied for a long time. But there is still no simple model that could be used by engineers to calculate the frictional force for a given pair of materials in contact. It is now widely accepted that friction results from complex interactions between contacting bodies. This includes the adhesion at the points of contact due to the molecular interaction, the effects of surface asperity deformation, and the plastic deformation of a weaker material. Figure 3.11 shows the schematic of normal contact and slope contact of asperities. The former tends to develop adhesions, while the latter can develop both adhesion and asperity deformation in the horizontal direction. A number of factors, such as the mechanical and physical-chemical properties of the materials in contact, surface topography, operational and environment conditions, and system dynamics determine the dominant components in the friction process. At a

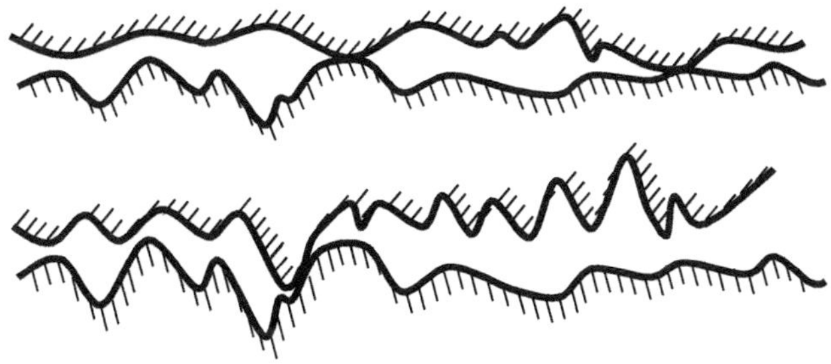

Figure 3.11 The normal contact and the slope contact of asperities

fundamental level, there are several major factors that determine the friction of dry solids:

(i) Real area of contact
(ii) Shear strength of the adhesive junctions formed at the points of real contact
(iii) Breakaway of these junctions during relative motion
(iv) System dynamics

Friction properties of a given material/component are not its intrinsic properties, they depend on many factors related to a specific applications. Quantitative values for friction in the forms of friction coefficient depend on the following basic groups of parameters:

(i) Friction pair—i.e., the components and their relevant properties of mechanical, physical, and chemical aspects and geometry
(ii) Interface—i.e., the formation and breakaway of contact junctions on multiple scales
(iii) Operational variables—i.e., load (pressure), speed, and time
(iv) Environmental variables—i.e., temperature and humidity
(v) System dynamics, the mutual interaction of the system's components, and their time-dependent variables (noting that the interface is variable depending on time due to wear and layer-forming)

Moreover, the coefficient of friction is complicated by other factors, such as scales. For instance, for a given interface, the coefficient of friction could be inversely proportional to the normal load. However, when load downs to the micro- to nano-Newton level, the coefficient of friction could have different features.

Friction is always associated with energy dissipation and conversion.

Mechanical energy is transformed within the real area of contact, mainly through elastic deformation and hysteresis, plastic deformation, plough, and adhesion.

Dissipation of mechanical energy takes place mainly through thermal dissipation (heat); storage within the bulk of the body by the generation of defects and cracks; and emissions like acoustic emission and thermal generation.

For typical contact interface, the friction is given by

$$F = A[\alpha\tau_a + (1 - \alpha)\tau_l], \tag{3.61}$$

in which τ_a, τ_l are the average shear stresses of the dry contact and of the liquid film, respectively; and α is the fraction of dry area. The average shear stress of the liquid film is contributed by both viscosity of the liquid and the meniscus effect:

$$\tau_l = \frac{\eta_l V}{h} + \tau_m, \tag{3.62}$$

in which η_l is the dynamic viscosity of the liquid, V is the relative sliding velocity, h is the liquid film thickness, and τ_m is the shear stress due to the meniscus effect. In boundary lubrication conditions or when the interfaces are exposed to humid environments, the meniscus and even the viscous effects could be significant.

In the next section, we discuss dry friction due to adhesion. We have shown that the area of real contact between two bodies is usually a very small fraction of the apparent geometrical area of contact. However, the friction force is determined almost entirely by the area of real contact.

Considering two dry surfaces, the upper body moves parallel to the lower body at velocity v. The friction force is given by

$$F = \tau_a A_r, \tag{3.63}$$

where A_r is the real contact area and τ_a is shear stress.

One of the most important components of friction originates from the formation and breakaway of interfacial adhesive bonds. Extensive theoretical and experimental studies have explained the nature of adhesive interaction. The main emphasis is on the electronic structure of the bodies in frictional contact. From a theoretical point of view, attractive forces within the contact zone include all those forces that contribute to the cohesive strength of a solid, such as metallic, covalent, and ionic short-range forces, as well as secondary van der Waals bonds, which are classified as long-range forces. The interfacial adhesion is as natural as the cohesion that determines the bulk strength of materials.

The coefficient of friction (cof) due to adhesion can be approximated as the ratio of the interfacial shear strength of the adhesive junctions to the yield strength of the asperity material:

$$\mu_a = \frac{F_a}{P} = \frac{A_r \tau_a}{P} \approx \frac{\tau_a}{p_r}, \quad [3.64]$$

in which p_r is the mean real pressure. Substituting A_r from Eq. [3.29], we obtain

$$\mu_a \approx \frac{3.2\tau_a}{E^*(\sigma_p/R_p)^{1/2}}. \quad [3.65]$$

For single asperity cases, it can be approximated as

$$\mu_a \propto \frac{1}{P^{1/3}}. \quad [3.66]$$

For plastic contact, it is

$$\mu_a \approx \frac{\tau_a}{H}. \quad [3.67]$$

In an elastic contact case, like a diamondlike coated disk against a ceramic slider, μ_a decreases with an increase in roughness. In the plastic contact case, μ_a could be independent of roughness in a moderate range. It can be seen that μ_a tends to be high at very smooth surfaces due to the increase in the real area of contact, and also tends to be high at very rough surfaces due to interlocking.

Evaluation of μ_a needs to quantify τ_a, which can be estimated by a limited analysis. Generally, the shear strength τ_a cannot substantially exceed the bulk shear strength for plastic contacts.

For most engineering materials, this ratio τ_a/p_r is on the order of 0.2, and therefore, the dry friction coefficient could be on the same order of magnitude. In the case of clean metals, where the junction growth is most likely to take place, the adhesion component of friction may increase 10–100 times.

The presence of any type of surface oxide layer, lubricant, or water film preventing the formation of the adhesive junction can dramatically reduce the magnitude of the adhesion component of friction. The real area of contact could be much larger than that is from the deformation due to applied load, this is because of the work of adhesion. Assuming that the attack angle or the slope of the asperity is θ, the modification of Eq. [3.64] with adhesion work can be derived by letting the work done by the normal force be equal to the work done in deformation and the change in surface energy. Using this equilibrium, this simple model can be supplemented by the surface energy of the contacting components, the friction coefficient is given by

$$\mu_a = \frac{\tau_a}{p_r}\left[1 - \frac{2w\tan\theta}{p_r}\right]^{-1}, \qquad [3.68]$$

where w is the surface energy.

Next, we discuss the plastic effect. Suppose that the material has a yield stress σ_y. If the initial area of contact is small, the asperities in contact will yield and deform plastically in compression. The area of contact will grow as these asperities are squashed and others come into contact. Eventually, the normal force $P = A_r\sigma_y$, and then the true area of contact will be

$$A_r = P/\sigma_y. \qquad [3.69]$$

When the sliding asperities pass each other, the frictional force F^s relates to average shear stress $\tau = F_s/A_r$ acting on the asperities. If the contact at the asperities is intimate and the asperities have deformed extensively, the asperities may have to shear plastically in order to slide. In this case, $\tau = \tau_a$, in which τ_a is the yield stress in shear. Since the yield stress in tension and compression is $\sigma_y = 2\tau_a$ in terms of the Tresca criterion,

$$\tau_a = \sigma_y/2. \qquad [3.70]$$

Therefore,

$$F_s = A_r\sigma_y/2. \qquad [3.71]$$

Considering that $A_r = P/\sigma_y$, we have $F_s = 0.5P$. The approximation of the friction law for asperity deformation yields a friction coefficient of 0.5. This is close to the value found for dry solids, including metals, ceramics, and glasses.

More typically, solids like metals oxidize to some extent in air and form a thin oxide layer on their surfaces. The interface between two oxide surfaces may slide at stresses less than τ_a of the metal, leading, therefore, to $\mu_a < 0.5$.

Note that the formation of a strong junction bond needs time to develop asperity creep deformation. Once sliding occurs, the alternative asperity attach and detach and the oscillations in normal direction make the time insufficient to achieve full atom-to-atom bonding over the entire contact area. Therefore, the coefficient of friction could decrease in the sliding process, which is one of the reasons that kinetic friction is less than dynamic friction.

Actual values of cof may range from as high as 5 to as low as 0.05, depending on the materials brought into contact and the conditions of their surfaces. For pure metals in contact with the local contact areas, junctions between asperities grow as sliding occurs due to extensive plastic deformation around the asperities. This growth in junction area leads to extensive bonding and cof >5. This phenomenon is part of the basis of "friction welding" of metals and plastics. If surfaces are rapidly rubbed against each other, the heat generated can melt low-melting materials such as the resin in bowed strings.

For metals with clean surfaces, the interfacial shear force can be approximated by [110]

$$\tau_s = c_0 + c_1 p + c_2 p^2, \quad [3.72]$$

in which c_0, c_1, c_2 are material constants and p is normal pressure.

This discussion applies to sliding contacts between the metals, where A_r is small but σ_y and τ_a are large so that cof is large. For other kinds of sliding pairs, the situation depends on the real contact area A_r, σ_y, and τ_a.

For ceramics on ceramics (e.g., slider-disk interfaces in hard disk drives), the hardness and σ_y are high, the contact area is small, the tendency to bond across the contact is small, and τ_a is small. Thus, cof is small.

Rigid polymer films on harder materials like Teflon [polytetrafluoroethylene (PTFE)] on metals yield a small coefficient of friction. This is because the polymer chains are oriented along the direction of the sliding and only weak van der Waals bonds between chains need to be broken.

In addition to adhesion, elastic and plastic deformation in continuum interfaces and the fracture process help develop friction. Considering the mode of failure due to crack propagation, the friction due to the fracture of an adhesion junction is derived:

$$F = \frac{c\sigma_{12}\delta_c}{n^2(PH)^{1/2}}, \quad [3.73]$$

in which c is coefficient, σ_{12} is the interfacial tensile strength, P is normal load, δ_c is the critical crack opening displacement, n is the work-hardening factor, and H is the hardness.

In the next section, we discuss dry friction that occurs due to deformation (slope). Besides the microscope interface interaction of smooth surfaces where asperity normal contact dominated, macroscopic interaction of asperity of rough surfaces could occur. Consequently, asperity deformation and ploughing of soft material by hard material usually occur.

In this sliding process, the mechanical energy is dissipated through the deformations of the contacting bodies. The slip-line (lines of maximum shear stress) model of continuum mechanics can be used to analyze the deformation of the single-surface asperity. Figure 3.12 shows a slip-line field-based deformation model of friction based on Prandtl two-dimensional stress analysis. The material in the region ABCDE flows downward and outward as the hard material moves forward. Three distinct regions of plastically deformed material may develop in these regimes. The flow shear stress of the material defines the maximum shear stress that can be developed in these regions [5,15]. The angle ψ can be chosen so that the velocities of elements of material on the free surface, contact surface, and boundary of the rigid plastic material are consistent. The coefficient of friction is given by

$$\mu_d = \lambda \tan\left\{ \arcsin\left[\frac{\sqrt{2}(2+\psi)}{4(1+\psi)} \right] \right\}, \qquad [3.74]$$

where λ is the portion of the plastically supported load, and it is a function of elastic modulus and hardness.

The proportion of load supported by the plastically deformed regions is a complicated function of the ratio of the hardness to the elastic modulus. For completely plastic asperity contact and an asperity slope of 45°, the

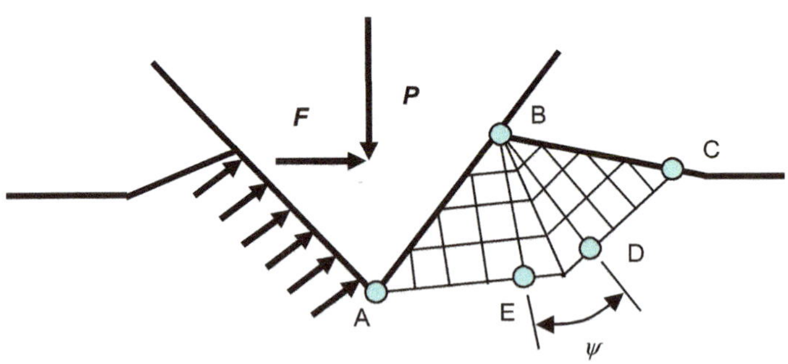

Figure 3.12 Schematic of a slip-line

coefficient of friction is 1.0. It decreases to 0.55 for an asperity slope approaching zero.

Another approach to this problem is to assume that the frictional work performed is equal to the work of the plastic deformation during steady-state sliding. This energy-based plastic deformation model of friction gives the following expression:

$$\mu_d = \frac{A_r}{P}\tau_m\left\{1 - 2\frac{\ln\left(1+\frac{\tau_s}{\tau_m}\right) - \left(\frac{\tau_s}{\tau_m}\right)}{\ln\left[1-\left(\frac{\tau_s}{\tau_m}\right)^2\right]}\right\}, \qquad [3.75]$$

where A_r is the real area of contact, τ_m denotes the ultimate shear strength of a material, and τ_s is the average interfacial shear strength.

The problem of relating friction to surface topography in most cases requires the determination of the real area of contact and studying the shear stress. If one surface is harder than the other, the asperities of the hard one could penetrate the soft one. It leads to grooves if the shear strength is exceeded. Ploughing not only increases friction, it also creates wear particles which change friction as well. To maintain the ploughing motion, a force is required and it may constitute a major component of the overall frictional force. The schematic of the ploughing of soft material by a hard conical asperity is shown in Figure 3.13a. Assuming that the effective slope is Θ, the formula for estimating the coefficient of friction in this case is as follows:

$$\mu_p = \frac{2}{\pi}\tan\Theta. \qquad [3.76]$$

For most engineering materials, the slope angle of asperity (the roughness angle) is very small (5−10 degrees), and the ploughing component of friction is correspondingly very small as well. A typical case

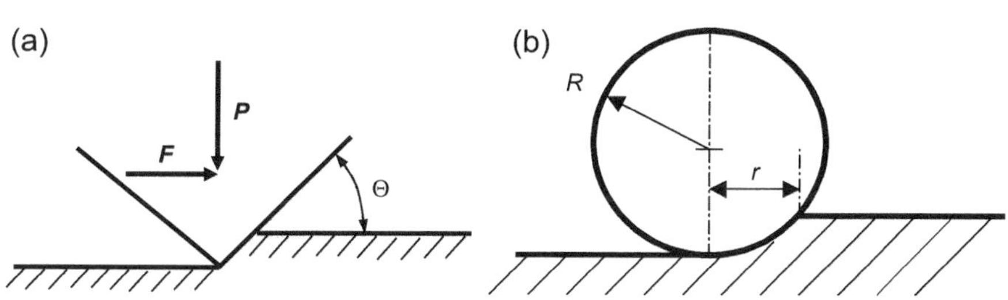

Figure 3.13 Ploughings of hard conical and sphere asperity against soft elastic substrates

is that the effective slope is less than 10 degrees, and the coefficient of friction μ_p is about 0.05 or less. In elastic contact, μ_p is often assumed to be negligibly small.

The frictional force due to ploughing is very sensitive to the ratio of the radius of curvature of the asperity to the depth of penetration. Considering a spherical asperity of radius R in contact with a softer body (as shown in Figure 3.13b), r is the size of the ploughing area that is proportional to the depth of penetration. The coefficient of friction is given by

$$\mu_p = \frac{2}{\pi}\left\{\left(\frac{R}{r}\right)^2 \sin^{-1}\left(\frac{r}{R}\right) - \left[\left(\frac{R}{r}\right)^2 - 1\right]^{1/2}\right\}. \qquad [3.77]$$

The coefficient of friction increases rapidly with an increase of r/R, and it increases as the sphere digs deeper.

Moreover, ploughing brittle material is associated with the microcracking process, a model developed in terms of fracture mechanics to quantify the effects of fracture toughness, elastic modulus, and hardness on cof is given as

$$\mu = c\frac{K_k^2}{E(HP)^{1/2}}, \qquad [3.78]$$

in which K_k is the fracture toughness, E is the elastic modulus, and H is the hardness.

For rough hard interfaces or hard interface-involved particles, the ploughing could be the dominant factor for energy dissipation. Plastic materials could undergo plastic deformation. Viscoelastic material could exhibit a hysteresis process. During the sliding, the softer material is pressed, and then the stress is released moving forward and contact continues. Each time the contact portion is stressed and energy is stored, when progressing to the next contact portion, most of the energy is released as the stress is removed. Only a small portion is dissipated as heat when hysteresis decreases.

Friction is a complex function that is dependent on a large variety of parameters, such as sliding speed, temperature, normal load, humidity, surface topography, and surface films. To further elaborate on the friction mechanism, we now consider energy dissipation during friction. In engineering interfaces, several friction mechanisms could coexist and interact with each other in a complicated way.

In general, frictional work is dissipated at two different locations within the contact zone. The first location is the interfacial region, characterized by high rates of energy dissipation and usually associated with an adhesion mechanism of friction. The other one involves the bulk of the body and the larger volume of the material subjected to deformation, where the rates of energy dissipation are much lower. Energy dissipation during ploughing and asperity deformations takes place in this second location. The distinction between these two locations is artificial and only for the purposes of simplifying a very complex problem. Actually, in describing KJL modeling, we illustrate the basic energy balance equation, which indicates the interaction aspect from the contact point of view. The various processes can be briefly characterized as follows:

(i) Plastic deformations and microcutting
(ii) Viscoelastic deformations leading to fatigue cracking and tearing, and subsequently to subsurface excessive heating and damage
(iii) True sliding at the interface leading to excessive heating and thus creating the conditions favorable for chemical degradation
(iv) Interfacial shear creating transferred films

During sliding contact, part of the kinetic energy produces waves and oscillations in the bodies, and part of it leads to the plastic and elastic deformation of asperity tips. Some energy expends through viscous dissipation, and the rest through adhesion, fracture, chemical reactions, and photoemissions. The distribution of energy conversion through this process varies for different applications. Each of these processes provides a mechanism for converting the original kinetic energy to an interim one in the form of vibration and sound, deformation energy, surface energy, tribochemical energy, and other tribo-emissions. In the end, part of the initial energy remains stored as potential energy, and part of it converts to thermal energy, eventually dissipating to the surroundings. Thus, friction can be viewed as a combination of processes that transforms ordered kinetic energy into either potential energy or a disorderedly thermalized state of kinetic energy. It then follows that the friction force can be considered as a combination of forces that resist motion during each of the following energy conversion processes:

$$F_f = F_{adhesion} + F_{elastic\ deformation} + F_{plastic\ deformation} + F_{viscousity} + F_{fracture} + F_{random}. \quad [3.79]$$

The schematic depiction illustrates the contributions to friction on different scales.

Characterizing this process affecting friction needs different models from macroscales to microscales or even smaller scales, as the various events take place on these scales. For the integration of the contribution from these scales, expertise is required from different disciplines of science and engineering. Usually, a continuum scale is suitable for modeling engineering problems. At the atomic scale, the problems relating to the dissipation and oscillation of atoms should be examined.

Many of the current efforts to model friction start at the continuum scale, relating surface roughness to friction. The obvious mechanisms that contribute to friction in such models include elastic/plastic contacts, viscous dissipation, fracture, and adhesion. Each of these processes develops at each true contact region between the surfaces. In addition, the true contacts take place between asperities on the surfaces or on particles between them.

3.3.2.2 Friction Transitions and Wear

The previous discussion was about interfaces with clean surfaces. However, engineering surfaces in real life are not likely to be clean. Oxide films could form on the surface of metal and other engineering materials, creating a boundary layer whose thickness could vary from an atomic thickness to a fraction of a micron. The boundary layers usually have lower shear strength than the original solid. For many engineering surface, the moisture or other chemical vapor could absorb on the surface to form adsorbed layer.

The transfer film forms more readily on roughened surfaces and the transfer film can exist in a solid state and in a low viscosity or fluid state. The formed film on the interface has a substantial effect on friction.

Friction is always associated with wear. They always coexist, and both are the result of one solid rubbing against another and interacting with each other. Wear is the effect of friction on material surfaces that rub together. It occurs as a result of interaction between two contacting surfaces, and it can be classified as the type of fatigue, adhesion, chemical wear, corrosion, and abrasion. Wear is usually associated with the loss of material from contracting bodies in relative motion. The formation and removal of interface films due to wear could lead to substantial variation in friction. Like friction, wear is controlled by the properties of the material, the environmental and operating conditions, and the geometry of the contacting bodies. Adhesive wear is invariably associated with the formation of adhesive junctions at the interface. Archard's classical wear model states that the rate of wear, $\dot{V}$, is proportional to the pressure, p, which is given by

$$\dot{V} = k\frac{pL}{3H}, \quad [3.80]$$

where k is the wear coefficient, L is the sliding distance, and H is the hardness of the softer material having the contact. Two materials exhibiting the same friction coefficient can exhibit quite different wear rates because the energy is partitioned differently between and within the materials.

Friction is intimately related to adhesion, interlock, and wear, which forms highly nonequilibrium processes occurring at multiple levels from the nanomolecular level to the micromolecular and macromolecular levels. It leads to what happens at the macroscopic level. The interface can be smooth or rough, hard or soft, elastic, viscoelastic or plastic, brittle or ductile, dry or wet (including lubricated), and of very different physics, chemistries, and thermal dynamics.

Next, we consider the energy balance associated with friction and wear. The energy that is transformed as a consequence of frictional contact can be stored in the friction system or dissipated in a number of different ways. If E_f is the energy resulting from sliding contact, E_{out} is the energy leaving the friction system, and E_{st} is the energy remaining in the friction system, then we have

$$E_f = E_{st} + E_{out}. \quad [3.81]$$

For example, mechanical energy from sliding can be converted to heat, vibrations, and sound or to material deformation, or new surfaces can be created by fracturing. Likewise, it can be stored in the material as the energy associated with microstructural defects.

The typical engineering surfaces are covered by adsorbed film that consists of water, oxygen, or even oil. For metal surfaces, they usually have oxide layers. Sliding surfaces are always progressive during the sliding process, due to the friction and wear processes. The changes in the surface also affect friction. For interfaces, after a certain period of sliding, the friction could level off to a steady state. That period of sliding is called the *break-in* or *run-in period*. After the break-in period, the cof could increase to a higher plateau or a lower plateau, exhibiting an *s*-shaped curve or a *Z*-shaped curve. The break-in process is associated with the peeling off of high asperities, surface polish being more matchable, original film breakage and wearing off, and new stable film forming. For some materials under specified limit operations, the severe wear could onset, which could give rise to a salient increase in friction.

Another associated problem involves temperature. When two surfaces slide over each other, the heat produced (Q) is proportional to the sliding velocity and friction force:

$$Q = \int_{A_r} V \tau dA_r, \qquad [3.82]$$

in which A_r is the actual area of contact, τ is the shear stress, and V is velocity. Q depends on area of contact A_r or the shear stress and velocity, or indirectly depends on normal load and coefficient of friction. The temperature increase on the surface depends on the thermomechanical properties of the two contacting bodies as well as Q.

The surface temperature generated in contact areas has a major influence on wear, material properties, and material degradation. The friction process converts mechanical energy primarily into thermal energy, which results in a temperature increase.

The friction intensity may not be sufficiently large to cause a substantial temperature increase on the body, but it could be large enough to cause a substantial temperature increase on the surface. The flash temperature is defined as the temperature increase in the contact area above the bulk temperature of the solid as a result of friction energy dissipation. The surface temperature increase can influence local surface geometry through thermal expansion, causing high spots on the surface that concentrate the load and lead to severe local wear. The temperature level can lead to physical and chemical changes in the surface layers. These changes can lead to substantial transitions in friction mechanisms and wear phenomena. Consider a simple formulation for the mean flash temperature in a circular area of real contact of diameter *2a*. The friction energy is assumed to be uniformly distributed over the contact region. For a stationary heat source, consider one of the sliding components (component 1), the mean temperature increase above the bulk solid temperature is

$$\Delta T = \frac{Q}{4ak_1}, \qquad [3.83]$$

where Q is the rate of frictional heat supplied to component 1 (Nm/s), k_1 is the thermal conductivity of body 1 (W/m, in degrees Celsius), and a is the radius of the circular contact area (m).

3.3.2.3 Static Friction, Hysteresis, Time and Displacement Dependence

In this section, we discuss static friction. For contact surfaces, the break force required to initiate motion could be lower than, higher than, or even equal to the force needed to maintain the surface sliding in the subsequent relative motion. Therefore, the coefficient of static friction could be less than, greater than, or equal to the coefficient of kinetic friction. Higher static friction usually appears for clean surfaces. For these cases, the static friction is usually a function of dwell time. The static friction increasing with time for dry interfaces is considered due to the interface bonding from the interaction of the atoms on the mating surface and the plastic flow and creep of interface asperities under load. The time-dependent static friction model usually can be represented as

$$\mu(t) = \mu_{\infty} - (\mu_{\infty} - \mu_0)\exp(-\alpha t_s), \quad [3.84]$$

where μ_0 is the initial value of the coefficient of friction, μ_{∞} is the limit value of static friction at long-term time, t_s is rest time, α is constant. Other models based on power law have also been developed [86].

In the next section, we present the hysteresis property of friction occurring in some systems. In acceleration and deceleration processes of certain system, friction versus velocity curves may not be identical. There usually is some delay to forming a hysteresis loop, as shown in Figure 3.14. The friction exhibits the memory feature. The state variable models have been developed to quantify this type of characteristics, which has critical applications, as in rock mechanics [87–88]. In solid mechanics, the constitutive relations have been used effectively at capturing steady-state

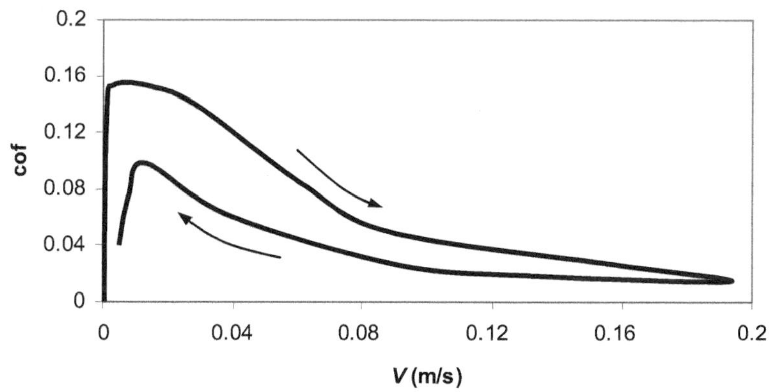

Figure 3.14 Hysteresis of a friction–velocity curve

and certain transient effects in a wide variety of materials with interfaces having micron-scale roughness. The approach involves expressing the friction in terms of the instantaneous slip speed at the interface and one or more state variables, for which phenomenological evolution equations are also introduced.

One of the underlying assumptions here is that the interfacial area is large enough to be self-averaging, so that a mean-field-like state variable is sufficient to capture the collective dependence of microscopic degrees of freedom on the dynamical variables, including time, displacement, and slip speed, which characterize the motion.

For instance, earthquakes have been considered to be fault-line stick-slip events. Recently, the relevant models have been extended to other materials, including lubricated metal, Teflon on metal, glass, and plastics. The state variable models contain a displacement-dependent feature. It not only adopts a steady-state velocity-dependent property incorporating the Stribeck curve (which will be illustrated in the next section of this chapter), but it also assumes an instantaneous velocity-dependent reflecting the fact that the instant change in velocity results in an instantaneous change in friction. Moreover, it assumes an evolutionary characteristic of sliding distance-dependent, which reflects the phenomena that following a sudden change in velocity, a steady-state friction curve is approached through an exponential decay over characteristic sliding distances.

Various phenomenological models offer such features, where the coefficients of each model are fitted to experiment data and then used to describe a wide range of observed frictional behaviors, such as the dilation of a liquid under shear and the transition between stick-slip (regular or chaotic) and smooth sliding friction. However, most state variables in models are unable to be quantitatively related to the properties of real physical systems.

In many applications, displacement-dependent static friction needs to be characterized. The static friction is actually a constraint, and it is associated with an elastic or plastic deformation under traction. The displacement before overcoming static friction can be modeled with an equivalent spring, k, which is a function of asperity, normal force, material elasticity, and surface energy. The presliding displacement has long been studied in many engineering problems, such as the microslip in magnetic recording tape. To elaborate on this point, let us consider the predisplacement of a data recording tape wrapping on a cartridge roller with tension before slip motion. The coefficient of friction between tape and roller is assumed to be μ. The friction causes a change in tape tension, thereby causing the elastic tape to extend and contract over the contact section on the roller. The resulted relative motion s is a microslip called elastic creep. The event has been experimentally observed for a tape belt/roller interface in a magnetic tape recording system. The presliding displacement results in wear, as well as wear debris built up on the roller, which affects real cof [82].

The microslip is the elongation of a belt due to tension before tape belt—roller relative rigid slippage, calculated as follows:

$$s = \left(T_{high} - T_{low}\right)/k, \qquad [3.85]$$

in which T_{high}, T_{low} are the tight and slack side tension, respectively; and k is the belt longitudinal stiffness. The contact zone is composed of the traction zone (in which the tension is a constant) and the microslip zone. Here, θ_s and θ_t are the slip zone and traction zone, respectively. The microslip occurs only in the slip zone, and the Euler formula can be used to approximate the relationship:

$$T_{high} = T_{low}e^{\mu\theta_s}, \qquad [3.86]$$

where μ is the coefficient of friction. The schematic representation of the belt—roller interface, in which the elastic tape belt undergoes microslippage during the transition of tension in the wrap section from T_{low} to T_{high}, is shown in Figure 3.15.

3.3.2.4 Effects of Environmental and Operational Conditions on Friction

In this section, we first discuss the temperature-dependent feature of friction. The friction of a certain interface is highly humidity-dependent. In a

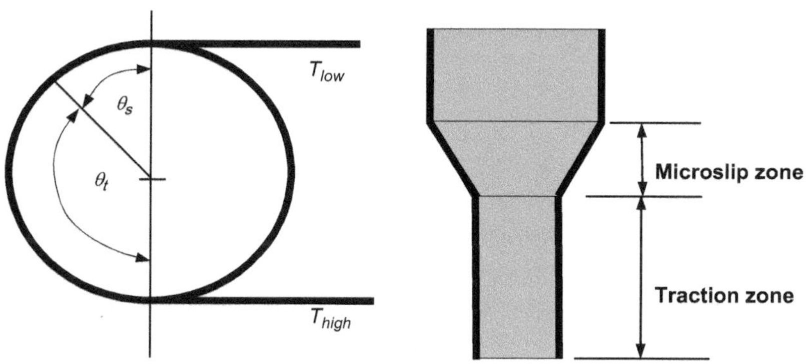

Figure 3.15 Microslip of a magnetic recording tape belt

humid environment, the amount of condensate water presenting at the interface increases with the increase in relative humidity. The adsorbed water film thickness on a diamondlike carbon-coated magnetic disk can be approximated as follows:

$$h = h_1(RH) + h_2 e^{\alpha(RH-1)}, \qquad [3.87]$$

where RH is the relative humidity ranging from 0 to 1, and h_1, h_2, and α are constants. For instance, some experiments have shown that the coefficient of friction of lubricated disk increases rapidly when relative humidity is above a critical value (60%), and this critical humidity is dependent upon the interface roughness. Above the critical limit, static friction could increase up to five times, whereas kinetic friction could only increase slightly. The static friction of some automotive brake-pad pairs under low speed/low control pressure increases in high-moisture air due to the increase of adhesion from water menisci formed at the contact junctions.

Load and velocity have strong effects on friction. The oxide films are likely produced on the surface of metal and alloys. This usually gives rise to lower friction than clean surface under low loads. When load increases, ploughing could occur, film could get penetrated, and higher friction could be given.

When a load is very high, the friction gets decreased due to the larger quantity of wear debris and the surface roughing. The coefficient of friction usually decreases with an increase in velocity. High-speed motion tends to reduce the formation of asperity junction of the interface. High speeds tend to generate heat, softening of asperity, and oxidation of film. On the other hand, softening could increase ploughing and high shear rates lead to a smaller real contact area, which makes the friction higher at high speeds. The real velocity dependence of friction depends on many competing factors. However, for the wet interface, the coefficient of friction usually decreases with an increase in velocity when velocity is under a certain threshold, which will be discussed in the next section.

The load and speed dependence of the friction of ceramics could be related to the changes in the chemical surface films and the extent of fracture. The cof is low at low loads, but it increases rapidly after the brittle fracture occurs. A decrease in friction with sliding velocity is attributed to the increase in the interface temperature, which promotes the formation of tribo-chemical films.

In many cases, the friction—velocity peak occurs because of the competing effects of strain rate dependence of shear strength and reduced contact area due to viscoelastic stiffening of some materials.

Friction is generally system vibration—dependent. The dependence of friction on vibrations and these interactions will be discussed further in the following chapters of this book.

3.3.3 Wet Friction

3.3.3.1 Stribeck Curve

Previously, we have discussed the interface with dry friction surfaces or surface friction without liquid involved. For many real engineering interfaces, water and moisture are mediated, and for lubrication system, grease and oil, as well as fatty acid lubricants, could be applied to the surfaces to allow them to slide easily over each other. It is well known that there may be four regimes of lubrication effect in an interface with fluid: (i) static friction, (ii) boundary lubrication, (iii) mixed lubrication (partial fluid and partial solid), and (iv) full fluid lubrication. Figure 3.16 gives the schematic of a generalized Stribeck curve showing the possible regimes.

Consider the contact-sliding interface with liquid film. To discuss the four possible lubrication regimes, let us define a ratio to quantify the relative

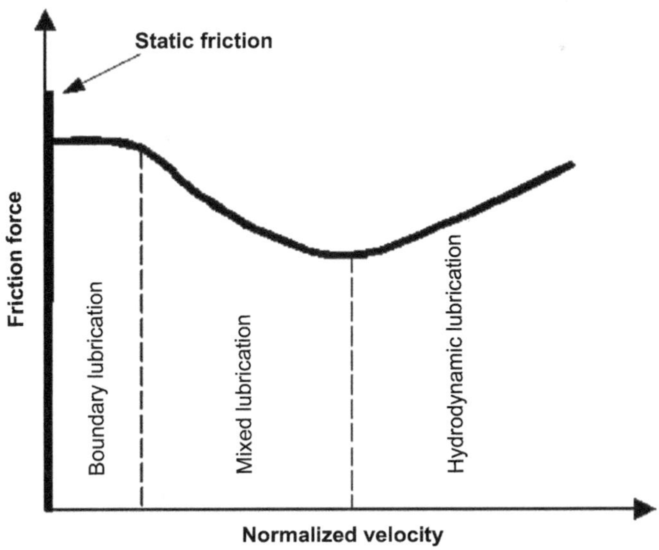

Figure 3.16 Schematic of a generalized Stribeck curve

thickness and the roughness of liquid. Here, λ is defined as the ratio of the mean liquid film thickness at interface to the combined roughness of both surfaces:

$$\lambda = h_0/R_q, \quad [3.88]$$

in which h_0 is the relative thickness of liquid, and R_q is the combined average rms roughness of the two surfaces.

Static friction is the tangential force that must be overcome to initiate the sliding of one object over another, while *kinetic friction* is the tangential force that must be overcome to maintain sliding. Strictly speaking, static friction is not a friction force, but a threshold of static force. Its presence implies that the objects have locked together into a local energy minimum that must be overcome before motion by the external force.

Different mechanisms have been suggested for static friction, such as geometric interlocking, elastic instabilities, interlocking mediated by so-called third bodies, plastic deformation, and crack propagation, just to name a few. All these mechanisms may well be applicable, although each one applies under different circumstances. Basically, the first regime does not depend on velocity. The static contact allows the asperity junctions to deform elastically and plastically and yields static bonding. A traction force is needed to break away the bond, or static friction. There is usually a pre-sliding displacement associated with the breakaway process. The ratio of the breakaway force to normal force is the coefficient of friction. The contacts are compliant in both the normal and tangential directions. The junction in the surfaces of the continuum system is like an equivalent spring k. Under the traction, the presliding displacement is linearly proportional to the applied force until it reaches a critical value equal to the static friction, and then breakaway occurs. The transition to sliding is actually not abrupt. Besides the solid—solid bond, meniscus due to liquid mediation may also contribute to the static friction.

The second possible regime is boundary lubrication, which occurs at a very low sliding velocity for $\lambda < 1$. The velocity is not sufficient to establish a film between the solid surfaces, and the liquid lubrication is not critical, which does not dominate friction. The applied load is actually carried by the surface asperities, and the friction depends upon the lubrication properties of the molecules on the surfaces. Coefficients of friction between 0.06 and 0.1 are typical when a low-shear-strength boundary film is present. If no such film is present, then coefficients ranging between 0.2 and 0.4 can be exhibited, even rising as high as 1.0 in some cases. In this regime, atomically flat surfaces could be separated by a few molecular layers of liquid; however, the behavior of the interface is qualitatively different from the bulk viscosity, which is traditionally associated with lubricant film. The interfacial material can pack into a solid-like structure due to its confinement and exhibit properties such as a finite yield stress and stick-slip. The boundary lubrication layer is basically solid-like. There is shearing in the liquid in boundary lubrication due to solid-to-solid contact. Boundary lubrication is basically a process of shearing in a solid, despite the liquid film involved.

The third possible regime is a mixed lubrication regime, which is approximately characterized by $1 \leq \lambda < 3$. In these contacts, part of the load is carried by the liquid, and part by the interacting asperities. The rubbing together of asperities can increase friction, and this can also be minimized by the compounds adhering to the surfaces or the tribolayer in the interface. The interface is partially supported by hydrodynamic force and partially by asperity contact force. Some liquid is expelled by pressure, but viscosity or wetting prevents all of the liquid from escaping, and thus a film is formed. The friction process could be dominated by the interaction of liquid viscosity, motion speed, pressure, and contact geometry. For this case, the greater the viscosity is or the higher the motion velocity is, the thicker the fluid film will be, and therefore the lower the cof. The surface roughness, asperity size, and orientation have essential impacts on the characteristics of formed films. In some other cases, the friction process could be determined by the tribolayer, speed, pressure, and contact geometry. In this case, the surfaces are so close that the liquid viscosity is relatively insignificant. It is the liquid's physical and chemical interaction with surfaces that dominates the friction. The mixed lubrication usually has liquid film thickness in the range of 30 nm to 3 μm. In tribology, additives are usually used to provide the desired properties, together with base liquid that needs low surface tension and a low contact angle. The additives are usually adsorbed on the surface or react with it to form monolayers with low shear strength, therefore reducing friction as the bottom line to protect the surface wear.

The possible fourth regime is full fluid lubrication, where solid-to-solid contact is eliminated. The process is governed by either the hydrodynamic or elasto-hydrodynamic lubrication, characterized by $3 < \lambda < 10$. In this situation, the surfaces are kept apart by a pressurized fluid. The clearance space is much larger than the average surface roughness, and therefore the surfaces can be considered smooth. The pressurization of the fluid is usually achieved by external means in hydrostatic bearings, but it is accomplished in hydrodynamic contacts by the relative motion and geometry of the surfaces.

3.3.3.2 Unsteady Liquid-Mediated Friction

In this section, we only discuss the first three possible regimes of friction and lubrication in the interface, which are the ones most likely associated with friction dynamics problems in engineering.

For liquid-mediated interface, the static friction or stiction could exhibit strong time-dependence: static friction increases remarkably with rest time. The empirical model can also be represented by Eq. [3.84], the power law or divided formula can also be employed. In some cases, like stiction in the slider-disk interface of a conventional hard disk drive, static friction starts to increase rapidly after a certain rest time. The long-term friction after days dwell could be 10 times higher than the static friction without dwell or having short dwell. Moreover, the higher the acceleration, the higher the static friction due to viscous effects for wet friction is.

Figure 3.17 shows the recorded static friction and long-term static friction in an interface of a new disk drive during the contact start-stop (CSS) and dwell processes. Each CSS cycle consists of the same operation: 3-s acceleration, 3-s constant speed operation, 3-s deceleration, and 6-s short soaking with a slider parking on the disk statically. Long-term (i.e., 10 h) soaking was introduced every 1000 CSS cycles. Figure 3.18 shows the similar results for the same interface in the next 5000 CSS operations. It can be seen that the static friction remains the same feature after the "run-in."

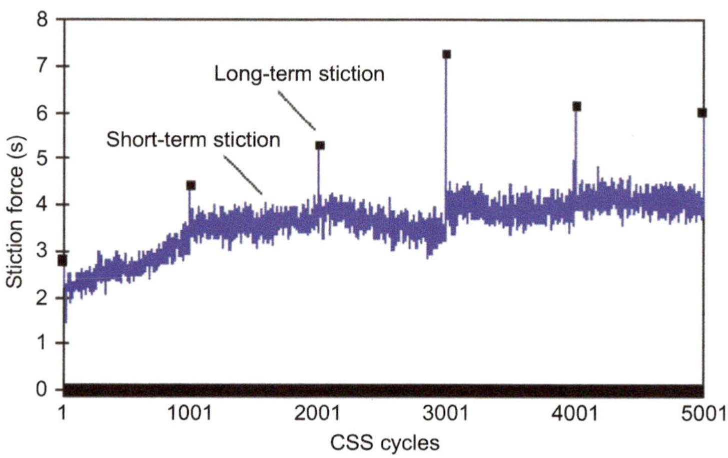

Figure 3.17 Static friction force versus CSS operation cycles for a new interface

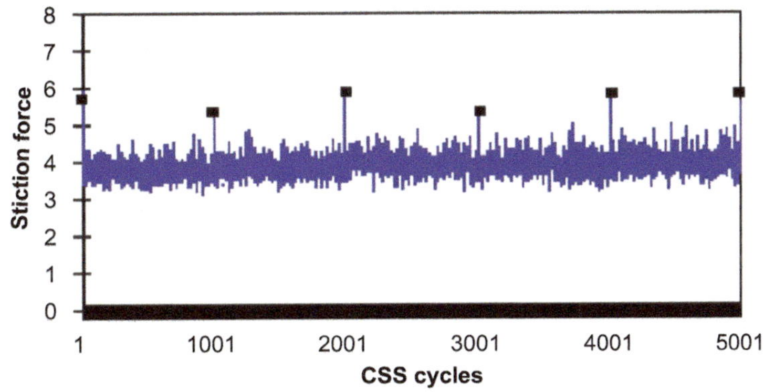

Figure 3.18 Static friction force versus CSS operation cycles for a conditioned interface

The ceramic slider and diamondlike, carbon-coated disk pairs are used in this interface. As shown in Figure 3.17, the long-term stiction is greater than the short-term stiction. This is because that the long-term soaking will lead to significant redistribution and accumulation of the interface lubricant, due to the microflow of lubricant around the toe-dipping regimes, in addition to the slider-disk asperity bond creep development.

Generally, there are several possible mechanisms allowing static friction force to increase with the contact time: the increasing of the area of real contact with the time of stationary contact; the redistribution and accumulation of the interface liquid, the microflow of lubricant around the toe-dipping regimes, and chain interdiffusion. Moreover, if the liquid film exists, dynamical phase transitions take place from the fluid state during slip to a solid state during stick whose formation is a nucleation process, then the static friction force could also increase with the time of stationary contact.

Figure 3.19 shows the measured static friction coefficient as a function of dwell time for different lubricants.

Many engineering wet interfaces and liquid-mediated interfaces exhibit unsteady mixed lubrication. Boundary and mixed lubrication is important for friction dynamics because they are usually associated with stiction and the negative slope of the friction versus velocity curve. For most wet interfaces, the measured static friction is almost always higher than kinetic friction, so stick-slip and negative damping could be a concern. In mixed lubrication regimes, friction is a function of velocity because the physical process of shear in the junction changes with velocity and the liquid film has hydrodynamic effects. Details of the velocity-dependent friction feature

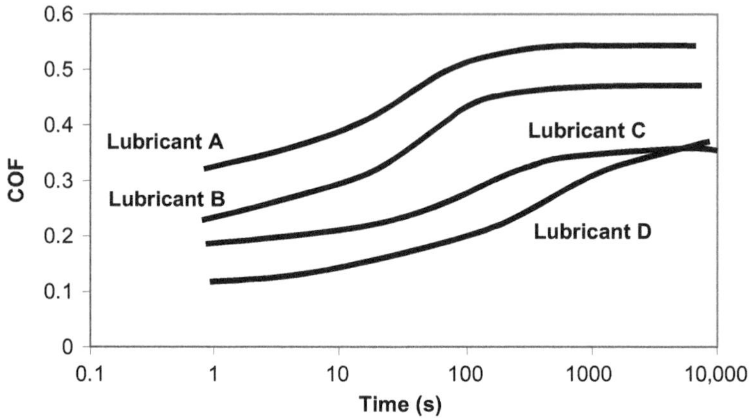

Figure 3.19 Static friction coefficient as a function of dwell time for different lubricants

depend upon the degree of boundary lubrication and the details of mixed lubrication.

Fluid friction arises from the shear of fluid film within contact, and can be represented as [100]

$$P_f = P_f\left(V, P, \eta_i, \alpha, h_0, E^*, r', T, b\right), \quad [3.89]$$

in which $V, P, \eta_i, \alpha, h_0, E^*, r', T$, and b are the velocity, load, fluid viscosity, pressure-viscosity coefficient, film thickness, effective elastic modulus, combined radius of curvature, temperature, and film length, respectively. On the other hand, the solid friction force can be written, in functional form, as

$$P_s = P_s\left(P, h_0, E^*, r', \sigma, b\right), \quad [3.90]$$

in which σ is the combined surface roughness. To model the total friction, the two forces are added and divided by the total normal load to acquire the friction coefficient:

$$\mu = \frac{P_f + P_s}{P} = \mu\left(V, P, \eta_i, \alpha, h_0, E^*, r', T, \sigma, \tau_s, b\right). \quad [3.91]$$

For mixed lubrication regimes, liquid viscosity could have an insignificant effect on the friction characteristics, and the interface load is dominated by asperity contacts. However, despite the fact that the liquid film does not control the friction, tribolayers could play a key role in distinguishing friction properties between wet friction and dry friction. This could be the case for many wet interfaces.

To acquire some insight, let us consider contact with hydrodynamic lubrication. The minimum film thickness for smooth surfaces with hydrodynamic lubrication can be calculated from the following formula:

$$h_0 = \frac{c\mu_0 RV}{P}, \qquad [3.92]$$

where c is constant, P is the normal load per unit width of the contact, R is the relative radius of curvature of the contacting surfaces, μ_0 is the lubricant viscosity at inlet conditions, and V is the relative surface velocity.

To estimate whether the interface has boundary lubrication, we can calculate the specific film thickness or the lambda ratio. For this case, the total normal load is shared between the asperity load and the film load.

Considering the particular lubricated concentrated contact, both the contacting asperities and the lubricating film contribute to supporting the load. Thus,

$$P = P_f + P_s, \qquad [3.93]$$

where P is the total load, P_l is the load supported by the liquid film, and P_s is the load supported by the contacting asperities. Load P_s, supported by the contacting asperities, results in the asperity pressure p_a, given by

$$p_a = \frac{4}{3}(\eta R_p \sigma_p) E^* (\sigma_p / R_p)^{1/2} F_{3/2}(d_e/\sigma), \qquad [3.94]$$

where $F_m(D)$ is a statistical function in the GW model of contact between two real surfaces, R is the relative radius of curvature of the contacting surfaces, E^* is the effective elastic modulus, N is the asperity density, r is the average radius of curvature at the peak of asperities, σ is the standard deviation of the peaks, and d_e is the equivalent separation between the mean height of the peaks and the flat, smooth surface. For the case $\lambda \approx 1$, the liquid pressure to total pressure is approximated by

$$p_l = \frac{1}{\lambda}\left(\frac{h_0}{h}\right)^{\alpha} p, \qquad [3.95]$$

where α is a constant, p is total pressure, h is the mean thickness of the film between two actual rough surfaces, and h_0 is the film thickness with smooth surfaces.

In liquid-mediated interfaces, there are strong friction delay-hysteresis phenomena. If there is a change in velocity, the corresponding change in friction has a delay. The friction memory can be attributed to the friction level lags a change in system state [86]. The hysteresis is the separation in friction levels during acceleration and deceleration.

An alternative to the state variable models described earlier in this chapter in Section 3.2.4.5 is the pure time lag model. In lubricated contacts, simple time delay better describes the effect,

$$F(t) = F_{vel}(\dot{x}(t - \Delta t)), \qquad [3.96]$$

in which $F(t)$ is the instantaneous friction force, $F_{vel}(\bullet)$ is friction as a function of steady-state velocity, and Δt is the lag parameter, which is the time by which friction lags a change in velocity. The lag increases with the increase of liquid viscosity and contact force.

3.3.3.3 Negative Slope of Friction—Velocity Curve

Wet friction occurs in many mechanical and automotive engineering applications. It is usually smaller than dry friction, but there are some results in which it is higher than dry friction. Moreover, the trend of the velocity-dependent feature of friction is usually determined by many factors and could have larger variations [161—172]. Next, we illustrate this by discussing several application cases in automotive engineering.

A typical case is the friction exhibited by friction material utilized as a torque converter in wet clutches. The converter is submerged in automatic transmission fluid (ATF). Friction-induced shudder is a major issue associated with the wet clutch used in automatic transmissions, which is due to velocity-dependent friction.

For automotive wet clutch interfaces with the same solid clutch friction material, the cof could have very different profiles when different automatic transmission fluids are used. Figure 3.20 plots the cof as a function of velocity for three different ATFs. A is considered to be favorable, while B and C may be susceptibly vulnerable to self-excited vibrations. The optimization of the fluid in question is achieved mainly by regulating the additives in the base liquid. On the other hand, a similar profile group could be attained by using the same fluid, but with different solid friction materials.

Figure 3.21 shows the friction—velocity curve of the clutch pair in dry friction testing and in testing with transmission fluids. It shows that the friction material produces a negative cof-velocity curve for both dry friction and lubrication with transmission fluids. At low sliding speeds, the coefficient of friction operating in transmission fluids is greater than that with dry friction.

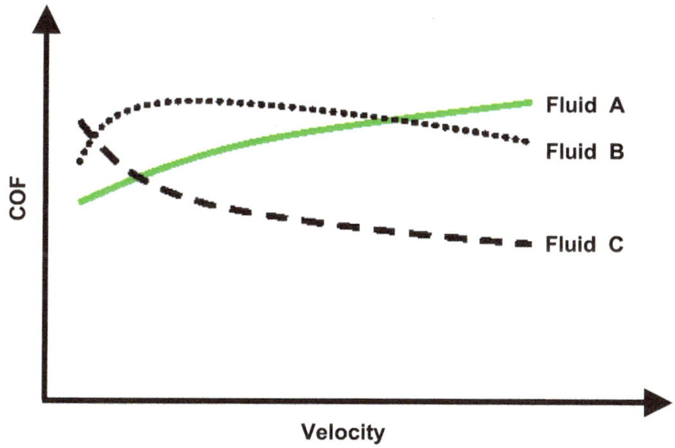

Figure 3.20 Schematic of friction–velocity curves of clutch friction pairs with three automatic transmission fluids

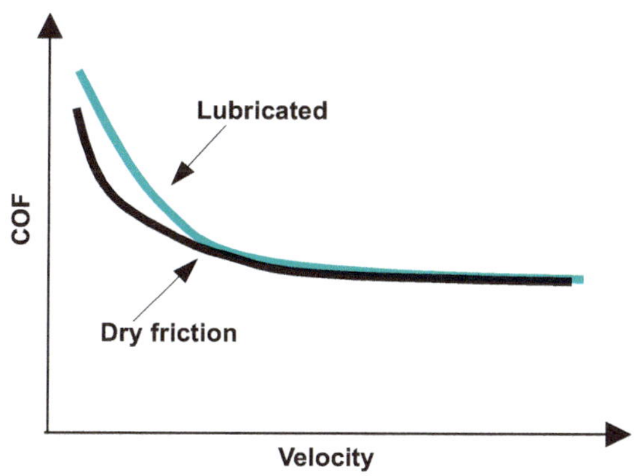

Figure 3.21 Dry and wet cof versus the velocity of a clutch interface

Another example is the complex friction features of an automotive brake friction material pair with water mediation, the wet cof could be higher than the dry cof, and the transition process is usually complex. The examples for the same friction couple are shown in Figure 3.22. During the first part of the test, the friction coefficient was measured with the brakes in a dry condition. Immediately thereafter, the brake disk was moistened with water. The results show that the adhesion force rises considerably when the brake disc is moistened. This may be caused by an increase in the adhesion forces between the lining and the disc due to water in the contact surface.

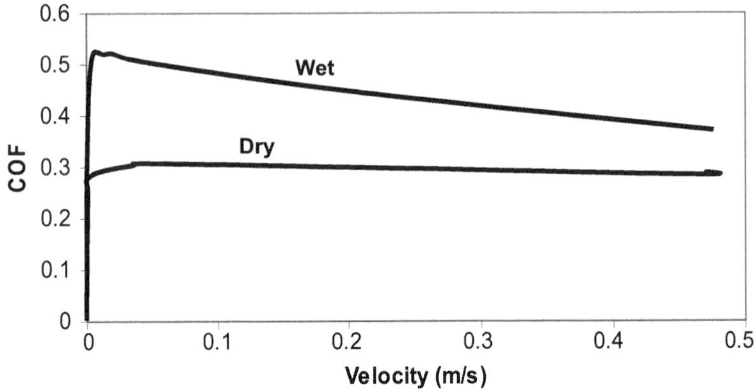

Figure 3.22 Dry and wet cof versus the velocity of a brake interface

In addition, the initial increase in the friction coefficient in all cases and the decrease in the friction coefficient during a further increase in the relative velocity are remarkable.

Another example is the temperature effects for the same friction couple. The cold wet cof is higher than the hot dry cof, and the hot dry cof is higher than the hot wet cof. All of them have a negative slope of the friction–velocity curve, as shown in Figure 3.23. Moreover, high relative air humidity could have obvious effects on the friction. Figure 3.24 shows the contrary effects of humidity on two different brake pairs. Figure 3.25 shows the typical friction versus slip velocity of tires under different road conditions.

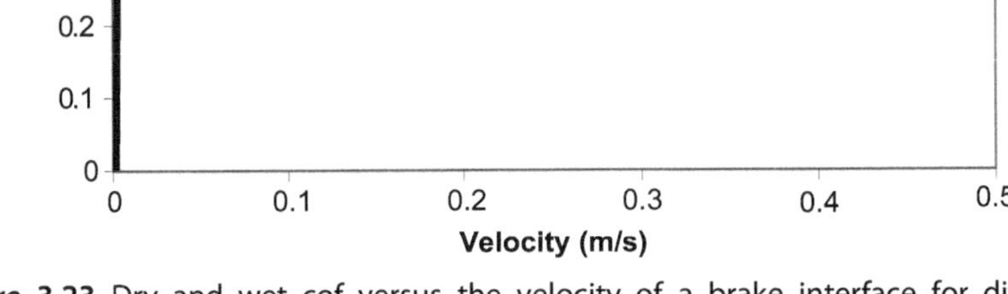

Figure 3.23 Dry and wet cof versus the velocity of a brake interface for different environmental temperatures

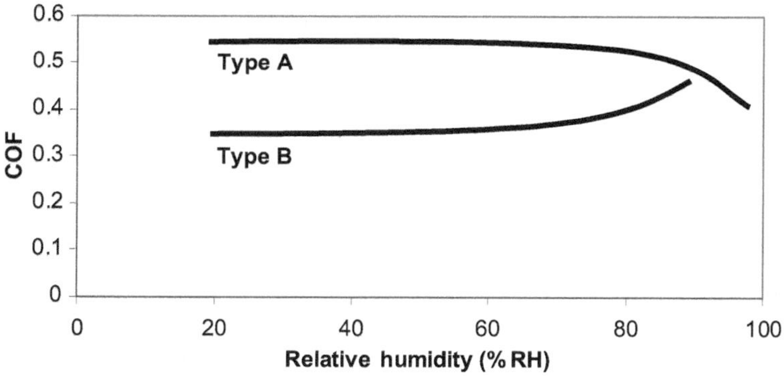

Figure 3.24 Cof versus relative air humidity for two different brake pads

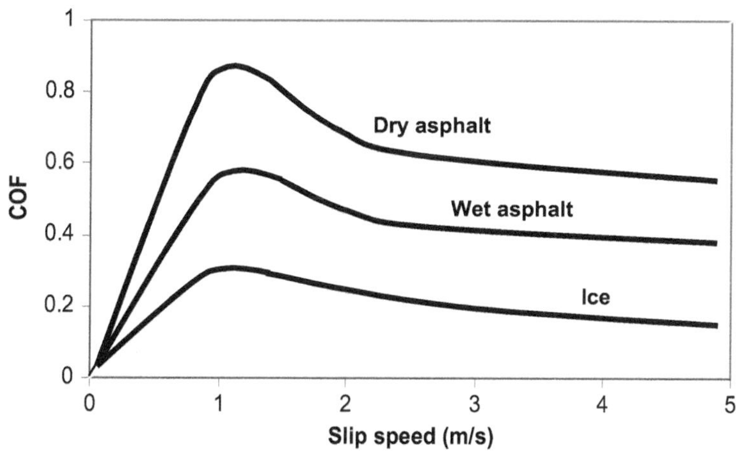

Figure 3.25 Cof versus slip velocity of tires in various road conditions

3.3.4 Friction Models

It is generally unlikely to study sliding friction dynamics directly using "first-principle" methods such as molecular dynamic calculations. It is also unlikely to study contact sliding dynamics directly using multiple physics-based models where Newton's force and motion equation, asperity contact mechanics, liquid film dynamics, and heat transfer are all included. These models are suitable for fundamental understanding and application interpretations, as well as in-depth investigation and research. However, the model is computationally expensive and too complicated to be implemented compared with empirical models, limiting its applications in engineering and science. For engineering applications, it is usually necessary to model friction and construct friction laws which can be used directly in Newton's equation of motion to predict the dynamic performance of the sliding macroscopic bodies [169–179].

It is usually feasible to construct empirical friction laws based on phenomenological observations to predict the sliding behavior of macroscopic bodies. This kind of model is based on macroscopic quantities observed, displacement, state variables, load, temperature, and the elements' history.

When we consider elastic continuum vibrations, the friction description ranges from Coulomb friction law to the microlevel surface tribological description. The former could be too simple to capture dynamic friction phenomena sufficiently; the latter could be too complex to be brought into the differential equation of dynamics motion. There have been many efforts in science and engineering communities to develop trade-off friction law for various problems, and numerous friction laws have been developed with phenomenological expressions.

The normal component of contact forces relating to the deformation of the contact areas can usually use Hertz theory for approximation. Let us consider an interface represented by spherical contact, with effective radii of curvature R. Based on Eqs. [3.9]–[3.12], the contact force $F_N(t)$ can be assumed as a function of elastic deformation in terms of normal displacement y_H,

$$F_N(t) = k_H y_H^{3/2}(t), \quad [3.97]$$

in which

$$k_H = 4E^* \sqrt{R}/3. \quad [3.98]$$

If the contact remains, this type of contact introduces nonlinear stiffness into the system. If the contact breaks, then a strong nonlinearity of vibro-impact will occur in the system. On the other hand, the coefficient of friction as a mathematical function of a variety of parameters can be developed by using varied fundamental formulations.

The most widely used friction law concerns the friction–velocity curve characterizing Stribeck features of real systems. The phenomenological expressions for a friction force versus slip velocity lead to qualitative results that help characterize the underlying physical phenomena. Polynomial and exponential forms of friction laws are used most commonly for capturing this kind of attribute. One of the exponential expressions takes the following form:

$$\mu(v_r) = \mathrm{sgn}(v_r)\left(1 - e^{-\beta_v|v_r|}\right)\left[1 + (f_r - 1)e^{-\alpha_v|v_r|}\right], \quad [3.99]$$

where v_r is slip speed and α_v and β_v are constants, with f_r representing the ratio of the static to kinetic friction coefficients.

Some applications require a more detailed and more accurate representation of friction to better predict system response and the development of instability. In these cases, in addition to describing the Stribeck curve, it is also necessary to quantify friction and coefficient of friction in terms of static friction transition, hysteresis, and presliding displacement (i.e., displacement that occurs just before a complete slip takes place). This presliding displacement is mainly due to lateral contact elasticity. The increase of static friction with time is mainly due to diffusion processes on the interface. The frictional lag is the effect of the friction force lagging in time behind changes in relative velocity, normal load, or both.

In the involved cases of oscillatory relative motion, particularly those with small amplitude, the modeling requires further details of the friction force in terms of its dependence on both displacement and velocity. Velocity reversal and the change of friction with displacement are important in many applications, where the friction force remains constant during sliding, but changes with displacement during each reversal of direction. The slope of the friction change with displacement can be treated as the equivalent linear tangential stiffness of the bonding force. Moreover, some modeling needs to take into account the static friction time-dependent characteristics.

The analysis of certain dynamic systems needs to represent friction as a function depending on pressure or temperature in addition to velocity [180–187].

A more general description of friction than Eq. [3.98] is given as follows to include Coulomb friction, viscous friction, static friction, and the Stribeck effect:

$$F = \begin{cases} F(v_r) & \text{if} \quad v_r \neq 0 \\ F_e & v_r = 0 \text{ and } |F_e| < F_s \, , \\ F_s\mathrm{sgn}(F_e) & v_r = 0 \text{ and } |F_e| \geq F_s \end{cases} \quad [3.100]$$

where $F(v_r)$ is an arbitrary function that captures the friction profile of the Stribeck curve. Different parameterizations have been proposed for $F(v_r)$, one of which is $F(v_r) = N\mu(v_r)$, in which N is normal force and $\mu(v_r)$ is given in Eq. [3.99]. A common form for the friction function is given by [180]

$$F(v_r) = f_c + (f_s - f_c)e^{-|v_r/v_s|^{\delta_s}} + f_v v_r, \qquad [3.101]$$

where f_s, f_c, and δ_s are constants characterizing static and dynamic friction, v_s is Stribeck velocity, and f_v is viscosity.

Another model is the hyperbolic model given by

$$F(v_r, x) = \begin{cases} F(x) = \min(|F_{ex}(x)|, F_s)\mathrm{sgn}[F_{ex}(x)], v_r = 0, \\ F(v_r) = -\dfrac{F_s \mathrm{sgn}(v_r)}{1 + \delta|v_r|}, v_r \neq 0 \end{cases}, \qquad [3.102]$$

in which $F_{ex} = kx$ is the externally applied force and is displacement-dependent. The friction force in the stick phase is limited by the maximum static friction force:

$$|F(x)| \leq F_s. \qquad [3.103]$$

The ADCK (which stands for "Armstrong-H'elouvry, Dupont, Canudas de Wit, and Karnopp") model is a more general model with more parameters [180–181]. It actually consists of two separate models with one for stiction and one for sliding. The friction is modeled as a stiff spring during stiction:

$$F_l(x) = -k_l x, \qquad [3.104]$$

where x is the presliding displacements and k_l is the equivalent stiffness. When sliding, the friction is modeled as follows, including Coulomb friction, viscous friction, and Stribeck friction:

$$F_f(\dot{x}, t) = \left[F_c + F_v|\dot{x}| + F_s(\gamma, t_2)\frac{1}{1 + (\dot{x}(t - \tau_L)/\dot{x}_s)^2}\right]\mathrm{sgn}(\dot{x}), \qquad [3.105]$$

in which

$$F_s(\gamma, t_2) = F_a + (F_\infty - F_a)\frac{t_2}{t_2 + \gamma}, \qquad [3.106]$$

in which F_f is friction force, F_c is Coulomb friction, F_v is viscous force, and F_s describes the varying friction level at breakaway or the static friction. The level of the static friction force F_s varies with dwell time t_2. Force F_a is the magnitude of the Stribeck friction at the end of the previous sliding period; force F_∞ is the magnitude of the Stribeck friction after dwell; γ is an empirical parameter characterizing time dependence of static friction; τ_L is time delay accounting for the desired frictional memory; and $\dot{x}_s$ is the characteristic velocity of Stribeck friction.

It is noted that the inherent characteristics of the velocity dependence of friction actually make it difficult to observe this phenomenon accurately in experiments. The friction is typically investigated via steady-state sliding experiments in friction testers whose response to sudden displacement input is elastic rather than viscous. A coefficient of friction of negative slope could eliminate the possibility of steady-state sliding, as the dynamics of the tester consists of negative damping and results in unstable oscillations. In other words, the apparent phenomena of a negative slope of friction versus speed in a steady-state sliding experiment could be the result of the interactions between the system dynamics and the complicated constitutive relationship for friction. The oscillation of the system could reduce interface friction, and system oscillation due to asperity is usually proportional to sliding speed. Thus, the friction speed dependence could be the result of interaction. Because of these considerations, rate and state friction models have been developed to simulate friction dynamics [188–189]. In these models, the dependence of friction on the relative velocity between the two bodies in contact is modeled using a differential equation. These models include the presliding displacement. The state variable friction models have been developed to quantify dynamic friction behavior, including Stribeck friction, rising static friction, and frictional memory.

A widely used nonlinear model that characterizes hysteretic behavior in mechanical systems is the Bouc-Wen model [190–195]. The model was developed to predict various hardening or softening and smoothly varying or nearly bilinear hysteretic behavior. This model was extended to characterize hysteretic behavior, as shown in Figure 3.26. The model is restricted to rate-independent hysteretic systems in which velocity-dependent damping effects are nearly negligible. The model has also extended to combine with the Coulomb friction and velocity-dependent damping in order to deal with the rate-dependent damping behavior of some systems like wire cable isolators. In the Bouc-Wen model, the interfacial force is assumed to consist of a linear part and a nonlinear part:

$$F(t) = kx(t) + k_n z(t), \qquad [3.107]$$

in which the friction force is denoted by $F(t)$, the excitation displacement is denoted by $x(t)$, and k is linear spring stiffness. The nonlinear force is characterized by $z(t)$. The constant k_n is used to scale the nonlinear force, so

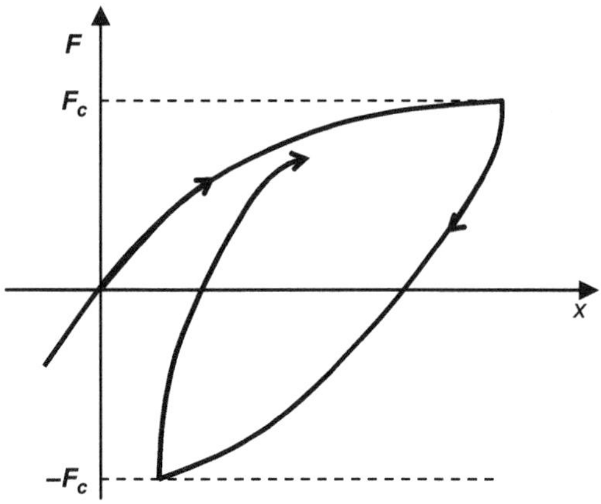

Figure 3.26 Friction force as a function of displacement

that here, $z(t)$ has the dimension of displacement. The Bouc-Wen model is represented by the following form:

$$\dot{z}(t) = \alpha\dot{x}(t) - \beta|\dot{x}(t)||z(t)|^{n-1}z(t) - \gamma\dot{x}(t)|z(t)|^n, \quad [3.108]$$

where the dot denotes the time derivative, and α, β, γ, and n are the model parameters to be determined. The parameters α, β, and γ and the exponent n can be estimated by parameter identification techniques based on experimental data.

In solid mechanics, various models have been developed to characterize hysteresis. One of such models is the Dahl model [196], which uses the stress–strain curve in classic solid mechanics as a starting point for the friction. The stress–strain curve is modeled using the following differential equation:

$$\frac{dF}{dx} = \sigma\left(1 - \frac{F}{F_c}\mathrm{sgn}(v_r)\right)^i, \quad [3.109]$$

where x is the displacement, F is the friction force, F_c is the Coulomb friction force, σ is the contact stiffness coefficient, v_r is slip speed, and the exponent i determines the shape of the stress–strain curve. Then the following friction model is introduced by including presliding displacement and incorporating tangential compliance:

$$F = \sigma z, \sigma > 0; \dot{z} = \dot{x}\left[1 - \frac{\sigma}{F_c}\mathrm{sgn}(\dot{x})z\right]^i, \quad [3.110]$$

where z specifies the state of strain in the frictional contact. The friction force is determined for all trajectories. The integer exponent i was used to characterize the transition rate of z to attain an optimal experimental fit. For simplification, the exponent i is usually assumed to be 1. The steady-state version of this model gives Coulomb friction. Dahl's model is a comparatively simple model capturing many phenomena like hysteresis, but it does not cover the Stribeck effect because the model assumes that the friction depends only on displacement.

The LuGre model extended the Dahl model to include the Stribeck effect [197–200]. The LuGre model can captures many aspects of friction, such as stick-slip motion. This model with continuous states is interpreted as an example of Prandtl's elasto-plastic material model, and has been used in many complicated engineering problems like tire friction modeling. The LuGre model is based on an internal variable z, which can be understood as the average deflection or tangential strain of the microscopic contact elements (i.e., the asperities or contact zones).

Moreover, some elasto-plastic models have been proposed to overcome the drawbacks of the LuGre model, which exhibits drift for arbitrarily small external forces [198]. Schematically, it uses a so-called bristle assumption to interpret friction [200]. Let us assume two proximity contact surfaces with contacting asperities extending from each. The asperities can be represented by small bristles. The friction between the two surfaces is assumed to be caused by many engaged bristles. The portion of friction contributed by each bristle is proportional to the strain of the bristle. When the strain exceeds a certain level, the bristle engagement or bond is broken. As such, the bristle behaves like stiff spring with damper, each giving rise to microscopic displacements and restore forces. If the displacement becomes too large, the engaged junctions break. When this breakaway occurs, macroscopic sliding (slipping) starts. The friction is thus modeled as the average deflection of the bristles. When a tangential force is applied, the bristles deflect like spring-damper. Denoting by z the average bristle deflection as the internal state, the model is given by

$$F = \sigma_0 z + \sigma_1 \dot{z} + \sigma_2 v_r, \sigma_i > 0, i = 0, 1, 2, \quad [3.111]$$

in which σ_0, σ_1, and σ_2 are Coulomb, damping, and viscous friction parameters for the tangential compliance, respectively; and $\dot{z}$ is given by

$$\dot{z} = v_r \left(1 - \frac{\sigma_0}{f_{ss}(v_r)} \mathrm{sgn}(v_r) z\right)^i, \quad [3.112]$$

where σ_0 can be considered as the stiffness of the bristles and σ_1 is the damping. One more extension of the models allowing a purely elastic regime is given by the following formulation [187,198]:

$$F(z, \dot{z}, v_r, w) = \sigma_0 z + \sigma_1 \dot{z} + \sigma_2 v_r + \sigma_3 w, \sigma_i > 0, i = 0, 1, 2, 3, \quad [3.113]$$

in which

$$\dot{z}(z, v_r) = v_r \left[1 - \alpha(z, v_r) \frac{\sigma_0 \mathrm{sgn}(v_r) z}{y_{ss}(v_r)} \right]^i, \quad [3.114]$$

where $\alpha(z, v_r)$ is an adhesion map that controls the rate of change of z to avoid drift.

The model defines the average bristle behavior as a first-order system, with z and $\dot{z}$ being interpreted as the mean bristle displacement and velocity, respectively; and v_r is the relative velocity between the two bodies in contact. The model suggests that the friction force F results from four components: an elastic term, an internal dissipation term, a viscosity term, and a noise term. The fourth component term $w(t)$ is assumed to be as a pseudorandom function of time series representing the random effects or uncertainties of interfaces like surface roughness. The auxiliary functions z_{ss} and α are defined as [30]

$$y_{ss}(v_r) = \left[f_c + (f_s - f_c) e^{-(v_r/v_s)^2} \right] \quad [3.115]$$

$$\alpha(v_r, z) = \begin{cases} \left.\begin{array}{ll} 0 & |z| < z_{ba} \\ \alpha_m(v_r, z) & z_{ba} < |z| < z_{\max}(v_r), \\ 1 & |z| > z_{\max}(v_r), \end{array}\right\} \text{ if } \mathrm{sgn}(v_r) = \mathrm{sgn}(z) \\ 0 \qquad\qquad\qquad\qquad\qquad \text{ if } \mathrm{sgn}(v_r) \neq \mathrm{sgn}(z), \end{cases} \quad [3.116]$$

in which the function $\alpha_m(v_r, z)$ is parameterized as

$$\alpha_m(v_r, z) = \frac{1}{2} \left[1 + \sin\left(\pi \frac{z - (z_{\max}(v_r) + z_{ba})/2}{z_{\max}(v_r) - z_{ba}} \right) \right], \quad [3.117]$$

which describes the transition between elastic and plastic behavior. The parameter z_{ba} defines the point where α starts to take nonzero values and is called a *breakaway displacement*.

In general, for the constant normal stress cases, the generalized friction model has the following form [181–184],

$$F_f(t) = f(V, x_1, x_2, \ldots, x_n) \quad [3.118]$$
$$\dot{x}_i = g_i(V, x_1, x_2, \ldots, x_n), i = 1, 2, \ldots, n,$$

in which x_i is a state variable. This model suggests that a sudden velocity change is unable to create a sudden change in the state, but it does change its time derivative. The state variables can be found in many physical interpretations in applications.

These models only consider the interfacial tangential direction force and deformation. Oden and Marins introduced a constitutive model for frictional interfaces that represents both the nonlinear normal compliance and sliding resistance of the interface [201–202]. The constitutive equations consist of two parts: a normal interface law and a friction law. For the normal response, the normal compliance of the interface is modeled as

$$\sigma_N = c_N a^{m_N} + b_N a^{I_N} \dot{a}, \text{ for } a \geq 0, \quad [3.119]$$

$$\sigma_N = 0, \text{ for } a < 0, \quad [3.120]$$

where σ_N is the normal stress, a is the normal approach on the interface, and $\dot{a}$ is its time derivative. The coefficients c_N,b_N, m_N, and I_N depend on the properties of the contacting surfaces and the materials of the two components. The second constitutive equation gives the friction law:

$$\text{When } a < 0, \text{then } \sigma_T = 0; \quad [3.121]$$

$$\text{When } a \geq 0, \text{then} |\sigma_T| \leq c_T a^{m_T}$$

$$\text{and} \begin{cases} |\sigma_T| < c_T a^{m_T}, \text{then } \dot{d} = 0 \\ |\sigma_T| = c_T a^{m_T}, \text{then } \dot{d} = -\lambda \sigma_T \quad (\lambda \geq 0) \end{cases}, \quad [3.122]$$

where $\dot{d}$ is the sliding velocity, calculated as the time derivative of sliding distance; and index T indicates a direction tangential to the contact surface. The friction force is a function of the normal approach of the two surfaces, which in turn depends on the normal force.

For a given surface profile and response curves, the parameters in Oden-Martins law can be determined and the static coefficient of friction can be estimated. The friction law was also modified to model velocity dependence, in which the coefficient of friction is given as a function of the relative tangential velocity at the interface:

$$\mu = \left| \frac{\sigma_T}{\sigma_N} \right| = \mu(v_T). \quad [3.123]$$

A further extension of the law is that the coefficient of friction depends on the normal load. The load dependence is often weak for hard materials but may be appreciable for relatively soft materials. For instance, Eq. [3.72] is a simplified friction model to characterize the relationship between coefficient of friction and interface contact pressure.

In engineering applications, data fitting–based empirical models have been used to quantify friction as the function of operation parameters like velocity, pressure, and temperature. If the involved database is very large, and the engagement data are directly gathered or indirectly derived from a variety of sources, the data can be simply treated as a large look-up table of friction under various conditions. They can be reduced to a compact mathematical representation through a system identification technique or using a typical static model. Static model is most commonly used, in which friction is measured as a function of applied pressure, slip speed, temperature, and time. A process parameter, denoted as L, may be added as another variable. Mathematically, friction can be written as

$$F = \text{function}\,(p, V, T, L, t). \qquad [3.124]$$

In practice, an empirical model is often coupled with an algebraic model.

Based upon certain system identification techniques and assuming a certain functional form $f_1, \ldots, f_k$ with a set of unknown parameters, $a_{1n}, \ldots, a_{in}, i = 1, \ldots, k$, we can express friction as

$$F = \sum_k f_i(p, V, T, L, t | a_{i1}, \ldots, a_{in}). \qquad [3.125]$$

It can be fitted to the measured data, and then it is solved for a friction coefficient set. Many techniques are available to prescribe this functional form and to determine its empirical parameters.

Generally, the modeled friction should be verified and validated. The friction is associated with system dynamics and interface geometry, mechanical, chemical, and physical properties.

Consider a dynamic mechanical/structural system consisting of a sliding interface, and assume that the system has dynamic load $P(t)$, as shown in Figure 3.27. The apparent velocity is V_0, and $\dot{x}(t)$ is vibration velocity for an arbitrary instant. The normal force is $N(t)$ and the tangential resistance force $F(t)$ is the friction force. The normal force is comprised of both the dynamic load due to the system and the normal component of any adhesive forces in the interface.

For a general dynamic system, $F(t)$ is obviously also an internal response or responsive force that depends on entire system dynamics [for example, through $N(t)$ and $\dot{x}(t)$], as well as the interface state and parameters (adhesion, wear, fracture, interlock, interface liquid films, and other factors). The interface affects system dynamics through $N(t)$ and $F(t)$.

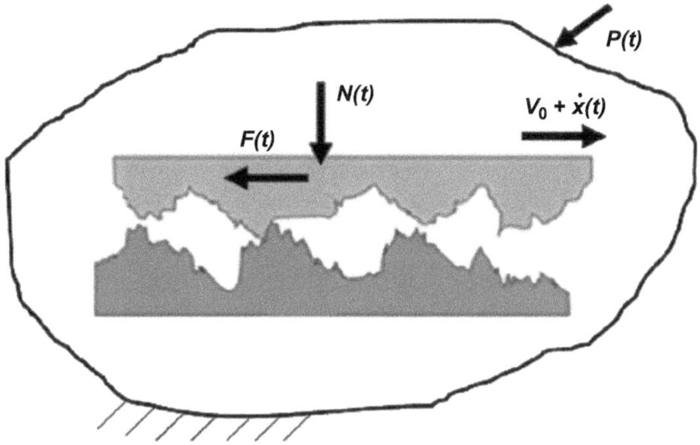

Figure 3.27 Schematic of a dynamical system consisting of a sliding interface

The interface forces are not fully determined in advance by interface parameters. In principle, the interface forces are time-varying responses of the dynamic system.

In most engineering applications, the concept of coefficient of friction $\mu = \overline{F}(t)/\overline{N}(t)$ has been used to simplify the analysis. In complex dynamics systems, the interface forces as a dynamic property or time-varying response of the dynamic system need to be quantified, verified, and validated based on a measured dynamics response. This is usually conducted by a model-based identification from measurements [203–212].

A comparatively new technique for system identification of friction and contact model is based on an artificial neural network [213–216], which is a tool for general pattern recognition. It empirically identifies the input-output relationships of an unknown system based upon available data. The artificial neural network is especially applicable for a nonlinear system-identification or system-modeling tool when the physical nature of a target system is not well understood. Next, we briefly illustrate the basic neural net structure and operations for the friction component modeling. The artificial neural network can be considered as a "black box" that accurately recreates system friction behaviors even if the system is indescribable with first principles. A two-layer feedforward neural net has the prescribed functional form as follows [213]:

$$y_k = f_2\left\{\sum_j W_{jk}\left[f_1\left(\sum_i W_{ij}x_i\right)\right]\right\}, \qquad [3.126]$$

where i is the index for input variables, j is the index for 1st layer neurons, and k is the index for output variables. Neural network weights (W_{ij}, W_{ik}) represent unknown model parameters. While a set of data flows through the net from left to right, a connector multiplies data by a weight. Then, a node carries out summation ($\sum$) and a simple linear or nonlinear operation (f_1,f_2) such as a sigmoid transformation. It is mathematically proven that this simple two-layer neural net is capable of capturing the input to output relationships of any continuous functions when connector weights are suitably selected. Thus, a key to construct a successful neural net friction model is to optimize a set of weights using the data from a target system. This process is often called *neural net training* and is analogous to determining coefficients in a conventional regression or system identification technique.

REFERENCES

[1] K.L. Johnson, Contact Mechanics, Cambridge University Press, Cambridge, 1985.
[2] F.P. Bowden, D. Tabor, Friction and Lubrication of Solids, Clarendon Press, Oxford, 1964.
[3] E. Rabinowicz (Ed.), Friction and Wear of Materials, second ed., Wiley, New York, 1995.
[4] B. Bhushan, Principles and Applications of Tribology, John Wiley & Sons, New York, 1999.
[5] T.A. Stolarski, Tribology in Machine Design, Butterworth Heinemann, Woburn, MA, 1990.
[6] P.J. Blau, Friction Science and Technology, Marcel Dekker, New York, 1986.
[7] J.N. Israelachvili, Intermolecular and Surface Forces, Academic Press, London, 1991.
[8] B.N.J. Persson, Sliding Friction: Physical Principles and Applications, Springer, Heidelberg, Germany, 1998.
[9] D.F. Moore, The Friction and Lubrication of Elastomers, Pergamon Press, Oxford, 1972.
[10] B. Bhushan (Ed.), Handbook of Micro/nanotribology, second ed., CRC, Boca Raton, FL, 1999.
[11] B. Bhushan, B.K. Gupta, Handbook of Tribology: Materials, Coatings, and Surface Treatments, McGraw Hill, New York, 1991.
[12] R. Wiesendanger, Scanning Probe Microscopy and Spectroscopy: Methods and Applications, Cambridge University Press, Cambridge, UK, 1994.
[13] A.W. Adamson (Ed.), Physical Chemistry of Surfaces, fifth ed., John Wiley, New York, 1990.
[14] G. Sheng, Q. Wang, Brake NVH Technology AMC report 82 (2002).
[15] A.C. Fischer-Cripps, Introduction to Contact Mechanics, Springer, Berlin, 1999.
[16] P.J. Blau, The significance and use of the friction coefficient, Trib. Intl. 34 (9) (2001) 585–591.
[17] B. Bhushan, J.N. Israelachvili, U. Landman, Nano tribology, wear, and lubrication at the atomic scale, Nature 374 (13) (1995) 607–616.
[18] G. He, M.H. Muser, M.O. Robbins, Adsorbs layers and the origin of static friction, Science 284 (5420) (1999) 1650–1652.
[19] E. Gerde, M. Marder, Friction and fracture, Nature 413 (2001) 285–288.

[20] K.C. Ludema, Friction, Wear, and Lubrication: A Textbook in Tribology, CRC Press, Boca Raton, FL, 1996.

[21] I.M. Hutchings, Tribology: Friction and Wear of Engineering Materials, CRC Press, Boca Raton, FL, 1992.

[22] I.L. Singer, H.M. Pollack, Fundamentals of Friction: Macroscopic and Microscopic Processes, Kluwer, Dordrecht, the Netherlands, 1992.

[23] J.B. Sokoloff, Theory of energy dissipation in sliding crystal surfaces, Phys. Rev. B 42 (1) (1990) 760–765.

[24] L.W. Bruch, M.W. Cole, E. Zaremba, Physical Adsorption: Forces and Phenomena, Clarendon Press, Oxford, UK, 1997.

[25] C.M. Mate, et al., Atomic-scale friction of a tungsten tip on a graphite surface, Phys. Rev. Lett. 59 (17) (1987) 1942–1945.

[26] J.A. Greenwood, J.B.P. Williamson, Contact of nominally flat surfaces, Proc. R. Soc. Lond. A 295 (1442) (1966) 300–319.

[27] K.L. Johnson, K. Kendall, A.D. Roberts, Surface energy and the contact of elastic solids, Proc. R. Soc. Lond A 324 (1558) (1971) 301–313.

[28] D.M. Tolstoi, Significance of the normal degree of freedom and natural normal vibrations in contact friction, Wear 10 (3) (1967) 199–213.

[29] B. Bhushan, Contact mechanics of rough surfaces in tribology: Single asperity contact, Appl. Mech. Rev. 49 (5) (1996) 275–298.

[30] B. Bhushan, Contact mechanics of rough surfaces in tribology: Multiple asperity contact, Tribol. Lett. 4 (1) (1998) 1–35.

[31] B. Bhushan, A. Majumdar, Elastic-plastic contact model of bifractal surfaces, Wear 153 (1) (1992) 53–64.

[32] W. Peng, B. Bhushan, A numerical three-dimensional model for the contact of layered elastic/plastic solids with rough surfaces by variational principle, ASME J. Tribol. 123 (2) (2001) 330–342.

[33] M. Yu, B. Bhushan, Contact analysis of three-dimensional rough surfaces under frictionless and frictional contact, Wear 200 (1) (1996) 265–280.

[34] S.K. Chilamakuri, B. Bhushan, Contact analysis of non-Gaussian random surfaces, Proc. Inst. Mech. Engrs. Part J: Journal of Engineering Tribology 212 (1) (1998) 19–32.

[35] L. Kogut, I. Etsion, A static friction model for elastic-plastic contacting rough surfaces, J. Tribol. 126 (1) (2004) 34–40.

[36] I. Etsion, I. Front, A model for static sealing performance of end face seals, Tribol. Trans. 37 (1) (1994) 111–119.

[37] W.R. Chang, I. Etsion, D.B. Bogy, Static friction coefficient model for metallic rough surfaces, J. Tribol. 110 (1) (1988) (1988) 57–63.

[38] W.R. Chang, I. Estion, D.B. Bogy, An elastic-plastic model for the contact of rough surfaces, J. Tribol. 109 (2) (1987) 257–263.

[39] L. Kogut, I. Etsion, A semi-analytical solution for the sliding inception of a spherical contact, J. Tribol. 125 (3) (2003) 499–506.

[40] W.R. Chang, I. Etsion, D.B. Bogy, Adhesion model for metallic rough surfaces, ASME J. Tribol. 110 (1) (1988) (1988) 50–56.

[41] L. Kogut, I. Etsion, Adhesion in elastic-plastic spherical micro-contact, J. Coll. Inter. Sci. 261 (2) (2003) 372–378.

[42] R.B. King, T.C. O'Sullivan, Sliding contact stresses in a two-dimensional layered elastic half-space, Int. J. Solids Structures 23 (5) (1987) 581–597.

[43] K. Komvopoulos, Elastic-plastic finite element analysis of indented layered media, J. Tribol. 111 (3) (1989) 430–439.

[44] W. Yan, K. Komvopoulos, Contact analysis of elastic–plastic fractal surfaces, J. Appl. Physics 84 (7) (1998) 3617–3624.

[45] K. Komvopoulos, D.H. Choi, Elastic finite element analysis of multiasperity contact, J. Tribol. 114 (4) (1992) 823–831.
[46] E.R. Kral, K. Komvopoulos, Three-dimensional finite element analysis of subsurface stress and strain fields due to sliding contact on an elastic-plastic layered medium, J. Tribol. 119 (2) (1997) 332–341.
[47] L.E. Malvern, Introduction to the Mechanics of a Continuous Medium, Prentice-Hall, New York, 1969.
[48] K. Mao, T. Bell, Y. Sun, Effect of sliding friction on contact stresses for multilayered elastic bodies with rough surfaces, J. Tribol. 119 (3) (1997) 476–480.
[49] T. Merriman, J. Kannel, Analyses of the role of surface roughness on contact stresses between elastic cylinders with and without soft surface coating, J. Tribol. 111 (1) (1989) 87–94.
[50] T. Nogi, T. Kato, Influence of a hard surface layer on the limit of elastic contact—Part I: Analysis using a real surface model, J. Tribol. 119 (3) (1997) 493–500.
[51] W. Peng, B. Bhushan, Modeling of surfaces with bimodal roughness distribution, Proc. Inst. Mech. Eng. Part J: J. Eng. Tribol. 214 (5) (2000) 459–470.
[52] I.A. Polonsky, L.M. Keer, A numerical method for solving rough contact problems based on the multi-level multi-summation and conjugate gradient techniques, Wear 231 (2) (1999) 206–219.
[53] I.A. Polonsky, L.M. Keer, Fast methods for solving rough contact problems: A comparative study, J. Tribol. 122 (1) (2000) 36–41.
[54] A.W. Bush, R.D. Gibson, G.P. Keogh, The limit of elastic deformation in the contact of rough surfaces, Mech. Res. Commun. 3 (3) (1976) 169–174.
[55] H.M. Pollock, Surfaces Forces and Adhesion. Fundamentals of Friction: Macroscopic and Microscopic Processes, Springer, Weteringschans, Netherlands, 1992, pp. 77–94.
[56] J.A. Ogilvy, Numerical simulation of elastic-plastic contact between anisotropic rough surfaces, J. Physics D: Appl. Phys. 25 (12) (1992) 1798–1809.
[57] Z. Rymuza, Energy concept of the coefficient of friction, Wear 199 (2) (1996) 187–196.
[58] P.R. Nayak, Random process model of rough surfaces, J. Tribol. 93 (3) (1971) 398–407.
[59] N. Yu, A.A. Polycarpou, Contact of rough surfaces with asymmetric distribution of asperity heights, J. Tribol. 124 (2) (2002) 367–376.
[60] G. Sheng, B. Liu, W. Hua, Discussion: "The stresses induced by dynamic load head-disk contacts and the parameter extrapolation for design", J. Tribol. 123 (3) (2001), 655–655.
[61] G. Sheng, Q.S. Chen, W. Hua, J. Li, An experimental study of dimple separations and head-disk impacts of negative pressure slider in unload process, IEEE Trans. Magnet. 37 (4) (2001) 1859–1862.
[62] G. Sheng, J. Zhang, An experimental and theoretical investigation of disk damage caused by head-disk impact in loading process, IEEE Trans. Magnet. 37 (4) (2001) 1863–1865.
[63] W. Hua, B. Liu, J. Li, G. Sheng, Further studies of unload process with a 9D model, IEEE Trans. Magnet. 37 (4) (2001) 1855–1858.
[64] G. Sheng, W. Hua, J. Zhang, Head-disk impact stress in dynamic loading process and the extrapolation of parameters for slider rounding and interface durability, J. Inform. Stor. Proc. Sys. 3 (3–4) (2001) 203–206.
[65] G. Sheng, B. Liu, W. Hua, An nonlinear dynamics theory for modeling slider-air bearings in hard disk drives, J. Appl. Phys. 87 (9) (2000) 6173–6175.
[66] G. Sheng, W. Hua, Q.S. Chen, B. Liu, Experimental and analytical study of head-disk impact and dynamics of negative pressure slider in unload process, J. Inform. Stor. Proc. Sys. 2 (2000) 281–292.

[67] W. Hua, B. Liu, G. Sheng, Y. Zhu, Investigations of disk surface roughness on the dynamic performance of proximity recording slider, J. Mag. Mag. Mater. 209 (2) (2000) 163–165.
[68] G. Sheng, B. Liu, W. Hua, Y. Zhu, Stable interface concept and design for nanometer spacing magnetic recording, J. Mag. Mag. Mater. 209 (1) (2000) 160–162.
[69] G. Sheng, B. Liu, W. Hua, Q. Chen, A theoretical model of acoustic emission sensing process and the experimental investigations for near-contact and contact interfaces in magnetic recording system, J. Appl. Phys. 85 (8) (1999) 5621.
[70] G. Sheng, B. Liu, A theoretical model of slider-disk interaction and acoustic emission sensing process for studying interface phenomena and estimating unknown parameters, Tribol. Lett. 6 (3–4) (1999) 233–239.
[71] G. Sheng, B. Liu, W. Hua, Structure and mechanics study of slider design for 5–15 nm head-disk spacing, IEICE Trans. Elec. 82 (12) (1999) 2125–2131.
[72] G. Sheng, B. Liu, W. Hua, A micro-machined dual slider-suspension for near-contact and contact recording, IEEE Trans. Mag. 35 (5) (1999) 2472–2474.
[73] B. Liu, et al., A dual-stage slider-suspension design for contact recording, J. Appl. Phys. 85 (8) (1999) 5609–5611.
[74] W. Hua, B. Liu, G. Sheng, Probability model and its application on the interaction of nanometer spaced slider-disk interface, IEICE Trans. Elec. 82 (12) (1999) 2139–2147.
[75] Z.M. Yuan, B. Liu, G. Sheng, Sensitivity study of 40 Gb/in2 recording system to the fluctuation of head-disk spacing, IEEE Trans. Magnet. 35 (5) (1999) 2241–2243.
[76] W. Hua, B. Liu, G. Sheng, Y.L. Zhu, Probability model for slider-disk interaction, J. Inform. Stor. Proc. Sys. 1 (3) (1999) 273–280.
[77] W. Hua, B. Liu, G. Sheng, Disk roughness and its influence on the performance of proximity recording sliders, IEEE Trans. Magnet. 35 (5) (1999) 2460–2462.
[78] B. Liu, et al., Study of clock head-disk interface failure mechanism in servo-writing process, IEEE Trans. Magnet. 34 (4) (1998) 1723–1725.
[79] J.I. McCool, The distribution of microcontact area, load, pressure, and flash temperature under the Greenwood–Williamson model, J. Tribol. 110 (1) (1988) 106–111.
[80] J.I. McCool, Extending the capability of the Greenwood–Williamson microcontact model, J. Tribol. 122 (3) (2000) 496–502.
[81] K. Deng, W.H. Ko, A study of static friction between silicon and silicon compounds, J. Micromech. Microeng. 2 (1) (1992) 14–20.
[82] D.P. Smith, Tribology of the belt-driven data tape cartridge, Tribol. Intl. 31 (8) (1998) 465–477.
[83] D.G. Alciatore, A.E. Traver, Multipulley belt drive mechanics: Creep theory vs shear theory, J. Mech. Des. 117 (4) (1995) 506–511.
[84] A.A. Polycarpou, A. Soom, Application of a two-dimensional model of continuous sliding friction to stick-slip, Wear 181/183 (1995) 32–41.
[85] P. Dupont, et al., Single-state elastoplastic friction models, IEEE Trans. Automotic Ctrl. 47 (5) (2002) 787–792.
[86] B. Armstrong-Helouvry, P. Dupont, C.C. De Wit, A survey of models, analysis tools, and compensation methods for the control of machines with friction, Automotica 30 (7) (1994) 1083–1138.
[87] J. Dieterich, Modelling of rock friction: 1. Experimental results and constitutive equations, J. Geophys. Res. 84 (B5) (1979) 2161–2168.
[88] A. Ruina, Slip instability and state variable friction laws, J. Geophys. Res. 88 (B12) (1983) 10359–10370.
[89] D.P. Hess, Interaction of vibration and friction at dry sliding contacts, in: A. Guran, F. Pfeiffer, K. Popp (Eds.), Dynamics with Friction, World Scientific Publishing, 1996, part II. New Jersey.

[90] M. Urbakh, et al., The nonlinear nature of friction, Nature 430 (6999) (2004) 525–528.
[91] Y. Zhao, D.M. Maietta, L. Chang, An asperity microcontact model incorporating the transition from elastic deformation to fully plastic flow, J. Tribol. 122 (1) (2000) 86–93.
[92] J.A. Greenwood, J.B.P. Williamson, Contact of nominally flat surfaces, Proc. Royal Soc. London A: Math., Phys., Eng. Sci. 295 (1442) (1966) 300–319.
[93] W.R. Chang, An elastic-plastic contact model for a rough surface with and ion-plated soft metallic coating, Wear 212 (2) (1997) 229–237.
[94] D. Berthe, M. Godet, A more general form of Reynolds equation—Application to rough surfaces, Wear 27 (3) (1974) 345–357.
[95] J.T. Oden, J.A.C. Martin, Models and computational methods for dynamic friction phenomena, Comp. Meth. Appl. Mech. Eng. 52 (1) (1985) 527–634.
[96] J.A.C. Martin, J.T. Oden, F.M.F. Simoes, A study of static and kinetic friction, Int. J. Engr. Sci. 28 (1) (1990) 29–92.
[97] W.W. Tworzydlo, et al., Computational micro- and macroscopic models of contact and friction: Formulation, approach and applications, Wear 220 (2) (1998) 113–140.
[98] A.A. Polycarpou, I. Etsion, Static friction of contacting real surfaces in the presence of sub-boundary lubrication, J. Tribol. 120 (2) (1998) 296–303.
[99] A.A. Polycarpou, I. Etsion, Comparison of the static friction subboundary lubrication model with experimental measurements on thin-film disks, Tribol. Trans. 41 (2) (1998) 217–224.
[100] A. Polycarpou, A. Soom, Modeling unsteady lubricated friction, in: A. Guran, F. Pfeiffer, K. Popp (Eds.), Dynamics with Friction, World Scientific Publishing, 1996, p. 197. New Jersey.
[101] P.J. Blau, The significance and use of the friction coefficient, Tribol. Intl. 34 (9) (2001) 585–591.
[102] L. Chang, A deterministic model for line—contact partial elasto—hydrodynamic lubrication, Tribol. Intl. 28 (2) (1995) 75–84.
[103] C.W. Wu, L.Q. Zheng, An average Reynolds equation for partial film lubrication with a contact factor, J. Tribol. 111 (1) (1989) 188–191.
[104] H. Christensen, K. Tonder, The hydrodynamic lubrication of rough journal bearings, J. Tribol. 95 (2) (1973) 166–172.
[105] S. Timoshenko, S. Woinowsky-Krieger (Eds.), Theory of Plates and Shells, second ed., McGraw-Hill, New York, 1987.
[106] Q. Wang, H.S. Cheng, A mixed lubrication model for journal bearings with a thin soft coating—Part I: Contact and lubrication analysis, Tribol. Trans. 38 (3) (1995) 654–662.
[107] J.J. Wu, The properties of asperities of real surfaces, J. Tribol. 123 (4) (2001) 872–883.
[108] X. Zhai, L. Chang, An engineering approach to deterministic modeling of mixed-film contacts, Tribol. Trans. 41 (3) (1998) 327–334.
[109] Y.W. Zhao, D.M. Maietta, L. Chang, An asperity microcontact model incorporating the transition from elastic deformation to fully plastic flow, J. Tribol. 122 (1) (2000) 86–93.
[110] H.A. Francis, Application of spherical indention mechanics to reversible and irreversible contact between rough surfaces, Wear 45 (2) (1977) 221–269.
[111] B.N.J. Persson, On the theory of rubber friction, Surface Science 401 (3) (1998) 445–454.
[112] B.N.J. Persson, Sliding friction, Surf. Sci. Rep. 33 (3) (1999) 83–119.
[113] L.E. Nielsen, R.F. Landel, Mechanical Properties of Polymers and Composites, Marcel Dekker, New York, 1991.

[114] N.K. Myshkin, M.I. Petrokovets, A.V. Kovalev, Tribology of polymers: Adhesion, friction, wear, and mass-transfer, Tribol. Intl. 38 (11) (2006) 910–921.
[115] K. Friedrich, Z. Lu, A.M. Hager, Recent advances in polymer composites' tribology, Wear 190 (2) (1995) 139–144.
[116] Z. Zhao, B. Bhushan, Effect of bonded-lubricant films on the tribological performance of magnetic thin -film rigid disks, Wear 202 (1) (1996) 50–59.
[117] H. Tian, T. Matsudaira, The role of relative humidity, surface roughness, and liquid buildup on static friction behavior of the head/disk interface, J. Tribol. 115 (1) (1993) 28–35.
[118] X. Tian, B. Bhushan, The micro-meniscus effect of a thin liquid film on the static friction of rough surfaces contact, J. Phys. D: Appl. Phys. 29 (1) (1996) 163–178.
[119] B. Bhushan, C.A. Kotwal, S.K. Chilamakuri, Kinetic meniscus model for prediction of rest stiction, J. Tribol. 120 (1) (1998) 42–50.
[120] C. Gao, X. Tian, B. Bhushan, A meniscus model for optimization of texturing and liquid lubrication of magnetic thin film rigid disks, Tribol. Trans. 38 (2) (1995) 201–212.
[121] Y. Li, F.E. Talke, A model for the effect of humidity on stiction of the head/disk interface, in: B. Bhushan (Ed.), Tribology and Mechanics of Magnetic Storage Systems, STLE, Park Ridge, IL, 1990, 79.
[122] N. Tas, et al., Stiction in surface micromachining, J. Micromech. Microeng. 6 (4) (1996) 385–397.
[123] C.H. Mastrangelo, C.H. Hsu, Mechanical stability and adhesion of microstructures under capillary forces—Part I: Basic theory, J. Microelectromech. Syst. 2 (1) (1993) 33–43.
[124] C.H. Mastrangelo, C.H. Hsu, Mechanical stability and adhesion of microstructures under capillary forces—Part II: Experiments, J. Microelectromech. Syst. 2 (1) (1993) 44–55.
[125] J.N. Israelachvili, D. Tabor, The measurement of van der Waals dispersion forces in the range 1.5 to 130 nm, Proc. R. Soc. A 331 (1584) (1972) 19–38.
[126] M.J. Matthewson, H.J. Mamin, Liquid mediated adhesion of ultra-flat solid surfaces, MRS Proc. 119 (87) (1988). Cambridge University Press.
[127] Y. Li, D. Trauner, F.E. Talke, Effect of humidity on stiction and friction of the head/disk interface, IEEE Trans. Magnetics 26 (5) (1990) 2487–2489.
[128] V. Raman, et al., The dependence of stiction and friction on roughness in thin film magnetic recording disks, J. Appl. Phys. 70 (3) (1991) 1826–1836.
[129] E.L. Decker, et al., Physics of contact angle measurement, Coll. Surf. A 156 (1) (1999) 177–189.
[130] R.V. Sedev, et al., Dynamic contact angles at low velocities, J. Coll. Inter. Sci. 159 (2) (1993) 392–399.
[131] D.Y. Kwok, C.J. Budziak, A.W. Neumann, Measurements of static and low-rate dynamic contact angles by means of an automated capillary rise technique, J. Coll. Inter. Sci. 173 (1) (1995) 143–150.
[132] C.J. Van Oss, Interfacial Forces in Aqueous Media, Marcel Dekker, New York, 1994.
[133] A.G. Egorov, K.G. Kornev, A.V. Neimark, Meniscus motion in a prewetted capillary, Physics of Fluids 15 (10) (2003) 3134–3143.
[134] K.G. Kornev, I.K. Shingareva, A.V. Neimark, Capillary condensation as a morphological transition, Adv. Coll. Inter. Sci. 96 (1) (2002) 143–167.
[135] V.M. Starov, Equilibrium and hysteresis contact angles, Adv. Coll. Inter. Sci. 39 (1992) 147–173.
[136] A. Hamraoui, et al., Can a dynamic contact angle be understood in terms of a friction coefficient? J. Coll. Inter. Sci. 226 (2) (2000) 199–204.
[137] L.M. Pismen, B.Y. Rubinstein, I. Bazhlekov, Spreading of a wetting film under the action of van der Waals forces, Phys. Fluids 12 (3) (2000) 480–483.

[138] C.H. Mastrangelo, C.H. Hsu, Mechanical stability and adhesion of microstructures under capillary forces, J. Microelectromech. Syst. 2 (1) (1993) 33–43.
[139] C.H. Mastrangelo, Adhesion-related failure mechanisms in micromechanical devices, Trib. Lett. 3 (3) (1997) 223–238.
[140] J. Valli, A review of adhesion test method for thin hard coatings, J. Vac. Sci. Tech. A 4 (6) (1986) 3007–3014.
[141] G. Sheng, et al., Design and analysis of MEMS-based slider-suspensions for high performance magnetic recording system, J. Micromech. Microeng. 10 (1) (2000) 64–71.
[142] Y. Zhu, et al., A study of interface dynamics for stiction-free slider and super smooth disk, J. Appl. Phys. 87 (9) (2000) 6149–6151.
[143] B. Liu, S.T. Lim, G. Sheng, Landing zone skew angle, long-term stiction, and contamination buildup on slider surfaces of disk drives, J. Appl. Phys. 81 (8) (1997) 5408–5410.
[144] B. Liu, G. Sheng, S.T. Lim, Meniscus force modeling and study on the fluctuation of stiction/friction force in CSS test process, IEEE Trans. Magnet. 33 (5) (1997) 3121–3123.
[145] B. Liu, et al., Tracking seeking, head-disk spacing variation, and head-disk interface failure, IEEE Trans. Magnet. 33 (5) (1997) 3136–3138.
[146] B. Liu, Q.S. Chen, G. Sheng, Skew angle and head-disk integration for Gb/in2 area density recording, Institute of Electronics, Information and Communication Engineers (Japan) Technical Report of IEICE (1996). MR96–39, 19.
[147] E.J. Berger, F. Sadeghi, C.M. Krousgrill, Analytical and numerical modeling of engagement of rough, permeable, grooved wet clutches, J. Tribol. 119 (1) (1997) (1997) 143–148.
[148] J.Y. Jang, M.M. Khonsari, Thermal characteristics of a wet clutch, J. Tribol. 121 (3) (1999) 610–617.
[149] S. Natsumeda, T. Miyoshi, Numerical simulation of engagement of paper based wet clutch facing, J. Tribol. 116 (2) (1994) 232–237.
[150] Y. Yang, R.C. Lam, T. Fujii, Prediction of torque response during the engagement of wet friction clutch, 1998. SAE Technical Paper 981097.
[151] N. Patir, H.S. Cheng, Average flow model for determining effects of three-dimensional roughness on partial hydrodynamic lubrication, J. Tribol. 100 (1) (1978) 12–17.
[152] B. Ganemi, R. Olsson, B. Lundström, Durability and friction characteristics of a novel high-performance four-wheel-drive automotive transmission fluid, 2000. SAE Technical Paper 2000–01–2907.
[153] R. Mäki, Limited slip wet clutch transmission fluid for AWD differentials; Part 1: System requirements and evaluation methods, 2003. SAE Technical Paper 2003–01–1980.
[154] Y. Yang, Modeling of heat transfer and fluid hydrodynamics for a multidisc wet clutch, 1995. SAE Technical Paper 950898.
[155] P. Zagrodzki, Numerical analysis of temperature fields and thermal stresses in the friction discs of a multidisc wet clutch, Wear 101 (3) (1985) 255–271.
[156] R. Mäki, Measurement and characterization of anti-shudder properties in wet clutch applications, 2005. SAE Technical Paper 2005–01–0878.
[157] J. Brecht, K. Schiffner, Influence of friction law on brake creep-groan, 2001. SAE Technical Paper 2001–01–3138.
[158] M. Eriksson, S. Jacobson, Friction behaviour and squeal generation of disc brakes at low speeds, Proc. Inst. Mech. Eng., Part D: J. Auto. Eng. 215 (12) (2001) 1245–1256.
[159] M. Eriksson, A. Lundquist, S. Jacobson, A study of the influence of humidity on the friction and squeal generation of automotive brake pads, Proceedings of the Institution of Mechanical Engineers, Part D: J. Auto. Eng. 215 (3) (2001) 329–342.

[160] H. Gao, G.C. Barber, Microcontact Model for Paper-Based Wet Friction Materials, J. Tribol. 124 (2) (2002) 414–419.
[161] G. Sheng, F.S. Cui, H. Zhen, Modeling and development of computer programe for friction induced instability analysis of HGA-Disk-cassette coupling system, SONY SRL Report-MCAP/169–2, 2000.
[162] G. Sheng, Q. Wan, Brake Noise Technology, CMA Consultancy Report, Anaheim, CA, 2002.
[163] G. Sheng, Q. Wan, Automotive Friction Noise Control Technology, CMA Consultancy Report, Anaheim, CA, 2002.
[164] G.G. Adams, S. Müftü, N.M. Azhar, A nanoscale multi-asperity contact and friction model, 2002 Proc. IMECE, New Orleans, LA, 2002.
[165] J.A. Hurtado, K.S. Kim, Scale effects in friction of single asperity contacts. Part I: From concurrent slip to single-dislocation-assisted slip, Proc. Roy. Soc. Lond. A 455 (1999) 3363–3384.
[166] J.A. Hurtado, K.S. Kim, Scale effects in friction in single asperity contacts. Part II: Multiple-dislocation-cooperated slip, Proc. Roy. Soc. Lond. A 455 (1999) 3385–3400.
[167] D. Maugis, Adhesion of spheres: The JKR-DMT transition using a Dugdale model, J. Coll. Inter. Sci. 150 (1) (1992) 243–269.
[168] K.L. Johnson, J.A. Greenwood, An adhesion map for the contact of elastic spheres, J. Coll. Inter. Sci. 192 (2) (1997) 326–333.
[169] A. Akay, Acoustics of friction, J. Acous. Soc. Amer. 111 (4) (2002) 1525–1548.
[170] A.A. Seireg, Friction-induced sound and vibration, in Friction and Lubrication in Mechanical Design, Marcel Dekker, New York, 1998.
[171] A. Guran, F. Pfeiffer, K. Popp, Dynamics with Friction: Modeling, Analysis, and Experiment, World Scientific Publishing, 1996. New Jersey.
[172] R.A. Ibrahim, Friction-induced vibration, chatter, squeal, and chaos. Part 1: Mechanics of contact and friction, Appl. Mech. Rev. 47 (7) (1994) 209–226.
[173] R.A. Ibrahim, Friction-induced vibration, chatter, squeal, and chaos Part 2: Dynamics and modeling, Appl. Mech. Rev. 47 (7) (1994) 227–253.
[174] A.H. Nayfeh, D.T. Mook, Nonlinear Oscillations, Wiley, New York, 1979.
[175] S. Timoshenko, D.H. Young, W. Weaver, Vibration Problems in Engineering, Wiley, New York, 1974.
[176] W. Soedel (Ed.), Vibrations of Shells and Plates, second ed., Marcel Dekker, New York, 1993.
[177] A.D. Pierce, Acoustics: An Introduction to Its Physical Principles and Applications, McGraw-Hill, New York, 1981.
[178] K.U. Ingard, Fundamentals of Waves and Oscillations, Cambridge University Press, Cambridge, UK, 1988.
[179] B. Feeny, A. Guran, N. Hinrichs, K. Popp, A historical review on dry friction and stick–slip phenomena, Appl. Mech. Rev. 51 (5) (1998) 321–341.
[180] L.C. Bo, D. Pavelescu, The friction-speed relation and its influence on the critical velocity of the stick-slip motion, Wear 82 (3) (1982) 277–289.
[181] D. Karnopp, Computer simulation of stick-slip friction in mechanical dynamic systems, ASME J. Dyn. Sys., Meas. Ctrl. 107 (1) (1985) 100–103.
[182] B. Armstrong-H'elouvry, P. Dupont, C.C. De Wit, A survey of models, analysis tools, and compensation methods for the control of machines with friction, Automatica 30 (7) (1994) 1083–1138.
[183] B. Armstrong-H'elouvry, Stick-slip and control in low-speed motion, IEEE Trans. Auto. Ctrl. 38 (10) (1993) 1483–1496.
[184] P.E. Dupont, The effect of friction on the forward dynamics problem, Int. J. Robotics Res. 12 (2) (1993) 164–179.

[185] D.P. Hess, A. Soom, Friction at a lubricated line contact operating at oscillating sliding velocities, J. Tribol. 112 (1) (1990) 147–152.
[186] A.A. Polycarpou, A. Soom, Application of a two-dimensional model of continuous sliding friction to stick-slip, Wear 181 (1995) 32–41.
[187] S. Serafin, The sound of friction: Real-time models, playability, and musical applications, PhD Thesis, Stanford University, 2004.
[188] J.H. Dieterich, Modelling of rock friction: 1. Experimental results and constitutive equations, J. Geophys. Res. 84 (B5) (1979) 2161–2168.
[189] J.R. Rice, A.L. Ruina, Stability of steady frictional slipping, J. Appl. Mech. 50 (2) (1983) 343–349.
[190] R. Bouc, Forced vibration of mechanical system with hysteresis, in: Proc. Fourth Conf, Nonlinear Oscill, Prague, 1967, p. 315.
[191] Y.K. Wen, Method of random vibration of hysteretic systems, J. Eng. Mech. 102 (2) (1976) 249–263.
[192] T.T. Baber, Y.K. Wen, Random vibration of hysteretic degrading systems, J. Eng. Mech. Div. 107 (6) (1981) 1069–1087.
[193] Y.Q. Ni, J.M. Ko, C.W. Wong, Identification of non-linear hysteretic isolators from periodic vibration tests, J. Sound Vib. 217 (4) (1998) 737–756.
[194] C.W. Wong, Y.Q. Ni, S.L. Lau, Steady-state oscillation of hysteretic differential model. I. Response analysis, J. Eng. Mech. 120 (11) (1994) 2271–2298.
[195] C.W. Wong, Y.Q. Ni, J.M. Ko, Steady-state oscillation of hysteretic differential model. II. Performance analysis, J. Eng. Mech. 120 (11) (1994) 2299–2325.
[196] P.R. Dahl, Solid friction damping of mechanical vibrations, AIAA J. 14 (12) (1976) 1675–1682.
[197] C.C. De Wit, et al., A new model for control of systems with friction, IEEE Trans. Auto. Ctrl. 40 (3) (1995) 419–425.
[198] P. Dupont, et al., Single-state elastoplastic friction models, IEEE Trans. Auto. Ctrl. 47 (5) (2002) 787–792.
[199] P. Dupont, B. Armstrong, V. Hayward, Elasto-plastic friction model: Contact compliance and stiction, 2000 ACC, Chicago, 2000.
[200] D.A. Haessig, B. Friedland, On the modeling and simulation of friction, J. Dyn. Sys., Meas., Ctrl. 113 (3) (1991) 354–362.
[201] J.T. Oden, J.A.C. Martins, Models and computational methods for dynamic friction phenomena, Comp. Meth. Appl. Mech. Eng. 52 (1) (1985) 527–634.
[202] J.A.C. Martins, J.T. Oden, F.M.F. Simoes, A study of static and kinetic friction, Intl. J. Eng. Sci. 28 (1) (1990) 29–92.
[203] C.T. Johnson, R.D. Lorenz, Experimental identification of friction and its compensation in precise, position controlled mechanisms, IEEE Trans. Indus. Appl. 28 (6) (1990) 1392–1398.
[204] J.H. Kim, et al., Identification and control of systems with friction using accelerated evolutionary programming, IEEE Ctrl. Sys. Mag. 16 (4) (1996) 38–47.
[205] R. Zglimbea, V. Finca, C. Marin, Identification of systems with friction via distributions using the simplified Dahl model, Int. J. Math. Model Meth. Appl. Sci. 2 (2008) 269–276.
[206] L. Ravanbod-Shirazi, A. Besançon-Voda, Friction identification using the Karnopp model, applied to an electropneumatic actuator, Proc. Instn. Mech. Engrs., Part I: J. Sys. Ctrl. Eng. 217 (2) (2003) 123–138.
[207] H.C. Meng, K.C. Ludema, Wear models and predictive equations: their form and content, Wear 181 (183) (1995) 443–457.
[208] S.D. Fassois, et al., Identification of pre-sliding friction dynamics, Chaos 14 (2) (2004) 420–430.

[209] J. Zhang, S. Bhattacharyya, N. Simaan, Model and Parameter Identification of Friction during Robotic Insertion of Cochlear-Implant Electrode Arrays, IEEE Press, 2009, 2009 IEEE International Conference on Robotics and Automation Chicago.

[210] S.K. Wang, J. Woodhouse, The frequency response of dynamic friction: A new view of sliding interfaces, J. Mech. Phys. Solids 59 (5) (2011) 1020–1036.

[211] G. Sheng, J. Xu, A parameter identification method for thermal flying-height control sliders, Microsys. Tech. 17 (9) (2011) 1409–1415.

[212] G. Sheng, Sensing and identification of nonlinear dynamics of slider with clearance in sub-5 nanometer regime (invited), 2011, http://dx.doi.org/10.1155/2011/282839. Adv. Tribol., ID 282839.

[213] Y. Fujii, et al., Review of wet friction component models for automatic transmission shift analysis, 2003. SAE Technical Paper 2003–01–1665.

[214] D. Aleksendric, D.C. Barton, Neural network prediction of disc brake performance, Tribol. Intl. 42 (7) (2009) 1074–1080.

[215] A. Senatore, et al., Experimental investigation and neural network prediction of brakes and clutch material frictional behaviour considering the sliding acceleration influence, Tribol. Intl. 44 (10) (2011) 1199–1207.

[216] G. Sheng, et al., Probing and diagnosis of slider–disk interactions in nanometer clearance regime using artificial neural network, Microsys. Technol. 18 (9–10) (2012) 1255–1259.

CHAPTER 4

Friction Dynamics of Vehicle Brake Systems

4.1 INTRODUCTION

The friction dynamics problems of brake systems have been the most challenging technical problems faced by the automotive industry. The prediction and control of the friction dynamics problems of brakes are problems where solutions have been elusive. This chapter presents the fundamental friction dynamics of brake systems. The time-varying, strong nonlinear and stochastic characteristics of the problem are emphasized. After briefly describing the typical brake system, several friction dynamics issues, including groan, judder, low-frequency squeal, and high-frequency squeal, are presented. The basic analytical and numerical approaches for brake friction dynamics are introduced, which include complex eigenvalue analysis and dynamic transient analysis using numerical approaches. The brake instability due to gyroscopic effects is illustrated. Finally, the perspectives of diagnosis and prognosis of friction dynamics are introduced.

4.2 BRAKE SYSTEMS AND FRICTION DYNAMICS

4.2.1 Brake Systems

Brake systems have been widely used in automobiles, motorcycles, rail trains, and aircrafts. In brake systems, friction is both a principal performance factor and a potential cause of undesirable noise and vibration. The structures and principles of the brake systems of different kinds of vehicles are analogous and similar. Friction dynamics has been one of the most challenging problems in the brake industry [1−36].

Friction dynamics related issues affect the reliability and quality of brake systems in many ways. For instance, brake squeal, an acoustic phenomenon caused by dynamic instability that occurs at one or more of the natural frequencies of the brake system, has been the most challenging issue in automotive brake systems, as it has been perceived to be equal to the quality of products by customers. In aircraft braking systems, friction-induced

Friction Dynamics
ISBN 978-0-08-100285-8
http://dx.doi.org/10.1016/B978-0-08-100285-8.00004-3

torque oscillations can lead to excessively high loads in the landing gear and brake structure, which results in passenger discomfort, component failure, or both, and thereby leading to warranty claims.

There are two major types of brakes: disc and drum. Disc brakes have been widely used in both heavy vehicles and passenger cars, despite the fact that drum brakes are predominantly used in trucks and buses. We focus our discussion on disc brakes.

As illustrated in Figure 4.1, a disc brake system consists of a rotor that can rotate about the wheel axis. The caliper assembly in brake system is mounted to the chassis, and the caliper housing can slide on the bracket. The pads can also slide on the anchor bracket, and the piston can slide inside the caliper housing. When hydraulic pressure is applied, the piston presses the pad against the rotor, thus generating braking torque through friction. The brake pad assemblies consist of a friction material that is mounted to a rigid-pad backing plate. The mounting can be achieved by using rivets or adhesives, or integrally molding the friction material to the backing plate. The two brake pad assemblies in a single disc brake are usually distinguished as inboard and outboard. These assemblies are mounted to the caliper or the caliper-mounting bracket in various ways. It is common to use vibration prevention measures like dampers (or shims) or

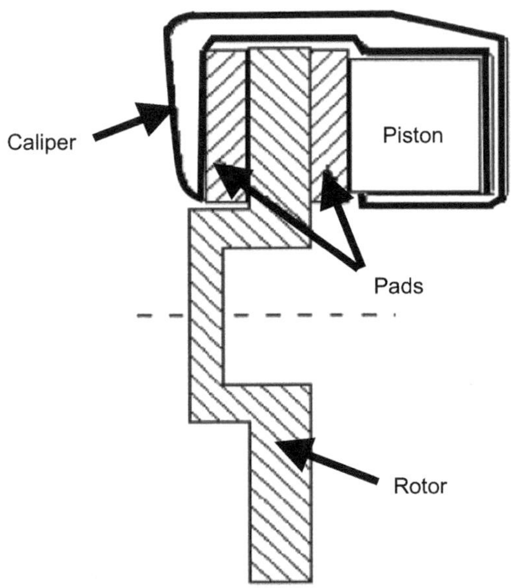

Figure 4.1 Schematic of an automotive disc brake system

damping substances between each of the pad backing plates and the caliper, and the caliper mounting bracket and piston. The rotors usually made of gray cast iron have dominated the market. Other materials like silicon carbide (SiC)—reinforced aluminum, carbon-SiC composites, and sintered carbon have started to emerge. Most pad friction materials are organic material that is based on a metal fiber-reinforced organic matrix. Other types of friction materials include metallic, semimetallic, and carbon-based material. Organic pads are a compound of a number of different materials (up to two dozen), with the function groups of a binder for adhesion, structural materials for enhancing strength, fillers for regulating cost and volume, and frictional additives for friction control. The binder or matrix is normally some kind of phenolic resin. The fibers include nonasbestos organic or resin-bonded metallic. The structural materials include metal, carbon, glass, and kevlar fibers that are added to the linings to give mechanical stability. Fillers are principally used to make the material less expensive and improve the main material. The frictional additives are added to control the coefficient of friction or to change the type of wear. Lubricants such as graphite are used to decrease the coefficient of friction. Abrasive particles are used to increase the coefficient of friction or to increase brake disc wear, for the purpose of removing iron oxides from the disc surface, thus providing more consistent braking properties.

4.2.2 Friction Dynamics of Brake Systems

Almost all of the power dissipated during a braking operation takes the form of heat. Only a very small portion usually affects dynamics. However, even this small amount can cause intolerable dynamic problems.

In braking, the rotor and the pads experience normal and tangential forces at their interface. These forces could interact and develop complicated vibrations.

Even assuming friction force to be constant, the tangential forces acting on the surfaces of a rotor could develop in-plane vibrations and the normal force can develop out-of-plane vibrations. A combination of in-plane and out-of-plane vibrations can change the contact area, thereby changing the presumed constant behavior of the friction force. Oscillatory normal force components always accompany the real oscillatory friction force, thus yielding bending waves and transverse vibrations in the rotor and pads.

Pad vibrations can easily travel to the brake caliper and beyond, causing caliper and other chassis subsystem resonances. Over the years, disc brake vibration and noise have been given various names, such as moan, groan, judder, squeak, squeal, that reflect the pattern of the vibrations and sound generated.

A simple definition is to classify the brake vibration and noise as low-frequency vibration and noise and squeal, depending on whether the frequency is lower or higher than about 1 kHz.

The low-frequency vibration and noise mainly consists of groan and judder. This category of noise is caused by friction material excitation, and the energy is transmitted as a vibrational response through the brake corner and couples with other chassis components.

Squeal is an annoying, usually close to a pure, high-pitched tone caused by dynamic instability that occurs at one or more of the natural frequencies of the brake system. The squeal includes low and high frequencies. Low-frequency squeal is generally classified as squeal with a narrow frequency bandwidth in a frequency range above 1000 Hz, yet below the first circumferential mode of the rotor. High-frequency brake squeal is defined as noise that contains the first circumferential mode of the rotor, which is produced by friction-induced excitation, usually involving coupled mode resonances with squeal occurring at frequencies above 5 kHz. One of the most troublesome types of noise is high-frequency brake squeal.

A typical groan has a spectrum ranging from 10 to 50 Hz, with harmonics extending to 500 Hz. It occurs at low speeds under moderate braking conditions. A groan due to stick-slip excites the low-frequency resonances of the brake system. In particular, resonances of the rigid-rotation mode of the caliper and the local suspension develop and radiate sound without the participation of the rotor.

Judder is due to rotational speed—dependent, continuous displacement pulsations between the rotor and pads, which could magnify themselves as low-frequency vibrations whose frequencies are integer multiples of the rotational speed of the wheel. It can transmit to the chassis and steering and manifest itself as vibration and harshness rather than noise. The conditions that lead to judder generally result from modulation and nonuniform friction force between the rotor and pads. This nonuniformity results from circumferential disc thickness variation and variation in surface finish.

A low-frequency squeal usually contains the first four nodal diameter vibrational modes of the rotor, which has nodal spacing larger than the pad length, and also involves the resonance of certain modes of the caliper bracket. On the other hand, a high-frequency squeal usually contains a higher nodal-diameter mode of the rotor, as well as the circumferential mode. The mode lock in a natural resonant mode due to velocity-dependent friction, or the mode coupling of two close different modes, could be the typical mechanisms of the squeal. The low-frequency noise is sometimes more likely attributed to the friction speed dependence (due to the difference between static and kinetic friction coefficients and the negative slope of the friction coefficient with respect to sliding speed).

In general, the squeal as dynamic instability could be due to the effect of mode lock-in or mode coupling or its combinations with the effects of negative friction slope, or other mechanisms such as sprag-slip, nonsmooth effects (involving friction reversal), parametric resonance, bouncing, and impulsive rubbing effects of friction. Squeal is usually high pitched, with a nearly fixed frequency that could be close to one of the natural frequencies of a component, or in between two natural frequencies of one or more components (coupling), or even their harmonics.

Squeal could exhibit nonstationary features due to the nonstationary nature of instantaneous friction and contact effects and give rise to limit cycles due to friction motion excitation. It usually has a fundamental frequency, which can be assumed to be the vibrational mode of the instantaneously formed system through friction/contact coupling. It is an instantaneous mode that is usually called *squeal mode*.

To provide a precise explanation to specific squeal phenomena, the classifications of some of these mechanisms are necessary. Which mechanism is dominant depends on the components, system/interface, operational conditions, environmental conditions, aging of the brake, and stochastic factors.

4.2.3 Groan and Judder

In the parlance of braking, friction-induced low-frequency vibration and noise are known as *judder, groan, moan,* and *howl,* which correspond to specific frequencies of around 10, 100, 500, and 1000 Hz, respectively.

The term *groan* denotes unpleasant high-intensity, low-frequency vibration (20—200 Hz) and noise generated by brake systems at very low speeds, either accelerating or decelerating, but more frequently in deceleration. Some brake run-in burnishing is usually necessary before groan occurs. The noise could happen throughout the entire stop process in the most severe cases. Usually, it is produced during the middle to end of the braking event. The motion of the brake corner caused by this is rigid body twisting of the caliper housing, caliper bracket, and knuckle.

While groan is rarely a symptom of a brake malfunction, it poses a serious customer satisfaction problem, which often results in costly warranty claims. The phenomenon may appear when the driver applies only light pressure on the brake and driving forces are acting on the vehicle. It occurs when there is the simultaneous application of torque to the wheel and the gradual release or increase of brake pressure. Eventually, the wheel torque load breaks the friction between pad and rotor, causing slippage and energy release. If the torque load is not large enough to maintain slippage, sustained stick-slip vibrations can occur, which transmits a low-frequency noise to the vehicle interior that is perceived as groan.

The root cause of groan is the stick-slip of the brake pads at the rotor surface due to friction coefficient velocity dependence, or a static coefficient of friction that is higher than the dynamic value. The effective brake friction material is likely to exhibit this type of characteristic.

The groan of low-frequency vibration involves rigid-body oscillations of the full assembly, which includes brake pads, caliper, and knuckle through which the system is mounted on the chassis. It is amplified and emitted by the chassis components (axle, suspension, or both).

The spectral content of the groan vibration shows a response at multiples of a certain fundamental frequency. The stick-slip vibration typically occurs at the first, second, or third order of the fundamental frequency. Figure 4.2 shows a typical case [37]. This condition is common in automatic transmissions since wheel torque is present at idle, but it can also occur with vehicles with manual transmission that are stopped on an incline.

Another type of low-frequency brake vibration and noise due to stick-slip is called *moan*; it is similar to groan in principle. It is typically a low-frequency, pure-tone noise of several hundred hertz. It can be considered

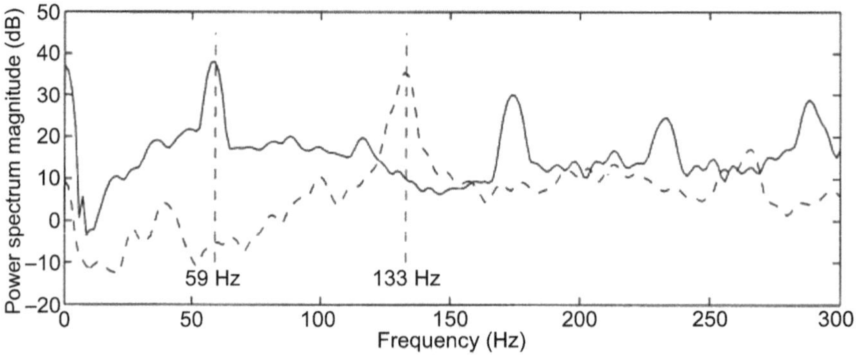

Figure 4.2 Power spectral density function; vertical acceleration measured on the front (solid line) and rear caliper (dotted line) during creep groan tonal phase; reference 1 m/s [37]

as an ultra-low-frequency "squeal" because it is similar to squeal in many ways. It could be either a system instability issue or resonance of certain low-frequency modes of brake assembly.

Brake groan/moan can be very difficult for a brake supplier to both detect and prevent because it is often dependent upon factors outside the direct control of the supplier, such as axle and suspension configuration. A successful investigation of this problem requires system-level analysis, including elements of the suspension and the foundation brake components.

Since it is the friction material characteristics that make the system prone to stick-slip behavior, most engineering efforts to date have concentrated on the characteristics between the friction material and the contacting surfaces and its effect on creep-groan vibration. Reducing the magnitude of the negative slope of the friction curve can significantly reduce the overall vibration levels. System mode frequencies usually do not shift, although the frequency of stick-slip vibration can jump to a different order.

Brake judder (roughness or shudder) can be roughly categorized into three major types. The first is due to disc thickness variation (DTV), in which the aggressiveness/wear of brake friction material generates DTV with brake temperature $< 100°C$. The second is corrosion-type brake judder, in which torque fluctuates due to corrosion-induced variation and surface damage. The third is hot judder, in which torque fluctuates due to disc rotor distortion caused by high temperature (brake temperature $>$ $200°C$). Other types include green roughness, new brake wet roughness, and water-soaked high-speed judder.

Brake judder causes vehicle vibration, jerkiness, or pulsation that is felt by the driver, passenger, or both, which is usually sensed through the steering wheel (shake or nibble), floorboard, dash panel, or car seat brake pedal during the stopping process. It is amplified by the resonance modes of the vehicle body (vibration transmission system) in relation to hydraulic pressure or torque fluctuations. It is created by repetitive variation in torque output from the brake, typically in the 5–60 Hz frequency range. Judder varies with wheel speeds of typically 1–2 pulsations per revolution, but can reach up to 10. Although all types have been observed to cause torque variation, the most serious case is caused by brake pressure fluctuation mainly due to DTV.

The uneven rotor wear caused by varying pad contact while driving at highway speeds is the factor influencing the creation of brake judder. There can be selective wear on either side of the rotor, or on both sides of the rotor. Typically, DTV in the range of 15—20 μm can create torque variation in the range of 50 Nm. This is normally sufficient to create vehicle judder. DTV is usually caused by uneven wear or the combined effect of corrosion of the disc rotor, which tends to induce a vibration called brake *cold judder.*

In brake systems, frictional heat is not uniformly distributed due to many reasons, such as nonuniform contact, imperfections in geometry, and thermal expansion. Moreover, the thermoelastic distortion due to frictional heating could further affect the contact pressure distribution and likely lead to thermoelastic instability, by which the contact load is concentrated on one or more small regions of the brake disc surface. These local contact regions reach very high temperatures and the passage of these hot spots moving under the brake pads can cause low-frequency vibration called *brake hot judder.* The nonuniform contact can also be generated by disc transverse runout. This type of judder occurs at higher orders of wheel rotation (6—12 have been typically noticed). The mechanism of this type of judder is considered to be the variation of the friction coefficient around the rotor. This variation is caused by chemical changes occurring to the friction couple when intense heat is generated at the friction interface. Visual observation of hot or blue spots around the rotor is common. Even after cooling, one can notice this effect by observing the color of the rotor surface. The order of vibration is related directly to the number of hot spots. The engineering development efforts for judder reduction or elimination include controlling characteristic parameters of the friction material, improving brake assembly and vehicle sensitivity.

4.2.4 Low-Frequency Squeal

Low-frequency squeal can be attributed to lock-in or mode coupling of brake corner components under frictional excitation. Many low-frequency squeals are due to the coupling of two or more modes of various structures, including calipers, pads, and rotors. As an example, one kind of low-frequency squeal is due to the modal coupling of the caliper and rotor. The friction variation between pads and rotor induces rotational vibration, which leads to longitudinal resonance of the caliper, which results in pressure variation on the pads. The pressure variation could further increase the friction vibration between pads and rotor. Thus, a feedback mechanism is established and the caliper system is self-exciting and magnifies itself, leading to squeal noise. Thus, separating the related vibrational mode of the rotor and caliper longitudinal mode can avoid mode coupling vibrations. This leads to the solution of decoupling the caliper and rotor modes by changing the rotor material from common gray cast iron to damped iron. The low-frequency squeal could be pad-induced, caliper bracket-induced, or both. Pad-induced vibration denotes one that is triggered by out-of-plane motions of the pad ends (pad-end flutter), which excites the out-of-plane vibration modes of the rotor. A typical case is pad-induced out-of-plane squeal (4–11 kHz), which are aligned with the out-of-plane modes of the rotor (15- and 16-in. brake corners with cast iron rotors). Another typical case could be the caliper bracket–induced squeal (2–6.5 kHz) for some floating caliper designs.

Low-frequency squeal could occur during low brake deceleration (for instance, at about 5 ft/s^2, vehicle speeds of 5–10 mph, and initial pad temperatures of 30 to 40°F). Low-frequency squeal could occur under special conditions, such as in very cold or high-humidity environment, and the occurrence rate is lower than for high-frequency squeal. This type of squeal could occur in the morning because of the overnight environmental conditions; in those cases, it is called *morning sickness/squeal.* Cold squeal tends to appear at low speeds in low temperatures.

4.2.5 High-Frequency Squeal

High-frequency squeal usually involves higher-order disc modes, with 5–10 nodal diameter mode. The squeal frequencies typically range between 5 and 15 kHz. Nodal spacing between the excited modes is normally comparable to or less than the length of the brake pads. For instance, a typical high-frequency squeal takes place in 30 revolutions per minute, 1000-kPa line pressure, and 150°C bulk rotor temperature. The system can squeal at a number of distinct frequencies.

An illustrative example [38] is shown in Figure 4.3. The out-of-plane modes of a typical rotor disc exist throughout the audible-frequency range (for instance, the second nodal diameter mode near 1 kHz and the 12 nodal diameter mode near 15 kHz). On the other hand, the modal density of in-plane modes is much lower. The first three circumferential (tangential) modes of the rotor disc consist of one, two, and three nodal diameters. A circumferential mode of a disc can be viewed as a compression wave in the disc circumference. From Figure 4.3, the frequency response function data of a rotor disc illustrates the strong relationship between the resonance frequency of the in-plane modes and the frequency of squeal. There is an

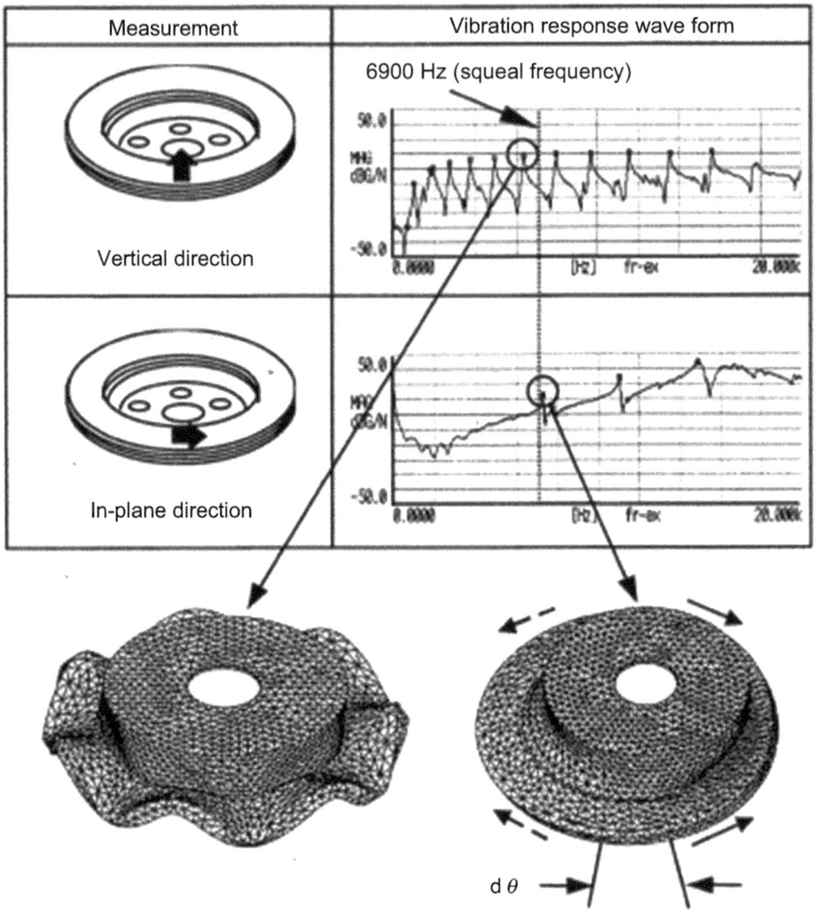

Figure 4.3 Rotor disc frequency response function of out-of-plane and in-plane modes and their modal shapes close to squeal mode (6900 Hz) [38]

in-plane primary eigenvalue at 7200 Hz that is close to the frequency of 6900 Hz, at which the brake squeal mainly occurred. The squealing frequencies are usually slightly less than the natural frequencies of the stationary rotor. Increased coefficient of friction between the rotor and pads of friction material increases the propensity for squeal.

In most cases of high-frequency squeal, the squeal frequency relates to an in-plane resonance frequency of the rotor disc, not an out-of-plane resonance. The frequencies at which high-frequency noise occurs are related to the in-plane modes of the rotor disc. Unlike the pure in-plane mode of the disc, the disc-hat shape of the rotor causes its in-plane mode to have out-of-plane components; thus, it can radiate noise. Sometimes the cross-coupling of an in-plane and out-of-plane mode of the rotor or an in-plane rotor and another mode of the brake system could occur.

The existing research illustrates that high-frequency squeals have good correlation with in-plane circumferential modes despite not being locked precisely in their frequencies, and are irrelevant with out-of-plane diametric modes, despite the out-of-plane modes usually providing more efficient sound radiation. It is also attributed to the velocity dependence of friction triggering the in-plane circumferential modes, which leads to out-of-plane vibration and sound radiation through normal-force coupling. In some cases, the direct in-plane and out-of-plane mode coupling could occur and lead to squeal.

Basically, high-frequency squeal is a result of the excitation of the in-plane modes of the rotor. The basic type of tangential rotor modes is compression, whose primary controlling parameter is the disc diameter. For the family of rotors used in automotive disc brakes (15- and 16-in. brake corners with cast iron rotors), the first three in-plane modes are in the range of 6—7, 9—11, 14—16 kHz, respectively.

For instance, a high-frequency brake squeal was characterized as an intermittent squeal that occurs once or twice per revolution of the brake rotor for a short time. Varied types of squeal events tend to occur at different physical locations on the brake rotor. The amplitude of vibration of squealing brakes is on the order of microns, and the vibration could be a standing wave or a traveling wave.

Mode lock-in of in-plane modes, or mode coupling between in-plane and out-of-plane modes, or simply rubbing excitation is the inherent mechanism for certain brakes to generate high-frequency squeal. The occurrence of squeal needs the right friction, pressure, and temperature to trigger mode lock-in or couple. The proper combination of geometric/structural/material parameters and assembly configurations, as well as operational conditions, gives rise to the proper threshold for squeal occurrence. So far, there is no approach to determine the squeal threshold or which mode of the components of a disc brake system can be used to infer a squeal propensity. The time-dependent nature of friction properties, manufacturing tolerances with respect to flatness and parallelism of rotor surfaces, and variability of material properties render the quantitative prediction of squeal elusive. A recommended approach to deal with squeal is to look at the entire brake corner as an interactive system and address all these interactions instead of sticking to only one component, sole factor, or identical mechanism.

Random impulsive excitation combined with self-excited systems yielding a nonstationary oscillation or instantaneous mode vibration constitutes the basic characteristics of brake noise.

Generally, the rotor is the noise radiator. The driving factor is the magnitude/slope of the friction force under specific operational and environmental conditions. The overall dynamic characteristics of the caliper/pad/rotor system determine squeal instability sensitivity. For low-frequency noise (groan and judder), the axle assembly subsystem as well as brake assembly may need to be taken into account.

4.3 NUMERICAL ANALYSIS OF FRICTION DYNAMICS OF BRAKES

Analytical and numerical approaches have been used to tackle the friction dynamics of brake systems. With these approaches, the design and prototype can be optimized to reduce the likelihood of vibration and noise. They also can help set up experiments and interpret experimental results, allowing for pinpointing the noise, vibration, harshness (NVH) failure root cause. Both the analytical/numerical and experimental approaches are indispensable tools for understanding and improving squeal issues in brake systems.

The basic analytical and numerical approaches can be classified into three categories: real modal analysis, complex eigenvalue analysis in the frequency domain, and transient analysis in the time domain. These three methods have been used within the automotive industry to model and analyze brake friction dynamics.

4.3.1 Real Modal Analysis

Real modal analysis can be implemented readily using widely available computer-aided engineering tools for structural dynamics analysis. The real modal method analyzes the natural frequencies and modes of the rotor, pads, caliper, and anchor bracket components. The obtained natural frequencies of the components are then checked for alignment with and coupling to other components. If the components had close natural frequencies, the propensity of instability or squeal through coupling would be considered.

This method can provide a quick approach to investigating the vibrations of components. The noncrossing of the eigenvalue is considered to be linked to the symmetric properties of the eigenvalue functions. By using the basic concept of modal lock-in or coupling, it simply assumes that dynamic instability or squeal could be triggered either when a natural mode is locked in through velocity-dependent friction, major components have close modal frequencies, or certain rotor tangential and normal vibration frequencies are aligned with each other.

4.3.2 Complex Eigenvalue Analysis

To take into account more parameters in dynamics analysis, such as friction or damping, complex eigenvalue analysis and transient analysis have been used [39−61].

In complex eigenvalue analysis, friction effects between the pads and rotor are treated as the excitation mechanism. In the dynamic governing equation of mass, damping, and structural stiffness, even the negative damping forces caused by friction can be introduced to evaluate the instability of brake systems.

Next, we present complex modal analysis, which has been widely used as an analytical tool for solving the system with mode-coupling instability [62−72]. Consider a multiple-degree-of-freedom system consisting of subsystems 1 and 2. They are connected by a contact sliding interface with coefficient of friction μ. The equation of motion of the system can be expressed as

$$[M]\{\ddot{u}\} + [C]\{\dot{u}\} + [K]\{u\} = \{F_f\}, \qquad [4.1]$$

where $[M]$, $[C]$, and $[K]$ are the mass, viscous damping, and stiffness matrices for the nonfriction system, respectively; $\{u\}$ is the displacement vector; and $\{F_f\}$ is the friction force vector between the two subsystems. The friction system consists of an interface with contact and is connected in the normal direction, but not in the tangential direction. The tangential friction is modeled as a force:

$$\{F_f\} = \mu(\{N_{static}\} + \{N_{dynamic}\}), \qquad [4.2]$$

where $\{N_{static}\}$, $\{N_{dynamic}\}$ are the static and dynamic normal forces, respectively. To solve the eigenvalue problem, static force is removed from the equation of motion. The dynamic normal force is caused by the vibration of the subsystem 1 and subsystem 2, and is represented by

$$\{N_{dynamic}\} = K_s(\{u_{N1}\} + \{u_{N2}\}), \qquad [4.3]$$

where $\{u_{N1}\}$ and $\{u_{N2}\}$ denote the displacements of the interface in the normal direction, and K_s is the local contact stiffness. Hence, the system equation of motion becomes

$$[M]\{\ddot{u}\} + [C]\{\dot{u}\} + [K]\{u\} = -\mu K_s[K_f]\{u\}. \qquad [4.4]$$

The matrix $[K_f]$ is the effective stiffness due to friction in the interface. It is a nonsymmetric matrix that couples the relative normal displacement with the tangential force. If the system has no damping and friction, it is degraded as

$$[M]\{\ddot{u}\} + [K]\{u\} = \{0\}. \qquad [4.5]$$

The modal domain transformation can be obtained from the equation as

$$\{u\} = [\psi]\{\gamma\},$$

where $[\psi]$ is the modal matrix of the system equation and $\{\gamma\}$ is the modal coordinate vector. With the modal transformation, the following relationship can be established:

$$[\psi]^T[M][\psi] = [I] \qquad [4.6]$$

$$[\psi]^T[K][\psi] = \begin{bmatrix} \omega_1^2 & 0 & \cdots & 0 \\ 0 & \omega_2^2 & & \vdots \\ \vdots & & \ddots & 0 \\ 0 & \cdots & 0 & \omega_n^2 \end{bmatrix}, \qquad [4.7]$$

where ω_i is the ith natural frequency of Eq. [4.5]. From Eqs. [4.6]–[4.7] and [4.4], the complex eigenvalue equation can be expressed as

$$\left\{ s^2 \begin{bmatrix} 1 & 0 & \cdots & 0 \\ 0 & 1 & & \vdots \\ \vdots & & \ddots & 0 \\ 0 & \cdots & 0 & 1 \end{bmatrix} + s \begin{bmatrix} 2\xi\omega_1 & 0 & \cdots & 0 \\ 0 & 2\xi\omega_2 & & \vdots \\ \vdots & & \ddots & 0 \\ 0 & \cdots & 0 & 2\xi\omega_n \end{bmatrix} \right. \qquad [4.8]$$

$$\left. + \begin{bmatrix} \omega_1^2 & 0 & \cdots & 0 \\ 0 & \omega_2^2 & & \vdots \\ \vdots & & \ddots & 0 \\ 0 & \cdots & 0 & \omega_n^2 \end{bmatrix} \right\} \{\gamma\} + \mu K_s[\Lambda_f]\}\{\gamma\} = \{0\}.$$

A single-degree-of-freedom system with friction has periodic limit-cycle oscillation solutions. Similarly, there are limit-cycle oscillation solutions in multiple-degree-of-freedom systems. The linear complex mode analysis can be used to predict the onset of the limit cycle. Although it cannot predict the magnitude and characteristics of the limit cycle, it helps to estimate the stability margin or threshold.

Usually, the stability propensity of a friction dynamical system can be evaluated by using a two-dimensional plot of complex eigenvalues called a *root locus plot* or *stability chart,* in which the ordinate represents frequency and the abscissa represents the real part of a complex eigenvalue. The real part of a complex eigenvalue is sometimes used to evaluate the stability propensity. Any point on the left side of the stability chart is considered to be a stable mode, as the vibration magnitude corresponding to the point will decay with time. Any point on the right side of the chart represents an unstable mode, as the vibration magnitude of the point will grow as time progresses.

For the complex modes solved from Eq. [4.8], the complex roots can be plotted on a root locus diagram, as shown in Figure 4.4. Roots in the right half-plane have damping that produces growing exponential solutions that are unstable. Roots in the left half-plane have stable damping. In Figure 4.4, two typical roots are initially assumed at B_1 and B_2. Consider changing a parameter in the system like the coefficient of friction to allow the roots to come together at A. Then the parameter is further increased until the roots split apart. One root remains in the left half-plane while the other goes to the right half-plane. Moreover, the roots that lie along the curve B_1^* –A–B_2^* are symmetrically located about point A. This symmetry is retained for a finite parameter change. It means that the two modes B_1^* and B_2^* vibrate at the

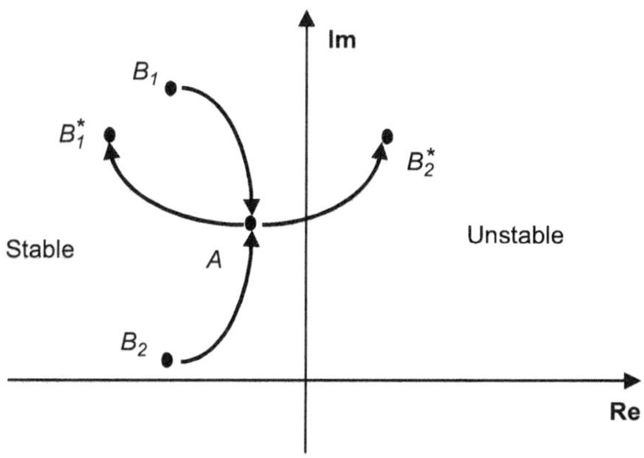

Figure 4.4 Root locus diagram: bifurcation roots

same frequency. Hence, they are coupled. These coupled motions exist within one system mode.

For this type of solution to be derived from Eq. [4.8], at least one of the roots must lie in the right half of the complex plane. The occurrence of a stable limit cycle due to system parameter change is the so-called *Hopf bifurcation*. During this process, two kinds of motion mode couple inside a complex mode, one of which has stable motion and the other has unstable motion. These motions exchange energy back and forth until energy is released and the system state changes. The noncrossing of the eigenvalue is linked to the symmetric properties of the eigenvalue functions. It is the self-excited oscillation. The excitation is from motion and interface friction, and a local change in interface parameter does not necessarily cause local or small effects on a global level stability for the system.

Next, we consider how to give a measure to evaluate the level of coupling for different systems. To this aim, the system complex eigenvalue is expressed as

$$[\Lambda_f] \equiv [\psi]^T [K_f][\psi] = K_s f([\psi_{Tr}], [\psi_{Nr}]). \qquad [4.9]$$

Obviously, the complex eigenvalue is dependent on the friction coefficient. The relevant equation can be solved directly for different values of friction to establish the relationship. However, it is instructive to develop an analytical expression that relates these two parameters. An approximation is obtained by a perturbation analysis with respect to the coefficient of friction to the second order. The complex eigenvalue of Eq. [4.9] can be expanded by a second-order series as

$$s_i^2 \approx -\omega_i^2 - \mu[\Lambda_f]_{ii} + \mu^2 \sum_{\substack{l=1 \\ l \neq i}}^{n} \frac{[\Lambda_f]_{il}[\Lambda_f]_{li}}{\left(\omega_l^2 - \omega_i^2\right)}. \qquad [4.10]$$

From the calculation of the real model in the fundamental frequency range of coupled mode, it usually results in

$$\omega_i^2 \gg \mu[\Lambda_f]_{ii}.$$

Hence, the diagonal terms of $[\Lambda_f]_{ii}$ can be neglected from the equation.

From Eq. [4.10], clearly the coupling strength between any two modes depends on the distance between the two modes and the product of the cross-coupling terms $[\Lambda_f]_{il}[\Lambda_f]_{li}$. Therefore, the coupling strength can be defined as

$$CS_{il} = \frac{[\Lambda_f]_{il}[\Lambda_f]_{li}}{\left(\omega_l^2 - \omega_i^2\right)}. \quad [4.11]$$

The coupling strength provides a measure of how fast the eigenvalues will move when the coefficient of friction increases from zero. It can be used to quantify the interaction between modes.

Eq. [4.10] is a closed-form solution that relates the complex eigenvalues with friction. The first- and second-order terms in Eq. [4.10] are the effect of friction. Eq. [4.10] shows the effect of friction on the complex eigenvalues of the system. It gives an indication of instability, which comes from the second-order term. It involves the dot products of relative normal and tangential displacements between two modes. A pair of modes that have a high dot product will either have a strong tendency to converge or diverge, depending on the sign function. Since it is known that instability is the result of the frequency convergence of the two modes, those pairs that have high second-order terms may have the potential to become unstable.

It is noted that the first-order term in Eq. [4.10] is typically negligible and has no effect on instability. Moreover, a stability index can be used for the estimation of instability propensity. Assuming the complex conjugate roots to be $\lambda = \sigma \pm i\omega$, the corresponding characteristic equation is

$$(\lambda - (\sigma + i\omega))(\lambda - (\sigma - i\omega)) = 0. \quad [4.12]$$

It can be simplified to the following:

$$\lambda^2 - 2\sigma\lambda + \sigma^2 + \omega^2 = 0.$$

The modal damping ratio is defined as

$$\varsigma = \frac{-\sigma}{\sqrt{\sigma^2 + \omega^2}}. \quad [4.13]$$

The modal damping ratio gives an index for an estimation of the stability margin.

The friction dynamical systems are characterized by a nonsymmetric stiffness matrix as illustrated previously. It could exhibit mass coupling as demonstrated by the system shown in Figure 4.5 [70–71].

In this model, subsystem 1 is represented by a three-degrees-of-freedom system on the top, and the subsystem 2 is modeled as a two-degree-of-freedom system on the bottom. Force P represents the applied force.

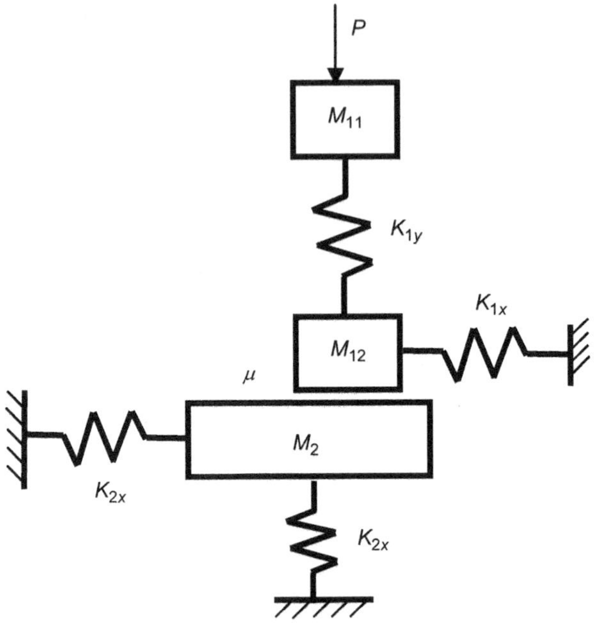

Figure 4.5 A system with a nonsymmetric mass matrix

Subsystem 2 travels horizontally at a constant speed V. The relative motion between subsystems 1 and 2 leads to interface friction forces. Let the interaction normal force be N and the friction force μN, where μ is the coefficient of friction. The system motion equation is given by

$$\begin{bmatrix} M_{11} & & & & \\ & M_{12} & & & \\ & & M_{12} & & \\ & & & M_2 & \\ & & & & M_2 \end{bmatrix} \begin{Bmatrix} \ddot{y}_{11} \\ \ddot{y}_{12} \\ \ddot{x}_1 \\ \ddot{y}_2 \\ \ddot{x}_2 \end{Bmatrix} + \begin{bmatrix} K_{1y} & -K_{1y} & & & \\ -K_{1y} & K_{1y} & & & \\ & & K_{1x} & & \\ & & & K_{2y} & \\ & & & & K_{2x} \end{bmatrix} \begin{Bmatrix} y_{11} \\ y_{12} \\ x_1 \\ y_2 \\ x_2 \end{Bmatrix} = \begin{Bmatrix} P \\ N \\ \mu N \\ -N \\ -\mu N \end{Bmatrix}. \quad [4.14]$$

In the right side of Eq. [4.14], except for the applied force P, all the other terms are internal forces that are to be determined by the response of the system. Defining the equilibrium state of the system as the quasi-static equilibrium state when subsystem 2 moves at a constant velocity V and the system does not vibrate, we are able to derive the equilibrium state from the following equations:

$$\begin{bmatrix} K_{1y} & -K_{1y} & & & \\ -K_{1y} & K_{1y} & & & \\ & & K_{1x} & & \\ & & & K_{2y} & \\ & & & & K_{2x} \end{bmatrix} \begin{Bmatrix} y_{110} \\ y_{120} \\ x_{10} \\ y_{20} \\ x_{20} \end{Bmatrix} = \begin{Bmatrix} P \\ N_0 \\ \mu N_0 \\ -N_0 \\ -\mu N_0 \end{Bmatrix}; \qquad [4.15]$$

let

$$\begin{Bmatrix} y_{11} \\ y_{12} \\ x_1 \\ y_2 \\ x_2 \end{Bmatrix} = \begin{Bmatrix} Y_{11} \\ Y_{12} \\ X_1 \\ Y_2 \\ X_2 \end{Bmatrix} + \begin{Bmatrix} y_{110} \\ y_{120} \\ x_{10} \\ y_{20} \\ x_{20} \end{Bmatrix} \text{ and } N(t) = N_0 + \widetilde{N}(t).$$

Now we are able to obtain the dynamic equations of the system about its equilibrium state as follows:

$$\begin{bmatrix} M_{11} & & & & \\ & M_{12} & & & \\ & & M_{12} & & \\ & & & M_2 & \\ & & & & M_2 \end{bmatrix} \begin{Bmatrix} \ddot{Y}_{11} \\ \ddot{Y}_{12} \\ \ddot{X}_1 \\ \ddot{Y}_2 \\ \ddot{X}_2 \end{Bmatrix} + \begin{bmatrix} K_{1y} & -K_{1y} & & & \\ -K_{1y} & K_{1y} & & & \\ & & K_{1x} & & \\ & & & K_{2y} & \\ & & & & K_{2x} \end{bmatrix} \begin{Bmatrix} Y_{11} \\ Y_{12} \\ X_1 \\ Y_2 \\ X_2 \end{Bmatrix}$$

$$= \begin{Bmatrix} 0 \\ N \\ \mu(N_0 + N) - \mu_V N_0 \\ -N \\ -\mu(N_0 + N) + \mu_V N_0 \end{Bmatrix}. \qquad [4.16]$$

From the second equation in Eq. [4.16], we express N as follows:

$$N = M_{22}\ddot{Y}_{12} - K_{2y}Y_{11} + K_{2y}Y_{12}.$$

On the other hand, the linear friction law is applied:

$$\mu = \mu_v - a\left(\dot{X}_2 - \dot{X}_1\right). \tag{4.17}$$

Then, we have the following:

$$\begin{aligned}
&\mu(N_0 + N) - \mu_v N_0 \\
&= \left[\mu_v - a\left(\dot{X}_2 - \dot{X}_1\right)\right]\left(N_0 + M_1\ddot{Y}_2 - K_{1y}Y_{11} + K_{1y}Y_{12}\right) - \mu_v N_0 \\
&= \mu_v M_1 \ddot{Y}_2 - aN_0\left(\dot{X}_2 - \dot{X}_1\right) - \mu_v K_{1y}(Y_{11} - Y_2) + O(X_2, Y_{11}, \ldots).
\end{aligned} \tag{4.18}$$

Substituting Eqs. [4.17] and [4.18] into Eq. [4.16], the linearized equations of the system can be derived.

We assume that the sliding contact interface does not have separation. Then we have the constraint condition $Y_2 = Y_{12}$. Finally, we obtain the motion equations of the system:

$$\begin{bmatrix} M_{11} & & & \\ & M_{12}+M_2 & & \\ & -\mu_V M_{12} & M_{12} & \\ & \mu_V M_{12} & & M_2 \end{bmatrix}\begin{Bmatrix} \ddot{Y}_{11} \\ \ddot{Y}_{12} \\ \ddot{X}_1 \\ \ddot{X}_2 \end{Bmatrix} + \begin{bmatrix} & & & \\ & & & \\ & & -aN_0 & aN_0 \\ & & aN_0 & -aN_0 \end{bmatrix}$$

$$\times\begin{Bmatrix} \dot{Y}_{11} \\ \dot{Y}_{12} \\ \dot{X}_1 \\ \dot{X}_2 \end{Bmatrix}\begin{bmatrix} K_{1y} & -K_{1y} & & \\ -K_{1y} & K_{1y}+K_{2y} & & \\ \mu_V K_{1y} & -\mu_V K_{1y} & K_{1x} & \\ -\mu_V K_{1y} & \mu_V K_{1y} & & K_{2x} \end{bmatrix}\begin{Bmatrix} Y_{11} \\ Y_{12} \\ X_1 \\ X_2 \end{Bmatrix}$$

$$= \begin{Bmatrix} 0 \\ 0 \\ 0 \\ 0 \end{Bmatrix}. \tag{4.19}$$

The friction coupling makes both the mass and stiffness matrices nonsymmetric, which causes the eigenvalues solutions to be complex.

Using the standard complex eigenvalue analysis algorithm, we may obtain the complex eigenvalues. Usually, the double-shift QR or QZ algorithm could have advantages for this type of problem. Based on obtained complex eigenvalues, the stability of the system can be further evaluated if any of the eigenvalues have negative damping characteristics.

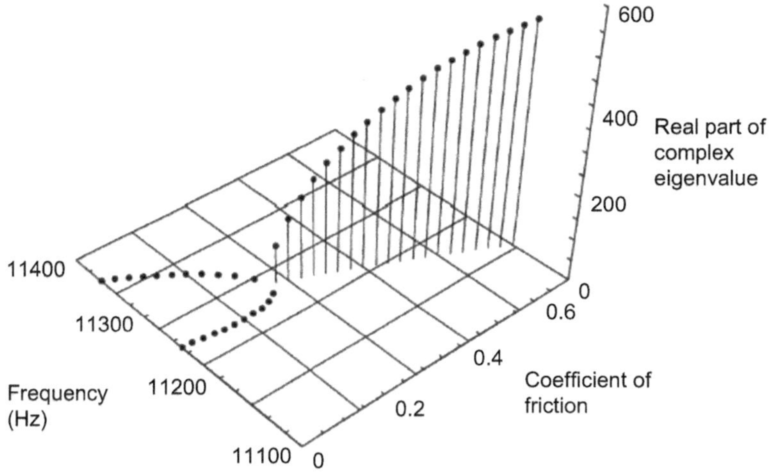

Figure 4.6 3D plot of eigenvalue [70–71]

Figure 4.6 shows a typical three-dimensional (3D) plot of an eigenvalue solution [70–71].

Next, we discuss the effect of damping. By properly adding damping to some part of the system, we could let the system tend to have more stability propensity. In terms of stability chart representation, for a typical multiple-degree-of-freedom system, Figure 4.7 depicts the effect of damping on system instability. It is actually a real example of an insulator in a brake system modeled by the finite element method. In the baseline model, no damping was assigned to any component. When damping was included in the insulator (viscoelastic layer), many of the real part of the complex

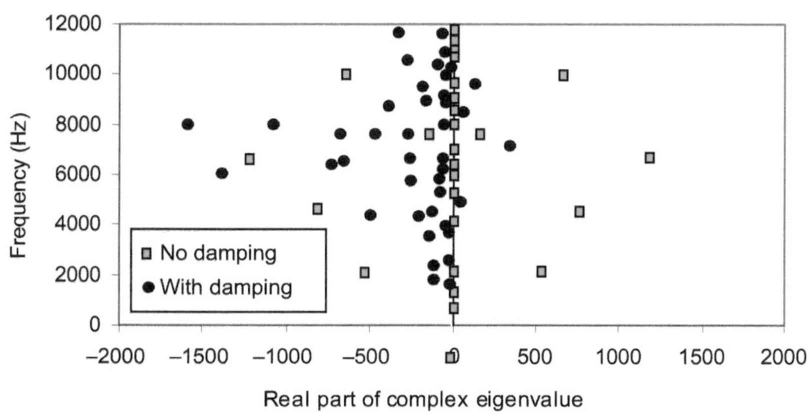

Figure 4.7 Effect of damping on eigenvalues

eigenvalue moves toward the negative end of the abscissa. Thus, some of the unstable modes become more stable.

A friction system could become unstable due to mode coupling with a high friction coefficient. Next, we discuss the effect of a nonconstant friction coefficient. Considering linear friction law of the form $\mu(v_r) = \mu_s - a_1 v_r$, Eq. [4.1] can be transferred to the following form:

$$\left[\overline{M}\right]\{\ddot{u}\} + \left[\overline{C}\right]\{\dot{u}\} + \left[\overline{K}\right]\{u\} = \{0\}. \qquad [4.20]$$

Note that in the motion equation, the damping matrix $\left[\overline{C}\right]$ is incorporated with the dynamic friction coefficient portion and the normal contact forces of the system. Both $\left[\overline{K}\right]$ and $\left[\overline{M}\right]$ matrices could be asymmetrical due to friction interaction. Through a transformation, Eq. [4.20] can be written as

$$[B]\{\dot{w}\} + [A]\{w\} = \{0\}, \qquad [4.21]$$

in which

$$[B] = \begin{bmatrix} [M] & 0 \\ 0 & [I] \end{bmatrix}, [A] = \begin{bmatrix} [C] & [K] \\ -[I] & 0 \end{bmatrix}, \{w\} = \begin{Bmatrix} u \\ \dot{u} \end{Bmatrix}.$$

Then we can get a generalized, first-order eigenvalue problem:

$$(A + \lambda B)\{w\} = 0, \qquad [4.22]$$

where λ is the complex eigenvalue and w is the right eigenvector.

The complex eigenvalue analysis shows that a negative slope of friction-velocity could render the system more unstable than constant friction. Figure 4.8 shows the effect of negative slope friction—velocity on

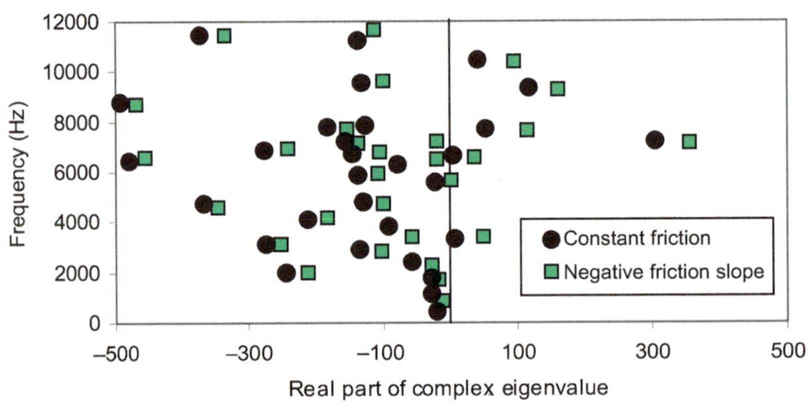

Figure 4.8 Effect of negative slope friction versus velocity on eigenvalues

eigenvalues. It is also based on an example of a brake system modeled by the finite element method. It shows that with the presence of negative slope friction, the real part of the eigenvalue moves toward the right side of the abscissa, thereby tending to be more unstable.

The treatment of the complex eigenvalue problem yields a linear stability analysis, which is a good approximation in the vicinity of an equilibrium point during steady contact sliding.

Although the results of complex eigenvalue analysis can be used to specify unstable modes theoretically, it is not feasible to directly use them to predict squeal occurrence. Actually, there are many unstable modes for a brake system, which are only the potential modes for squeal. Each component in a brake system has its own natural modes. Even for a typical rotor, there may be up to 80 modes below 20 kHz. Theoretically, many of them can be locked in as instability subject to velocity-dependent friction features. Moreover, the natural modes of the rotor, caliper, anchor, and pad could dynamically couple together, resulting in a series of coupled vibration modes, which are different from the free vibration modes of the individual component. The inclusion of friction coupling forces results in a stiffness matrix containing asymmetric coupling terms. This asymmetric coupling term could be the root cause of the brake squeal. Complex eigenvalue analysis usually uncovers many unstable modes, but only a very small number (or sometimes just one) correspond to real squeal, which is due to very specific operational and environmental conditions. This is partially because a minor change in the interface does not necessarily result in insignificant effects on the global vibration and noise level for the system. The small variations in operating and environmental conditions can lead to changes in the coefficient of friction, thereby causing differing squeal propensities or frequencies. The brake squeal usually exhibits a scholastic nature that is usually nonrepeatable. A brake may not always retain squeal for a given condition that is apparently unchanged. On the other hand, a squeal always exhibits sustainable and limited oscillation amplitudes, which suggests that the occurrence of a limit cycle due to system parameter change is due to Hopf bifurcation. The excitation is fundamentally due to motion and friction. The instability could be triggered by proper friction or friction properties, and the limit cycle oscillations could be maintained for a short duration. Overall, it exhibits a combined feature of motion of limit cycle and scholastics as well as nonstationary.

Usually, finite element models are used to model components with complicated geometry and are validated through component modal testing.

The complex eigenvalue analysis is used for brake noise analysis and reduction by identifying and reducing unstable modes. The dynamometer and vehicle tests are used to diagnose and validate squeal occurrence. The complex mode shapes are capable of depicting the noise-generating mechanisms, and the complex eigenvalues are able to depict the noise frequencies.

The real part of the complex eigenvalue λ_R has been used as a squeal factor in brake complex eigenvalue analysis [73–80]. From Eq. [4.10], the second-order term is the summation of cross-mode coupling between the *i*th mode and other modes. The complex eigenvalue trace is usually controlled by cross-mode coupling. Hence, a basic characteristic of brake squeal instability is that it is mainly due to cross-mode coupling. Generally, a few strong coupling mode pairs make the largest contribution to the summation. Brake squeal happens when the strong coupling mode pairs are close enough in frequency to provide a chance for the two modes to merge and become an unstable complex eigenvalue. Thus identification of mode pairs with high coupling could be an indication of the possibility for instability. There are two requirements for the cross-coupling term to have a high value:

1. Similarity of the relative normal and tangential shape. Eq. [4.9] shows that the coupling value is large if $\{\Psi_{N,r}\}$ and $\{\Psi_{T,r}\}$ have similar shapes for modes *i* and *l*. This means that either the normal force for mode i corresponds with a relative tangential displacement for mode *l* or the normal force for mode *l*, or both correspond with a relative tangential displacement for mode *i*. The coupling is highest when both conditions are true, but coupling can be high even when only one is true. Here, the normal force shape from mode *i* is similar to the relative tangential displacement shape of mode *l*. Thus another characteristic of brake squeal instability is that the strong coupling modes have similar normal and tangential mode shapes at the pad/rotor interface.
2. High normal force shape. In addition to mode shape similarity, it is believed that the shape for the normal force itself must be high. This is partly evident from Eq. [4.10] and also follows from the belief that out-of-plane bending modes of the rotor are the source for the radiated squeal noise. Logically, the bending modes of the rotor tend to compress the softer lining and create a high normal force. This leads to the assertion that a high normal force is required.

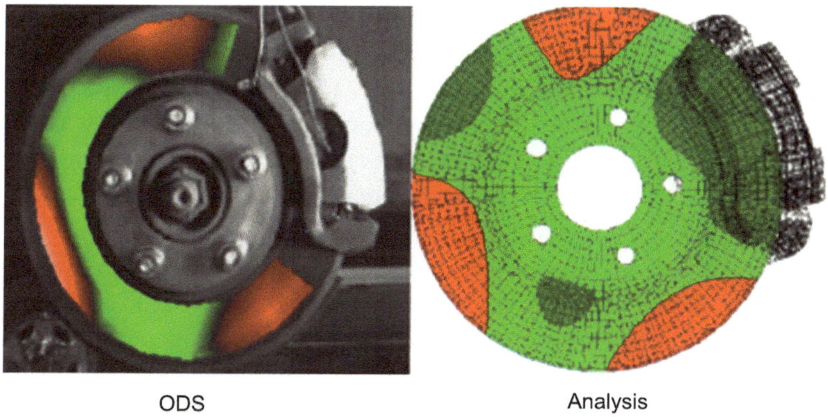

Figure 4.9 Rotor test and analysis correlation [81]

Generally, the complex eigenvalue resulting from numerical analysis needs experimental validation. As an example, Figure 4.9 shows the comparison of a tested operational deformation shape of a rotor and the corresponding mode shape by modeling [81]. Figure 4.10 shows the tested squeal frequency and the root locus diagram from complex eigenvalue analysis. The complex eigenvalue analysis could provide more than one unstable coupling mode. But in experiments, there are much smaller squeal modes, or even one mode. For instance, as shown in Figure 4.10a, laboratory testing shows squeal frequencies between 3500 and 3800 Hz. The

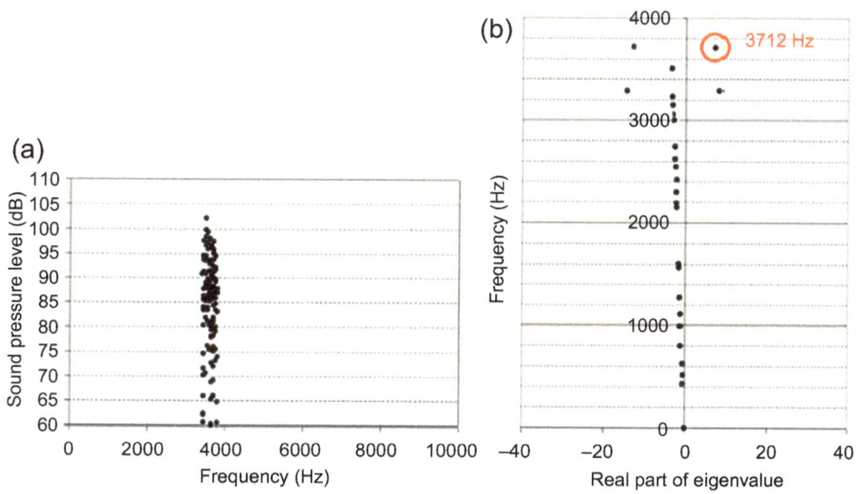

Figure 4.10 (a) Laboratory testing results; (b) analysis results [81]

analysis showed unstable frequencies at 3712 Hz and approximately 3300 Hz, as illustrated in Figure 4.10b. Only 3712 Hz correlates with the squeal frequencies observed during testing. The instability at 3300 Hz is an analysis overprediction. This is very common in complex eigenvalue analysis. Overpredictions like this prevent complex eigenvalue analysis from being a true predictive tool. Complex eigenvalue analysis is best used as an analysis tool after the overpredictions are clarified by test data.

Another limit of the complex eigenvalue approach is that the predictions of unstable modes are sensitive to slight changes in the system model. Since a small shift (i.e., less than 3%) in modal frequencies can make a significant difference in the occurrence of instability, the accurate prediction of instability is difficult. This kind of condition sensitivity is also partially the reason why squeal noise does not always occur with every brake application in practice. A further drawback of the method is that it does not provide any insight into the response amplitudes.

4.3.3 Dynamic Transient Analysis Using Numerical Approach

For general complex systems with friction, to obtain an analytical solution is difficult. The finite element model—based dynamic transient analysis is a feasible approach to quantitative prediction.

A multiple-degree-of-freedom system with friction exhibits many basic nonlinear characteristics, such as primary resonance, secondary resonance, combination resonance, bifurcation, and chaos. For multiple-degree-of-freedom friction systems, complex eigenvalue analysis offers an efficient way to estimate the instability associated with the primary resonance. But this predication is inaccurate and incomplete. To predict the vibration amplitude and the complex characteristics such as secondary resonance and combination resonance, numerical analysis of transient nonlinear dynamics is necessary.

Although it is approximate, linear stability analysis is more economical than full transient analysis using numerical approaches. However, for deeper understanding, we have to use numerical transient analysis by taking into account nonlinearity and friction law. For advanced analysis, the uncertainty should be quantified by using the methods for stochastic differential equations, such as using Monte Carlo approaches or Bayesian probabilistic methods. Figure 4.11 describes the conventional methodology used for the analysis of stability of friction dynamics [82].

Generally, one of the major difficulties in modeling brake friction dynamics is to bring different scale problems together. Effects on very small

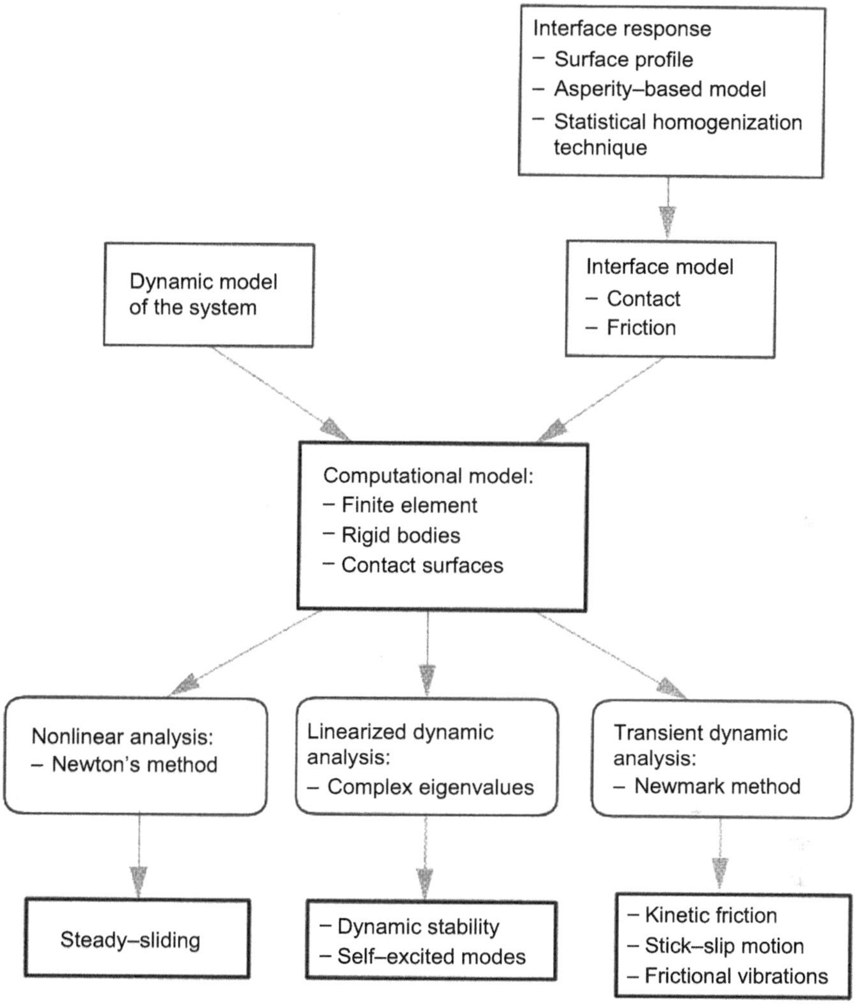

Figure 4.11 A Flowchart for the general approach to model friction-induced vibrations [82].

scales in length and time (i.e., microscopic contact phenomena and high-frequency vibrations) interact in critical ways with effects on larger scales (such as motions and dynamics of large brake substructures). Another major difficulty in modeling brake noise is to properly predict the invasive process like friction and wear over the life of the brake and quantify friction, which is multiple functions of many parameters. Moreover, there is difficulty in modeling a high-frequency natural mode of component due to its high sensitivity to minor structure and property uncertainty. The large-scale motion (braking operation) could lead to the substantial changes of interface on the small scale (surface), but the resultant small-scale change does not necessarily have no substantial effect on the large-scale quantities like frictions, vibrations, and squeal in the next operation. It is this unpredictable change in small scale that makes the accurate modeling and prediction elusive.

The modeling and analysis of brake vibration and noise are a challenging multidisciplinary problem in physics from the nano-, micro-, and macro-levels, involving contact mechanics, nonlinear dynamics, acoustics, and tribology. The problems are generally characterized by progressive interfaces due to environmental and operational effects, time-varying dynamic properties, spatial and time scales, wide frequency range from tens of hertz to 20 kHz.

The modeling and analysis of groan and judder are a challenging engineering problem involving interface friction properties, component structural vibrations, and system dynamics, including both brake system and suspensions. A number of theories have been formulated to explain the mechanisms of brake squeal, and numerous studies have tried with varied success to apply them to the dynamics of disc brakes. However, none of these models have attempted to include the effects at all the scales mentioned previously. This has led to models that capture some features of brake squeal well and ignore many others. Experimental studies also tend to have limited applicability, with their results pertaining only to one brake configuration or to one automobile type.

The experiments have been the most reliable method for investigating brake squeal. In fact, experiments have been so far considered to be the sole means for verifying any solution to brake squeal. It has been argued that the finite element method/lumped parameter models is as yet incapable of accurately modeling the complicated boundary conditions and friction mechanisms found in brake systems, and that the finite element method/lumped parameter models used to model squealing brakes simplify the problem to a degree that many of the important parameters are disregarded. Nevertheless, the finite element method has become an indispensable tool for modeling disc brake systems and providing new insights into the problem of brake squeal. There are many possible mechanisms leading to brake noise and vibrations, and there is no single mechanism that can comprehensively capture all of them and put into a unified model. Given the number of factors involved and the wide range of design options, a unified sole model intended to capture everything may be too complex to be useful. Therefore, the multilayer modeling and interactive analysis incorporating with testing could be an appropriate approach to achieve the analysis and diagnosis goal.

These might include the influence of the variable speed and pressure, the inclusion of thermal and wear effects, and the incorporation of more complex friction models and constitutive models for the pads of friction material, to ensure that all the important factors are covered. By interactive analysis, their interactions are likely to be uncovered and the root cause is identified. There are many approaches tried, which include the multilayer modeling to cover the following possible mechanisms: using finite element modeling for real modal analysis; using finite element modeling for complex eigenvalue analysis; using a continuum model or lumped parameter model to quantify complicated phenomena under special excitation like sprag-slip, nonsmooth effect, or impact effects; using proper friction law that is quantified in parameter space to quantify contact and friction; checking entire operational conditions, environmental conditions to extrapolate the aging conditions; checking uncertainty from structure and boundary conditions; conducting comprehensive testing; establishing the database of vibration, noise, and tribology parameters versus environmental and operational conditions and time; correlating multilayer analysis with testing results; eliminating the overpredictions and refining the analysis where the correct correlation and prediction have been attained; and updating the analysis and testing by interactive procedure.

In braking, both complex eigenvalue analysis and dynamic transient analysis have been widely employed to study friction dynamics issues. Modern CAE software tools enable transient analysis to be conducted to simulate multiple physics process in addition to dynamics; as such, the complex process such as thermal process, tribological process, and system parameter uncertainty have been integrated into finite element–based dynamics model in order to determine the dynamics level. Figure 4.12 shows an improved numerical methodology for studying disc brake squeal [83].

Next, we discuss how to incorporate nondiscontinuous friction law and present the basic numerical method for transient dynamic analysis. In many applications, the modeling of vibration systems with friction yields a set of differential equations with discontinuity. The standard methods to solve discontinuous differential equations consist of the smoothing method (normalization method) and switch method. The smoothing method replaces the discontinuous system by a smooth adjoint system. The smoothing

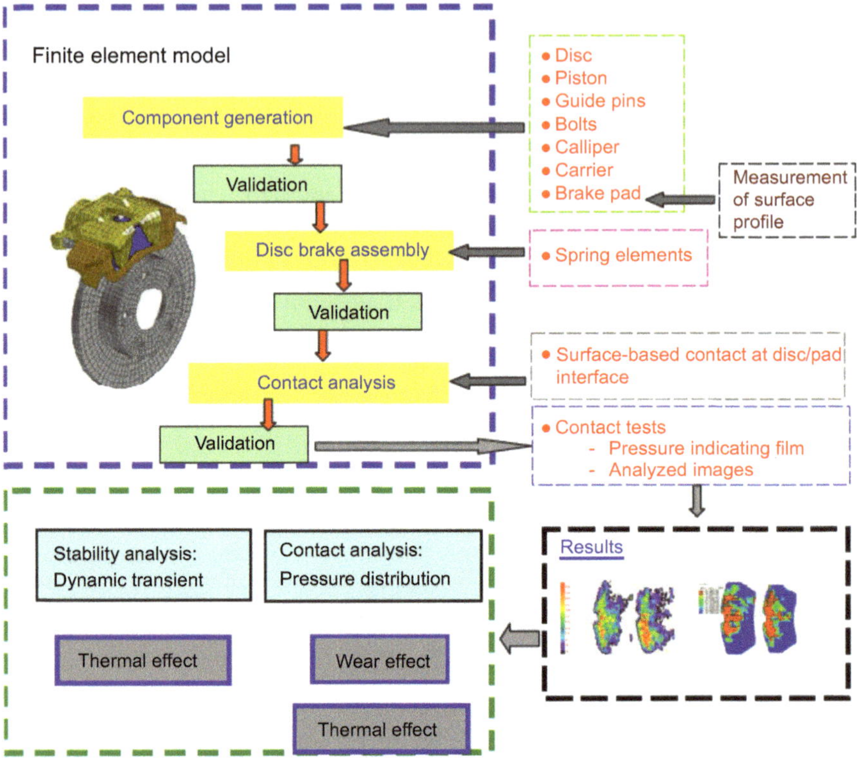

Figure 4.12 Overall simulation scheme [83]

method yields a system of ordinary, but stiff, differential equations and consequently leads to large computational times.

Consider the friction law $\mu(v_r) = \mu_0 \mathrm{sgn}(v_r)$. The transition at zero point is not smooth if static friction is not zero. It can be approximated by

$$\mu(v_r) = (2\mu_0/\pi)\tan^{-1}(\varepsilon v_r), \qquad [4.23]$$

in which a trigonometric function is used to smooth the transition. Sometimes this form could be a more realistic representation of a physical friction. Here, ε is a steepness parameter, and increasing it improves the approximation, especially for v_r values close to 0. Even for truly nondiscontinuous curves, the friction curve will be almost the same for large values of the steepness parameter, like $\varepsilon = 10^6$. However, a steep slope arises at $v_r = 0$ and is given by $-(2/\pi)\varepsilon\mu_0$. In the end, this causes a stiff differential equation, which is numerically costly to integrate.

The problems of the smoothing method led to the development of models that switch between different sets of equations by using alternate friction models for positive and negative velocities. The switch model starts from an initial state with a set of differential equations. After each time step, the state vector is inspected on a possible discontinuous transition within this time step. If meets an event, the integration process is paused, and an iteration procedure is used to find the accurate switching point. With the switching point, a new integration process is started with a modified set of differential equations and initial conditions at the switching point. An advantage of the switch model is that it is able to incorporate a variety of friction laws discussed in the previous section (for instance, the time-dependent static friction).

Next, we present the basic algorithms used for the transient dynamic analysis shown in Figure 4.11 (i.e., the Newmark method). The global equation of motion of friction system can be written as

$$\boldsymbol{M}\ddot{\mathbf{X}}(t) + \boldsymbol{C}\dot{\mathbf{X}}(t) + \boldsymbol{K}\mathbf{X}(t) = \mathbf{F}(t), \tag{4.24}$$

where $\boldsymbol{M}$, $\boldsymbol{C}$, and $\boldsymbol{K}$ are the mass, damping, and stiffness matrices of the system, respectively. The forcing term $\mathbf{F}(t)$ includes external force, contact force, and friction force. $\mathbf{X}(t)$ is the unknown displacement vector, and a dot denotes differentiation with respect to time t. Conventionally, the Runge-Kutta family and linear multistep method, such as the Newmark method, are used to solving this kind of problems. It has been demonstrated that the linear multistep method is more efficient for this type of problem. After balancing many requirements for simulating Eq. [4.24] e.g., low calculation cost, unconditional stability, and avoidance of nondissipation in all frequencies—the Newmark method with parameters $\delta > 1/2, \alpha > 1/4$ has been suggested and applied widely. The Newmark method is expressed as

$$\dot{\mathbf{X}}_{t+\Delta t} = \dot{\mathbf{X}}_t + \left[(1-\delta)\ddot{\mathbf{X}}_t + \delta\ddot{\mathbf{X}}_{t+\Delta t}\right]\Delta t \tag{4.25a}$$

$$\mathbf{X}_{t+\Delta t} = \mathbf{X}_t + \dot{\mathbf{X}}_t\Delta t + \left[(1/2-\alpha-\delta)\ddot{\mathbf{X}}_t + \alpha\ddot{\mathbf{X}}_{t+\Delta t}\right]\Delta t^2. \tag{4.25b}$$

The method has the following features:

1. Unconditional stability when applied to linear problems
2. No more than one set of implicit equations is to be solved at each step
3. Controllable algorithmic dissipation in the higher modes

The complex eigenvalue analysis could give sufficient estimation to linear instability analysis. However, in order to capture complicated interface parameters in which coefficient of friction is parameter dependent, time dependent, or having random features, nonlinear transient analysis is needed. For instance, the transient analysis of finite element models is needed for disc brake squeal analysis so as to feature a coefficient of friction depending on velocity, interface temperature field, and the interface pressure field. Transient analysis can deal with the system where an experimentally fitted law of contact force or stiffness between the pads and the rotor is involved.

Transient analysis has the promise of simulating the braking event real dynamics. However, it is difficult to relate detailed design iterations with transient analysis results. The solution time is also prohibitively long for a comprehensive modeling.

Transient analysis has been conducted by incorporating a nonlinear frictional contact mode into the finite element model. As an illustrative example [84], a constitutive model of the interface is applied to deal with an aircraft brake. In the investigations, the statistical information of the surface and the material properties of the bulk material are used to define the response of the surface to normal and tangential loading. A representative asperity is defined as having a tip curvature radius of $R_a = 5$ μm and height $h = 0.85$ μm. For this asperity, an elasto-plastic solution is obtained using an analytical solution to a spherical contact problem. The calculated response of a single asperity is then used in conjunction with a statistical homogenization procedure to produce constitutive characteristics of the interface. This includes normal pressure versus approach, real contact area versus approach, and friction force versus approach. The approach involves the coming together of the two surfaces from the point of initial contact of the tips of the tallest asperities. Two of these curves, normal pressure and frictional resistance, are shown in Figure 4.13.

Figure 4.14 shows the phase plots of the normal and tangential direction motions of a selected node, which lies on the face of the pad/stator in contact with the spinning surface directly underneath the inner edge of a piston friction.

Note that both oscillations are excited and grow in this case. The strength of instability can be quantitatively compared for different types of instabilities induced by either mode lock-in due to friction velocity dependence or modal coupling.

Due to the nonlinear effects inherent in this problem, the oscillations do not grow infinitely; rather, they develop into a limit cycle oscillation.

In transient analysis, the time dependence of the friction force between disc and pads can be readily incorporated into the system equation to investigate the generation of low- and high-frequency squeals [85–86].

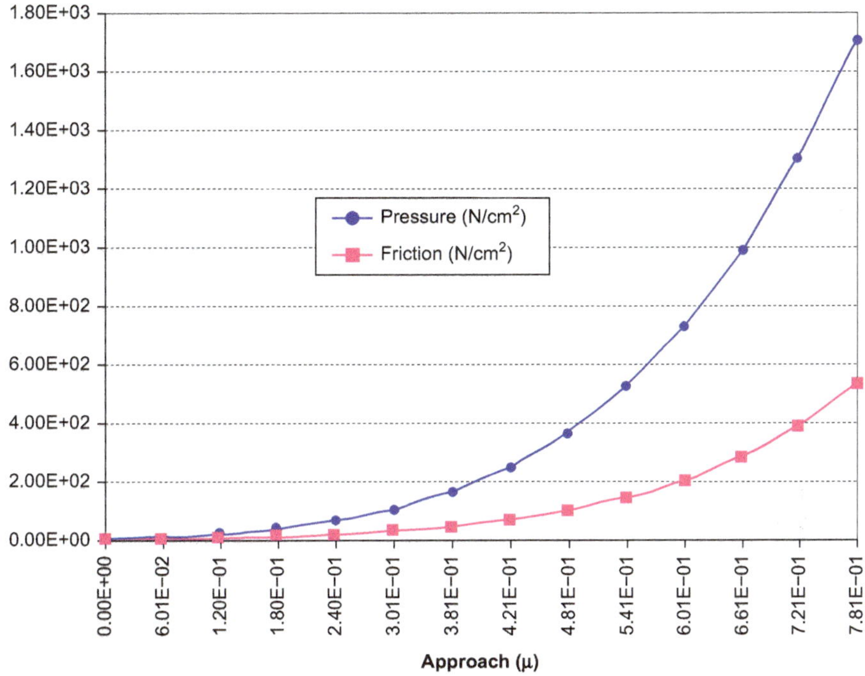

Figure 4.13 Normal pressure and frictional resistance as a function of the approach described in [84]

In [85], both the deterministic and random friction coefficient is considered. Figure 4.15 shows the friction coefficient—velocity relationship with a random component included. The numerical simulation with random friction helps to reveal a modulated intermittent response of both the disc and the pad. It is found that the frequency component corresponding to circumferential vibrations of the disc is always dominant in the case of deterministic friction. However, random friction creates impulsive forces that excite the transverse mode and has little effect on circumferential vibrations. Figure 4.16 shows a typical time—frequency spectrum of circumferential disc motion with asymmetric mode excited.

The results show that contact forces create modulated intermittent motions of both discs and pads. The intermittent motion consists of high-frequency oscillations in addition to a low-frequency component. The frequency content is changing with time due to the time dependency of contact forces as the relative velocity is changing with time. The time length of high- or low-frequency alternating motions is found to depend

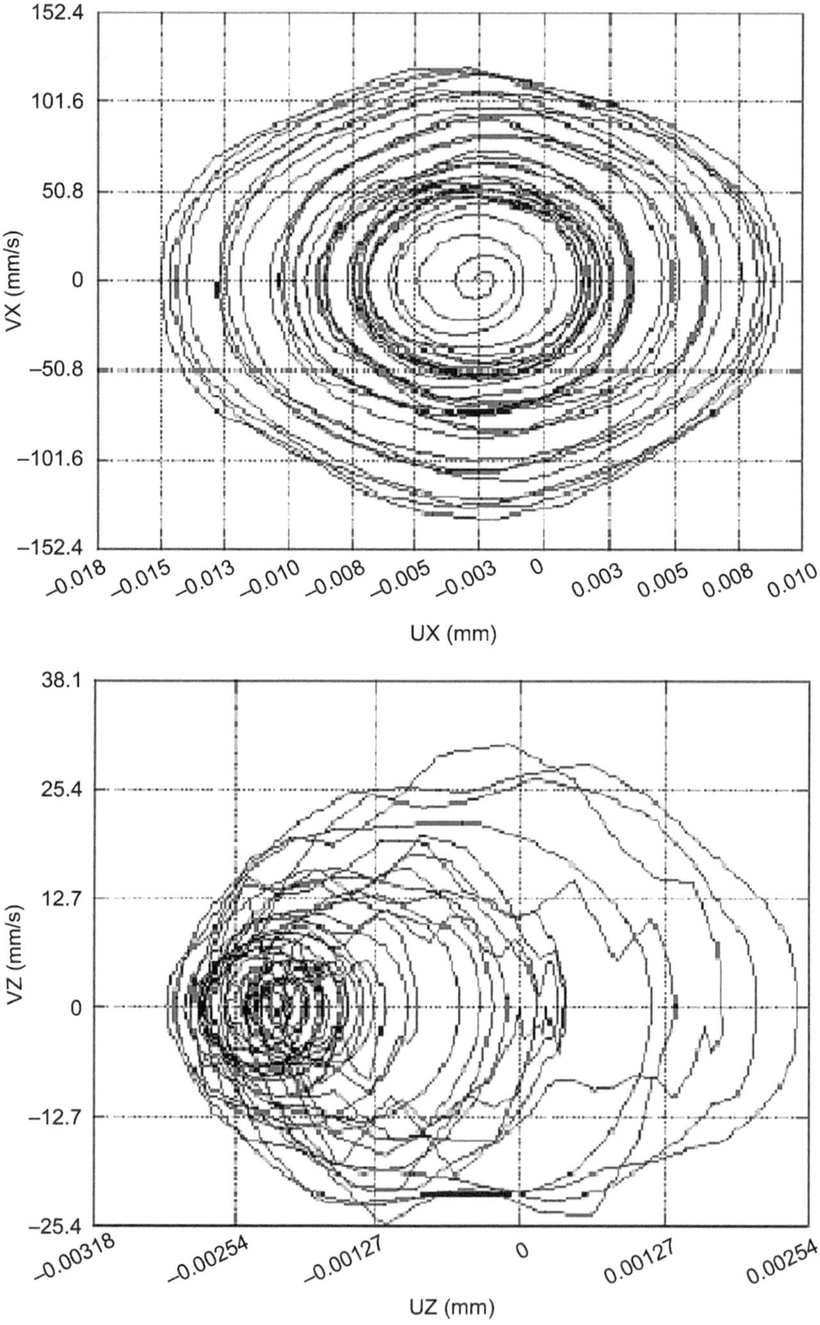

Figure 4.14 Phase plots of tangential and normal motions of selected node during unstable oscillations due to dynamic coupling [84]

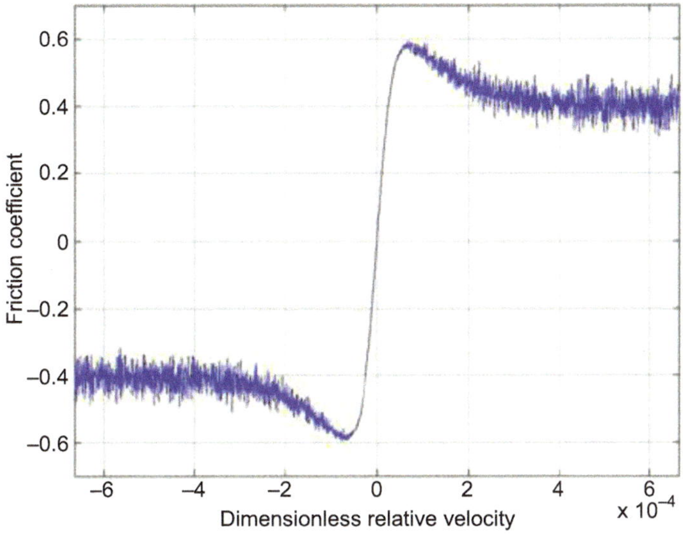

Figure 4.15 Friction coefficient–velocity relationship [85–86]

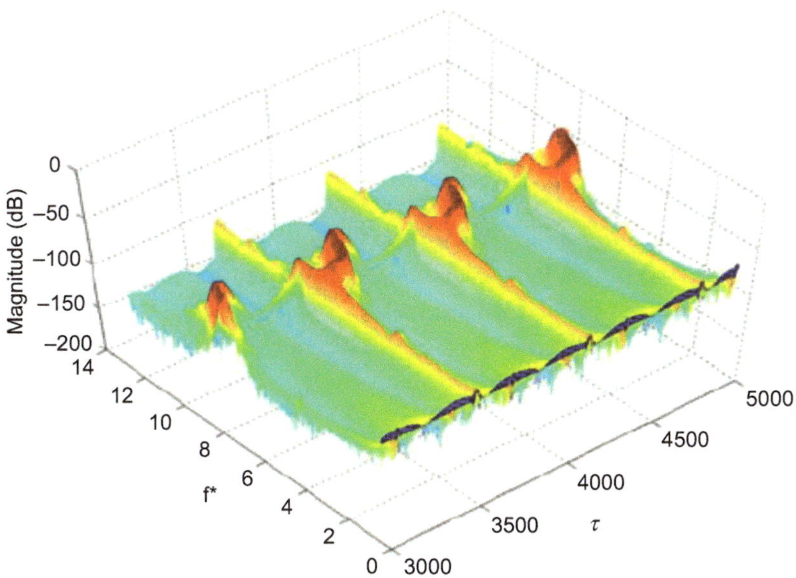

Figure 4.16 Time–frequency spectrum of circumferential disc motion: Asymmetric mode [85–86]

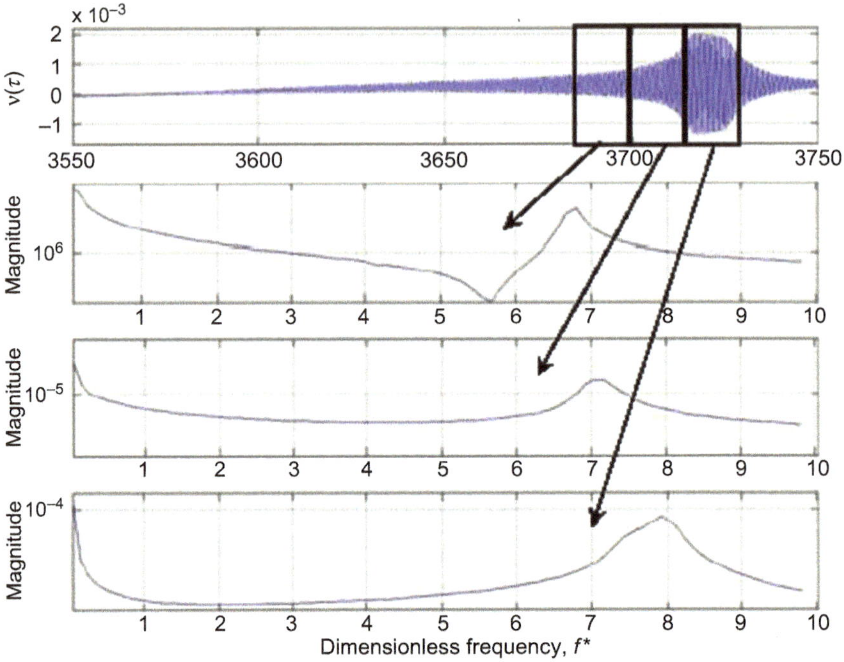

Figure 4.17 Magnification of time history for disc circumferential motion and fast Fourier transform (FFT) for different time sections [85–86]

on the system parameters, such as damping, brake force, and disc speed. Intermittent character of the response is also maintained in the case of random friction.

The time evolution of the high-frequency component is shown in Figure 4.17. It can be observed that oscillations with $f = 6.7$ dimensionless frequency are triggered first, and then both magnitude and frequency increase with time to the dominant peak of $f = 8.2$.

The time ranges of high- and low-frequency alternating zones are very much influenced by the brake force, damping, disc speed, and nonlinear characteristics of the friction coefficient. The unstable mode could remain at certain fixed-frequency components over the entire time domain. The frequency switching could occur for some unstable modes. The operational deflection shapes of rotor during the squeal event can be extracted from numerical time domain data. Moreover, many details, such as the areas with largest displacements on the pads during the squeal event, could be depicted.

4.3.4 Instability Due to Gyroscopic Effects

In this section, we present the instability of brake systems due to both friction and gyroscopic effects [87–94]. Consider the model shown in Figure 4.18, which consists of a rotating rigid disc with angular velocity Ω having thickness h and mass moment of inertia I, I, I_φ, with respect to the body's fixed b_1, b_2, b_3 frame [93]. The disc is hinged in a spherical joint in its center of mass and viscoelastically supported by rotational springs (rotational stiffness k_t) and rotational dampers (damping coefficient c_t), such that it can wobble while rotating with constant angular speed about its nominal rotation axis.

An inertial coordinate system is defined by the Cartesian coordinate system with unit vectors n_1, n_2, n_3. The disc is in contact with two pads with preload N_0, located at distance r from the center of the disc. The Coulomb friction law is assumed between the pad and the disc, with a constant coefficient of friction μ. The preload is assumed to be large enough not to cause loss of contact. The contact pad pairs supported by the spring (stiffness k) and the dampers (damping c) are in the n_1, n_2 plane and can move in the n_3 direction. The orientation of the disc is described by the second set of Euler angles q_1, q_2, q_3, and a_1, a_2, a_3, and b_1, b_2, b_3 are two intermediate coordinate

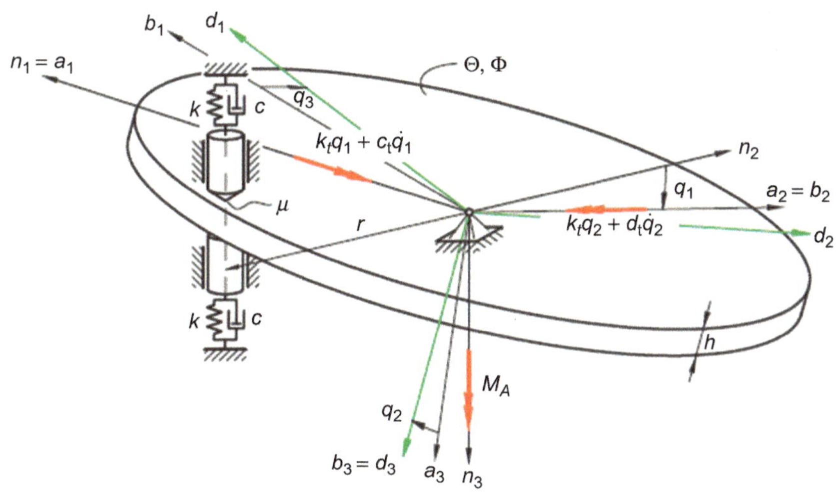

Figure 4.18 Schematic of wobbling disc in contact with spring and damper [94]

systems. By using Euler equations, the equations of motion of the system could be derived:

$$\begin{bmatrix} \Theta & 0 \\ 0 & \Theta \end{bmatrix}\begin{bmatrix} \ddot{q}_1 \\ \ddot{q}_2 \end{bmatrix} + \begin{bmatrix} \frac{1}{2}\mu N_0 \frac{h^2}{r\Omega} + 2dr^2 + d_t & \Phi\Omega \\ -\Phi\Omega - \mu dhr & d_t \end{bmatrix}\begin{bmatrix} \dot{q}_1 \\ \dot{q}_2 \end{bmatrix}$$
$$+ \begin{bmatrix} k_t + 2kr^2 + N_0 h & \frac{1}{2}\mu N_0 \frac{h^2}{r} \\ -\mu(khr + 2N_0 r) & k_t + (1+\mu^2)N_0 h \end{bmatrix}\begin{bmatrix} q_1 \\ q_2 \end{bmatrix} = \begin{bmatrix} 0 \\ 0 \end{bmatrix}. \quad [4.26]$$

The equations of motion contain gyroscopic terms. The asymmetric coupling terms in the equation makes the system likely to have self-excited type instability or mode coupling type instability.

The stability analysis of Eq. [4.26] is conducted [94] by carefully choosing realistic parameters to allow the stability to have a high sensitivity to those parameters:

$$h = 0.02\,\text{m}, k_t = 1.88 \times 10^7\ \text{Nm}; N_0 = 3.00\,\text{kN}, \Theta = 0.16\,\text{kg/m}^2,$$
$$r = 0.13\,\text{m}, k = 6.00 \times 10^6\,\text{N/m}, \Omega = 5\pi s^{-1}, \Phi = 2\Theta, \mu = 0.6,$$
$$c_t = 0.1\,\text{Nms}, c = 5.00\,\text{Ns/m}.$$

Eq. [4.26] could be solved for eigenvalues of the linearized system. Figure 4.19 shows the root locus of the eigenvalues for varying speeds of the disc. It shows that above a certain critical speed Ω_{crit}, there are eigenvalues with positive real parts, suggesting that the trivial solution becomes unstable.

Figure 4.20 shows the critical speed Ω_{crit} for varying stiffness k of the brake pads. It demonstrates that the critical speed increases for decreasing stiffness of the brake pads. Figure 4.21 shows the critical speed versus μ and N_0 with constant braking torque. The critical speed does not significantly depend on the friction coefficient m within this model. The figure also illustrates the influence of the combination of these two parameters on the critical speed. From this image, it is obvious that the stable zone could be maximized by a high normal force and a relatively low coefficient of friction.

4.3.5 Diagnosis and Prognosis

In this section, we offer perspectives on the diagnosis and prognosis of friction dynamics. In the automotive industry, friction-induced noise and

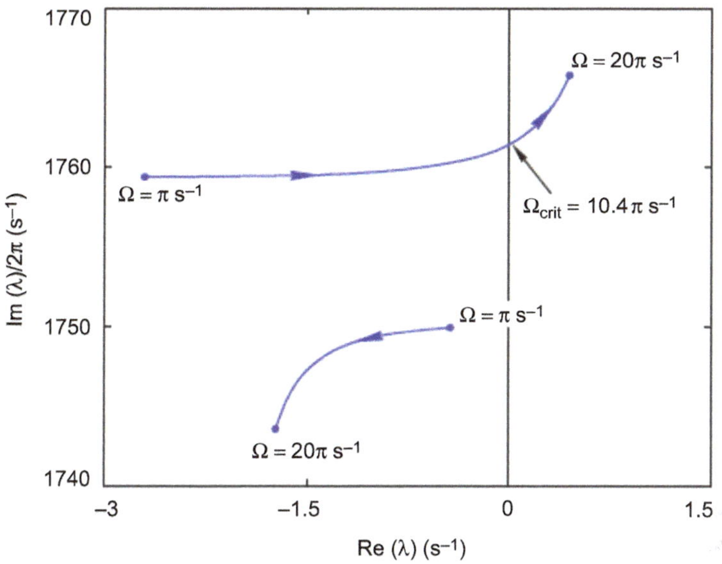

Figure 4.19 Root locus of the eigenvalues for varying Ω values (upper half-plane shown only) [94]

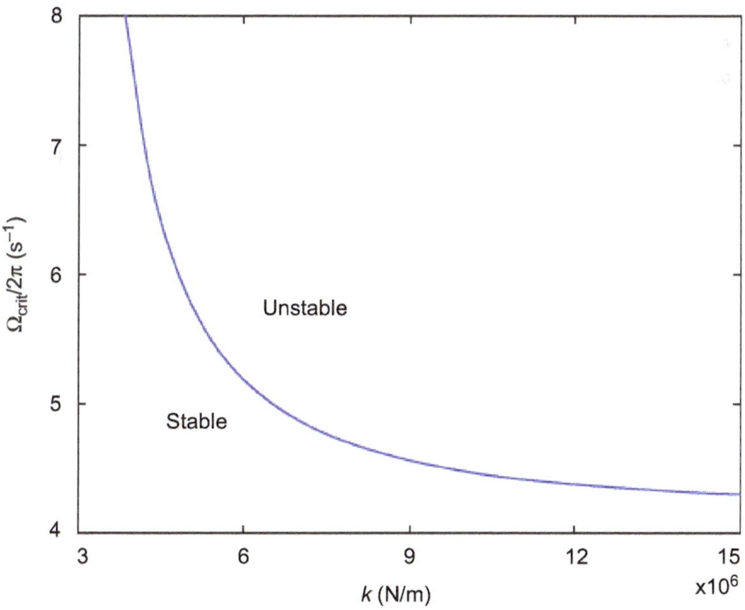

Figure 4.20 Critical speed versus stuffiness k [94]

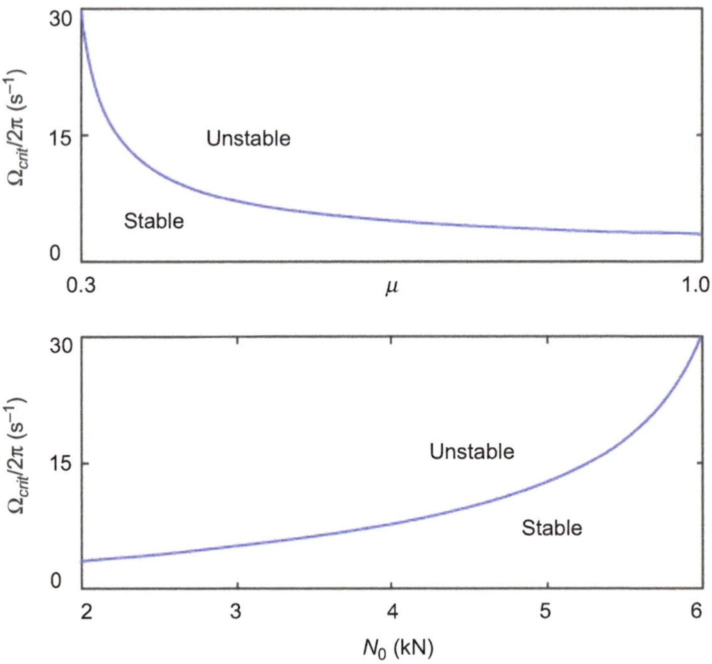

Figure 4.21 Critical speed versus μ and N_0 with constant braking torque [94]

vibrations have been a critical cost factor because of customer dissatisfaction. Even though efforts have been made over the last several decades to address this, friction dynamics remains surprisingly unpredictable and poorly understood. There are many unique dynamic phenomena and underlying mechanisms associated with friction dynamics. The most prominent ones are as follows: stick-slip due to difference between static and dynamic frictions; self-excited vibrations due to velocity-dependent friction; mode couplings; nonsmoothness effects; transient contact separation or vibro-impact effects; random excitation; impulsive excitation, sprag-slip; complex vibrations due to complicated friction laws such as rate-dependent friction; parametric vibrations of systems with variable normal force and friction; instability due to gyro-effects; vibrations due to nonlinear properties such as nonlinear contact stiffness; time-varying effects due to radical interface changes such as material generation and removal; uncertainties such as interface wear; and combinations and interactions of varied mechanisms. In general, friction-induced noise and vibrations have time-varying, nonlinear stochastic features.

For example, in the automotive brake industry, an extremely costly trial-and-error troubleshooting process has been applied for brake noise and vibration problems, despite the fact that numerical modeling and experiments are widely conducted. Most research has been directed toward either analytical and experimental studies of brake noise mechanisms or the prediction of brake noise propensity using finite element methods. Investigation into brake squeal has been conducted by various experimental and analytic methods [95–106].

Experimental methods, for all their advantages, are expensive mainly due to their hardware cost and long turnaround time for design iterations. Frequently, discoveries made on a particular type of brakes or on a particular type of vehicles are not transferable to other types of brakes. Product development is frequently carried out on a trial-and-error basis. There is also a limitation on the feasibility of the hardware implementation of ideas. A quietness margin for robust design is usually not found experimentally. Analytical or numerical modeling can simulate different structures, material compositions, and operational and environmental conditions of brakes. With these methods, noise performance and improvement measures can be examined conceptually before a prototype is made and tested. Theoretical results can also provide guidance to an experimental setup and help interpret experimental findings. The conventional numerical methods consist of complex eigenvalue analysis in the frequency domain and transient nonlinear analysis in the time domain.

Complex eigenvalue analysis may overpredict or underpredict the number of unstable vibration modes, and not all predicted unstable vibration modes will result in troublesome noise. These could be attributed to the inherently high sensitivity of the friction dynamics problems to the variation or the small changes of parameters, the uncertainty of the structure, materials and friction parameters, and the unforeseen effects of micro-level interface evaluations on friction. Brake components possess a large range of variations in material properties and sizes. The influence of these variations can be seen from measured and simulated frequencies on the system and component levels. The variations present great difficulties in validating a numerical model or a physical design. A robust theoretical model that takes into consideration the uncertainties due to material and size variations is needed.

Note that the repeatability of friction tests and brake noise tests is notoriously poor. Brake noise is a largely elusive phenomenon, partly due to the stochastic features of friction and the relevant time-varying factors such as interface profile, wear, and interface films. The theoretical stochastic models have been developed to address variability and uncertainty in brake noise [107–116].

A typical study [108] documented a robust design of a disc brake through structural optimization using the complex eigenvalue approach, considering the variation of friction coefficient, major elastic constants, and the effect of material wear-off. The evaluation of robustness can be conducted using commercial available software to give varied statistical measures [111]. Two studies [112–113] explored the sensitivity and uncertainty of predictions from a dynamics modal point of view. A method for efficiently estimating prediction error bounds is presented and validated using representative parametric uncertainties. It is illustrated that physical effects such as contact stiffness, damping, and friction laws can lead to extreme sensitivity. In [115], a design of experiment study of brake noise is presented. The experimental results were analyzed using statistical methods, including correlation analysis, semblance analysis, and joint recurrence quantification analysis. These results highlight the nonlinear character of brake squeal and indicate the potential of using nonlinear statistical analysis tools to analyze disc brake squeal.

In [116], the neural network method is used to predict disc brake performance, and the polynomial chaos expansions approach is used to tackle stochastic eigenvalue problems with applications to linear stability calculations for a simplified brake system in which the stability of a finite element model of a brake is investigated when its friction coefficient or the contact stiffness are modeled as random parameters. Results are compared to Monte Carlo simulations. The analysis of brake noise propensity is primarily based on linear approaches. On the other hand, it has been recognized that the brake noise and vibration involves a number of nonlinear phenomena such as bifurcation and chaos [117–124].

It is estimated that the onset of squeal happens in linear conditions and that nonlinearity defines the envelope of the vibration amplitudes. Nonlinear models are necessary to estimate the vibration amplitude after the onset of squeal and predict the complex modes and interactions. The transient nonlinear analysis is time-consuming and the results are difficult to interpret. How to include nonlinearity in the prediction of brake squeal propensity remains a challenge. In [120], test data obtained from a full brake system is investigated and the chaotic structures of noisy data from the squealing events are characterized by using recurrence plots. The lower-dimensional attractors are quantified by Lyapunov exponents. In [122], brake vibration data up to ultrasonic regimes are investigated using recurrence analysis and phase space reconstruction. It indicates that irregular vibration states of friction brakes are strongly dominated by intermittency phenomena that are dominated by low-dimensional irregular deterministic dynamics rather than by high-dimensional stochastic processes.

In [124], the investigation shows the complexity of the contributions of different harmonic components in transient friction-induced vibrations with the coexistence of multiunstable modes. Besides fundamental resonance, certain additional unstable modes can appear, which are not predicted by complex eigenvalues analysis. Besides numerical analysis, extensive testing of brake noise on dynamometers is required to ensure that the noise performance of brakes is acceptable. In particular, the ability of the complex eigenvalue method to predict brake squeal propensity is limited, and the numerical analysis in practice needs to be supplemented by extensive dynamometer tests. Both theoretical methods and experimental methods are needed to capture the underlying physics. These methods remain indispensable tools for understanding brake noise and for achieving improvement of noise performance [125–129].

Despite several decades of research where much progress has been made, the underlying mechanisms of noise and vibration generation are still not fully understood. The efficient and effective methodology for the prediction of brake noise has still been unavailable, and trial-and-error-based simple correlations of numerical modeling and test data have still been the major method used for brake noise analysis and prediction in the automotive industry. Even though there have been many studies on modeling, numerical simulation, and experiments, as well as qualitative correlation and approximate model updating based on linear modal analysis theory, research into model verification and validation with mathematical rigor remains rare. For instance, there has been a lack of research into critical aspects such as nonlinear system identification, stochastic system identification, and nonlinear stochastic system identification of brake noise and vibration.

Diagnosis is the early detection of faults in complex nonlinear stochastic systems, including classification and identification of the root causes of faults. On the other hand, prognosis means to predict the propensity of fault initiation and the evolution of fault occurrence conditions and to estimate the resulting uncertainty. The prognostics and health management and health monitoring technology have been fully developed by the professional community of electronics, aerospace, and civil engineering [130–131].

Similar technology can be applied to tackle friction-induced friction and noise problems in automotive systems. The diagnosis and prognosis framework could be developed for efficient and effective prediction of brake noise and vibrations due to its nonlinear and stochastic natures. Existing methods for brake noise troubleshooting and prediction tend to center around linear analysis and modal testing. These often fail to provide correct predictions due to the neglect of the following inherent characteristics of friction-induced vibrations: complex time-varying, nonlinear stochastic phenomena; high sensitivity of friction-induced vibrations to small changes in parameters; unavoidable variability and uncertainty of geometry/material/structures; and the significant effect of the unpredictable small-scale property's evolutions on macroscale dynamical phenomena. The existing nonlinear and stochastic analysis of brake noise and vibrations has been in the infancy stage.

The development of a diagnosis and prognosis offers solutions to overcome these limitations by comprehensively dealing with brake noise and vibration problems using nonlinear signal processing, phase space reconstruction, and nonlinear stochastic system identification in the context of computational intelligence techniques. A hierarchical multilevel strategy could be used in both diagnosis and prognosis. In a streamlined process, the measured vibration and noise signal could be processed by using nonlinear signal processing approaches to characterize specific nonlinear and nonstationary vibration features, and then the signal could be evaluated to extract possible chaotic properties and to characterize random behavior by using phase reconstruction and recurrence analysis. Nonlinear stochastic identifications could be subsequently followed by using measured signal, numerical model, and order-reduced models. These elements help capture the dynamics and tribological essentials of the brake system for reliable prediction. Finally, statistical pattern recognition algorithms could be used to integrate these elements and the data sets/knowledge accumulated in previous practices. This is expected to develop a combination of data-driven, physics model—based, and knowledge-based diagnostic and prognostic algorithms for automotive brake noise and vibrations, thus offering an unprecedented opportunity to fix brake noise and vibration problems effectively and efficiently in order to improve performance.

REFERENCES

[1] N.M. Kinkaid, O.M. O'Reilly, P. Papadopoulos, Automotive disc brake squeal, J. Sound Vib. 267 (1) (2003) 105—166.
[2] H. Ouyang, et al., Numerical analysis of automotive disk brake squeal: A review, Int. J. Vehicle Noise and Vib. 1 (3—4) (2005) 207—231.

[3] A. Papinniemi, et al., Brake squeal: A literature review, Appl. Acous. 63 (4) (2002) 391–400.
[4] F. Chen, C.A. Tan, R.L. Quaglia, Disk Brake Squeal, 2006. SAE International, PA.
[5] H. Misra, et al., Brake analysis and NVH optimization using MSC, in: NASTRAN, MSC Worldwide Automotive Conference, Atlanta, 1999.
[6] H. Nakata, et al., A new analysis approach for motorcycle brake squeal noise and its adaptation, 2001. SAE Technical Paper 2001–01–1850.
[7] T. Hamabe, et al., Study of a method for reducing drum brake squeal, 1999. SAE Technical Paper 1999–01–0144.
[8] S. Panier, P. Dufrénoyband, P. Brémondc, Infrared characterization of thermal gradients on disc brakes, Proc. SPIE 5073 (2003) 295–302.
[9] T.E. Rook, S. Kumar, Dynamic aircraft gear simulation using flexible multibody methods in Adams to guide component design and testing, in: ADAMS User Conference, Detroit, June, 2001.
[10] J.E. Mottershead, Vibration- and friction-induced instability in disks, Shock Vib. Dig. 30 (1) (1998) 14–31.
[11] J.E. Mottershead, S.N. Chan, Brake squeal—An analysis of symmetry and flutter instability, in: R.A. Ibrahim, A. Som (Eds.), Friction-Induced Vibration, Chatter, Squeal, and Chaos, DE-49, ASME, New York, 1992, p. 87.
[12] W. Österle, M. Griepentrog, T. Gross, I. Urban, chemical, and microstructural changes induced by friction and wear of brakes, Wear 251 (1) (2001) 1469–1476.
[13] H. Abendroth, Advances/progress in NVH brake test technology, 1998. SAE Technical Paper 982241.
[14] D.A. Crolla, A.M. Lang, Brake noise and vibration—The state of the art, in: D. Dowson, C.M. Taylor, M. Godet (Eds.), Vehicle Tribology, No. 18 in Tribology Series, 1991, pp. 165–174. Leeds, U.K.
[15] S. Yang, R.F. Gibson, Brake vibration and noise: Reviews, comments, and proposals, Intl. J. Mat. Prod. Technol. 12 (4–6) (1997) 496–513.
[16] M.G. Jacko, P.H.S. Tsang, S.K. Rhee, Automotive friction materials evolution during the past decade, Wear 100 (1) (1984) 503–515.
[17] J.O. Hultén, J. Flint, An assumed modes method approach to disc brake squeal analysis, 1999. SAE Technical Paper 1999-01-1335.
[18] J.O. Hultén, J. Flint, T. Nellemose, Mode shape of a squealing drum brake, 1997. SAE Technical Paper 972028.
[19] H. Ouyang, J.E. Mottershead, A bounded region of disc-brake vibration instability, J. Vib. Acous. 123 (4) (2001) 543–545.
[20] H. Ouyang, et al., A methodology for the determination of dynamic instabilities in a car disc brake, Intl. J. Veh. Des. 23 (3–4) (2000) 241–262.
[21] H. Ouyang, et al., Dynamic instabilities in a simple model of a car disc brake, 1999. SAE Technical Report 1999-01-3409.
[22] M. Nishiwaki, Generalized theory of brake noise, Proc. Inst. Mech. Eng., Part D: J. Auto. Eng. 207 (3) (1993) 195–202.
[23] J.E. Mottershead, S.N. Chan, Flutter instability of circular discs with frictional follower forces, J. Vib. Acous. 117 (1) (1995) 161–163.
[24] J.G. McDaniel, et al., Analysis of instabilities and power flow in brake systems with coupled rotor modes, 2001. SAE Technical Paper 01NVC-108.
[25] J. Flint, The effect of distributed parameters examined in a model for simulation of disc brake squeal, 2000. SAE Technical Paper 2000-01-2766.
[26] Y. Yuan, A study of the effects of negative friction-speed slope on brake squeal, in: Proc. 1995 ASME Des. Eng. Conf., Boston, 3 (Part A), 1995, pp. 1153–1162.
[27] A.J. Day, M. Tirovic, T.P. Newcomb, Thermal effects and pressure distributions in brakes, Proc. Inst. Mech. Eng., Part D: J. Auto. Eng. 205 (3) (1991) 199–205.

[28] M. Tirovic, A.J. Day, Disc brake interface pressure distributions, Proc. Inst. Mech. Eng., Part D: J. Auto. Eng. 205 (2) (1991) 137–146.
[29] Y.B. Yi, et al., Effect of geometry on thermoelastic instability in disk brakes and clutches, J. Tribol. 121 (4) (1999) 661–666.
[30] S. Du, J.W. Fash, Finite element analysis of frictionally-excited thermoelastic instability in a 3D annular disk, Intl. J. Veh. Des. 23 (3–4) (2000) 203–217.
[31] S. Du, et al., Finite element analysis of frictionally excited thermoelastic instability, J. Therm. Stress. 20 (2) (1997) 185–201.
[32] K. Lee, J.R. Barber, Frictionally excited thermoelastic instability in automotive brakes, J. Tribol. 115 (4) (1993) 607–614.
[33] P. Zagrodzki, et al., Nonlinear transient behavior of a sliding system with frictionally excited thermoelastic instability, J. Tribol. 123 (4) (2001) 699–708.
[34] J.R. Barber, Thermoelasticity and contact, J. Therm. Stress. 22 (4–5) (1999) 513–525.
[35] H. Abendroth, Advances in brake NVH test equipment, Auto. Eng. Intl. 107 (2) (1999) 60–63.
[36] F. Chen, J. Chern, J. Swayze, Modal coupling and its effect on brake squeal, 2002. SAE Technical Paper 2002-01-0922.
[37] M. Bettella, M.F. Harrison, R.S. Sharp, Investigation of automotive creep groan noise with a distributed-source excitation technique, J. Sound Vib. 255 (3) (2002) 531–547.
[38] H. Baba, T. Wada, T. Takgi, Study of reduction of brake squeal caused by in-plane vibration on rotor, 2001. SAE Technical Paper 2001–0103158.
[39] S. Dom, M. Riefe, T.S. Shi, Brake squeal noise testing and analysis correlation, 2003. SAE Technical Paper 2003-01-1616.
[40] M. Watany, et al., Abdel-Gawad, brake squeal generation, 1999. SAE Technical Paper 1999-01-1735.
[41] S. Mahajan, Y.K. Hu, K. Zhang, Vehicle disc brake squeal simulations and experiences, 1999. SAE Technical Paper 1999-01-1738.
[42] G.D. Liles, Analysis of disc brake squeal using finite element methods, 1989. SAE Technical Paper 891150.
[43] W. Liu, J. Pfeiffer, Reducing high-frequency disc brake squeal by pad shape optimization, 2000. SAE Technical Paper 2000-01-0447.
[44] T.A. Nosseir, et al., Significance of squeal in disk brake design, Intl. J. Veh. Des. 19 (1) (1998) 124–133.
[45] Y. Kumemura, et al., Analysis for reducing low-frequency squeal of a disc brake, 2001. SAE Technical Paper 2001-01-3137.
[46] Y. Denou, M. Nishiwaki, First-order analysis of low-frequency disk brake squeal, 2001. SAE Technical Paper 2001-01-3136.
[47] S.N. Chan, J.E. Mottershead, M.P. Cartmell, Parametric resonances at subcritical speeds in discs with rotating frictional loads, Proc. Inst. Mech. Eng., Part C: J. Mech. Eng. Sci. 208 (6) (1994) 417–425.
[48] H. Abendroth, A new approach to brake testing, 1985. SAE Technical Paper 850080.
[49] H.A. Sherif, On the design of anti-squeal friction pads for disc brakes, 1991. SAE Technical Paper 910575.
[50] H. Ouyang, J.E. Mottershead, Unstable travelling waves in the friction-induced vibration of discs, J. Sound Vib. 248 (4) (2001) 768–779.
[51] H. Ouyang, et al., Friction-induced parametric resonances in discs: effect of a negative friction-velocity relationship, J. Sound Vib. 209 (2) (1998) 251–264.
[52] J.E. Mottershead, et al., Parametric resonances in an annular disc, with a rotating system of distributed mass and elasticity; and the effects of friction and damping, Roy. Soc. Lond. Proc. Series A: Math. Phys. Eng. Sci. 453 (1956) (1997) 1–19.

[53] W.V. Nack, Brake squeal analysis by finite elements, Intl. J. Veh. Des. 23 (3–4) (2000) 263–275.
[54] P. Blaschke, M. Tan, A. Wang, On the analysis of brake squeal propensity using finite elements, 2000. SAE Technical Paper 2000-01-2765.
[55] F. Moirot, A. Nehme, Q.S. Ngyuen, Numerical simulation to detect low-frequency squeal propensity, 2000. SAE Technical Paper 2000-01-2767.
[56] M. Chargin, L. Dunne, D. Herting, Nonlinear dynamics of brake squeal, Fin. Ele. Anal. Des. 28 (1) (1997) 69–82.
[57] Y.K. Hu, S. Mahajan, K. Zhang, Brake squeal DOE using nonlinear transient analysis, 1999. SAE Technical Paper 1999-01-1737.
[58] Y.K. Hu, L.I. Nagy, Brake squeal analysis by using nonlinear transient finite element method, 1997. SAE Technical Paper 971510.
[59] L. Lee, et al., A systematic approach to brake squeal simulation using MacNeal method, 2002. SAE Technical Paper 2002-01-2610.
[60] L. Lee, et al., Direct finite element analysis on disc brake squeal using the able algorithm, 2003. SAE Technical Paper 2003-01-3350.
[61] A.R. Daudi, Effect of brake plate and fin joint on rotor noise damping, 1999. SAE Technical Paper, 1999-01-0145.
[62] N. Hoffmann, et al., A minimal model for studying properties of the mode-coupling type instability in friction induced oscillations, Mech. Res. Comm. 29 (4) (2002) 197–205.
[63] P.E. Dupont, D. Bapna, Stability of sliding frictional surfaces with varying normal force, J. Vib. Acous. 116 (1994), 237–237.
[64] P. Blaschke, M. Tan, A. Wang, On the analysis of brake squeal propensity using finite element method, 2000. SAE Technical Paper 2000-01-2765.
[65] H. Nakata, et al., A new analysis approach for motor cycle brake squeal noise and its adaptation, 2001. SAE Technical Paper 2001-1-1850.
[66] R.A.L. Rorrer, J.C. Brown, Friction-induced vibration of oscillating multi-degree of freedom polymeric sliding systems, Trib. Intl. 33 (1) (2000) 21–28.
[67] C.H. Chung, J.M. Donley, Mode coupling phenomenon of brake squeal dynamics, 2003. SAE 2003-01-1624.
[68] C.H. Chung, et al., A new analysis method for brake squeal. Part I: Theory for modal domain formulation and stability analysis, 2001. SAE Technical Paper 2001-01-1600.
[69] H. Lee, N. Kikuchi, Sensitivity analysis of complex eigen solutions for brake noise, 2003. SAE Technical Paper 2003-01-1626.
[70] J. Flint, The effect of distributed parameters examined in a model for simulation of disc brake squeal, 2000. SAE Technical Paper 2000-01-2766.
[71] J. Flint, J. Hultén, Lining-deformation induced modal coupling squeal generator in a distributed parameter disk brake model, J. Sound Vib. 254 (1) (2002) 1–21.
[72] H. Ouyang, et al., A methodology for the determination of dynamic instabilities in a car disc brake, Int. J. Veh. Des. 23 (3–4) (2000) 241–262.
[73] H. Abendroth, B. Wernitz, The integrated test concept: Dyno-vehicle, performance-noise, 2000. SAE Technical Paper 2000-01-2774.
[74] M. Reeves, et al., A study of brake disc modal behavior during squeal generation using high-speed electronic speckle pattern interferometry and near-field sound pressure measurements, Proc. Inst. Mech. Eng., Part D: J. Auto. Eng. 214 (3) (2000) 285–296.
[75] N. Talbot, J.D. Fieldhouse, Animations of a disc brake generating noise, 2001. SAE Technical Paper 2001-01-3126.
[76] J.D. Fieldhouse, P. Newcomb, The application of holographic interferometry to the study of disc brake noise, 1993. SAE Technical Paper 930805.

[77] J.D. Fieldhouse, M. Rennison, An investigation of low-frequency drum brake noise, 1998. SAE Technical Paper 982250.
[78] J.D. Fieldhouse, A proposal to predict the noise frequency of a disc brake based on the friction pair interface geometry, 1999. SAE Technical Paper 1999-01-3403.
[79] H.V. Chowdhary, A.K. Bajaj, C.M. Krousgill, An analytical approach to model disc brake system for squeal prediction, Proc. ASME DETC, Pittsburgh, 2001.
[80] M. Matsuzaki, T. Izumihara, Brake noise caused by longitudinal vibration of the disc rotor, 1993. SAE Technical Paper 930804.
[81] P.S. Shaw, M.A. Riehle, S.W. Kung, Mechanical Properties of Friction Materials and the Effect on Brake System Stability, 2003. SAE Technical Paper 2003-01-1619.
[82] W.W. Tworzydlo, et al., Friction-induced oscillations of a pin-on-disk slider: analytical and experimental studies, Wear 236 (1) (1999) 9−23.
[83] A.R. AbuBakar, H. Ouyang, A prediction methodology of disk brake squeal using complex eigenvalue analysis, Intl. J. Veh. Des. 46 (4) (2008) 416−435.
[84] O.N. Hamzeh, et al., Analysis of friction-induced instabilities in a simplified aircraft brake, 1999. SAE Technical Paper 1999-01-3404.
[85] D.M. Beloiu, R.A. Ibrahim, Disc-pad interaction related to brake squeal, 2004. SAE Technical Paper 2004-01-0826.
[86] R.A. Ibrahim, et al., Experimental investigation of friction-induced noise in disc brake systems, Intl. J. Veh. Des. 23 (3−4) (2000) 218−240.
[87] H. Ouyang, J.E. Mottershead, Friction-induced parametric resonances in disc: Effect of a negative friction−velocity relationship, J. Sound Vib. 209 (2) (1998) 251−264.
[88] H. Ouyang, et al., Friction-induced vibration of an elastic slider on a vibrating disc, Intl. J. Mech. Sci. 41 (3) (1999) 325−336.
[89] J. Kang, Squeal analysis of gyroscopic disc brake system based on finite element method, Intl. J. Mech. Sci. 51 (4) (2009) 284−294.
[90] A. Fidlin, O. Drozdetskaya, B. Waltersberger, On the minimal model for the low-frequency wobbling instability of friction discs, Eur. J. Mech. A/Sol. 30 (5) (2011) 665−672.
[91] O.N. Kirillov, Sensitivity analysis of gyroscopic and circulatory systems prone to dissipation-induced instabilities, Proc. MOVIC 2008, Int. Conf. Motion Vib. Ctrl. (Nov 5, 2008) 15−18. Munich, Germany.
[92] B. Hervé, et al., Friction-induced mode coupling model including damping and gyroscopic effects, in: Proc. 2007 IMAC-XXV Conf. Structural Dynamics, 2007, 68−29.
[93] B. Hervé, et al., Analysis of squeal noise and mode coupling instabilities including damping and gyroscopic effects, Eur. J. Mech. A/Sol. 27 (2) (2008) 141−160.
[94] U. von Wagner, D. Hochlenert, P. Hagedorn, Minimal models for disk brake squeal, J. Sound Vib. 302 (3) (2007) 527−539.
[95] A.R. Abu-Bakar, H. Ouyang, Recent studies of car disc brake squeal, in: N.W. Benjamin (Ed.), New Research on Acoustics, Nova Science Publishers, Inc., 2008, pp. 159−198. Long Beach, CA.
[96] A.A.Y. Mohammed, I.A.R. Rahim, Analysing disc brake squeal: Review and summary, Intl. J. Sci. Tech. 2 (4) (2013) 67−72.
[97] O. Giannini, A. Akay, F. Massi, Experimental analysis of brake squeal noise on a laboratory brake setup, J. Sound Vib. 292 (1) (2006) 1−20.
[98] A. Akay, et al., Disc brake squeal characterization through simplified test rigs, Mech. Sys. Sig. Proc. 23 (8) (2009) 2590−2607.
[99] J. Dreyer, et al., Effect of disc-pad contact modification on the brake judder source using a simplified elasto-kinematic model, 2013. SAE Technical Paper 2013-01-1907.

[100] J. Liette, J. Dreyer, R. Singh, Dynamic characterization of the rectangular piston seal in a disc-caliper braking system using analytical and experimental methods, Proc. Inst. Mech. Eng., Part D: J. Auto. Eng. 226 (12) (2013) 1613−1629.
[101] C. Duan, R. Singh, Analysis of the brake judder problem by employing a simplified source-path-receiver model, Proc. Inst. Mech. Eng., Part D: J. Auto. Eng. 225 (2) (2011) 141−149.
[102] Y. Dai, T.C. Lim, Suppression of brake squeal noise applying finite element brake and pad model enhanced by spectral-based assurance criteria, Appl. Acous. 69 (3) (2008) 196−214.
[103] F. Cantone, F. Massi, Numerical investigation into the squeal instability: effect of damping, Mech. Sys. Sig. Proc. 25 (5) (2011) 1727−1737.
[104] F. Massi, L. Baillet, A. Culla, Structural modifications for squeal noise reduction: Numerical and experimental validation, Intl. J. Veh. Des. 51 (1−2) (2009) 168−189.
[105] A. Culla, F. Massi, Uncertainty model for contact instability prediction, J. Acous. Soc. Amer. 126 (3) (2009) 1111−1119.
[106] P. Grange, et al., Brake squeal analysis by coupling spectral linearization and modal identification methods, Mech. Sys. Sig. Proc. 23 (8) (2009) 2575−2589.
[107] R.S. Langley, A unified approach to probabilistic and possibility analysis of uncertain systems, J. Eng. Mech. 126 (11) (2000) 1163−1172.
[108] F. Chen, et al., On automotive disc brake squeal. Part IV: Reduction and prevention, 2003. SAE Technical Paper 2003-01-3345.
[109] S.L. Qiao, R.A. Ibrahim, Stochastic dynamics of systems with friction-induced vibration, J. Sound Vib. 223 (1) (1999) 115−140.
[110] I. Elishakoff, Y. Ren, Finite Element Methods for Structures with Large Stochastic Variations, Oxford University Press, Oxford, UK, 2003.
[111] R.F. Nunes, et al., Robustness evaluation of brake systems concerned to squeal noise problem, Weimarer Optimierungs und Stochastiktage 4 (2009).
[112] T. Butlin, J. Woodhouse, Sensitivity of friction-induced vibration in idealized systems, J. Sound Vib. 319 (1) (2009) 182−198. Munich, Germany.
[113] T. Butlin, J. Woodhouse, Friction-induced vibration: Quantifying sensitivity and uncertainty, J. Sound Vib. 329 (5) (2010) 509−526.
[114] S. Oberst, J.C.S. Lai, Statistical analysis of brake squeal noise, J. Sound Vib. 330 (12) (2011) 2978−2994.
[115] D. Aleksendric, D.C. Barton, Neural network prediction of disc brake performance, Tribol. Intl. 42 (7) (2009) 1074−1080.
[116] E. Sarrouy, O. Dessomb, J.J. Sinou, Piecewise polynomial chaos expansion with an application to brake squeal of a linear brake system, J. Sound Vib. 332 (3) (2013) 577−594.
[117] K. Popp, P. Stelter, Stick-slip vibration and chaos, Philos. Trans. Royal Soc. Lond. A 332 (1642) (1990) 89−105.
[118] H. Hetzler, D. Schwarzer, W. Seemann, Analytical investigation of steady-state stability and Hopf-bifurcations occurring in sliding friction oscillators with application to low-frequency disc brake noise, Comm. Nonlin. Sci. Num. Sim. 12 (1) (2007) 83−99.
[119] F.H. Yang, W. Zhang, J. Wang, Sliding bifurcations and chaos induced by dry friction in a braking system, Chaos, Solitons, Fractals 40 (3) (2009) 1060−1075.
[120] S. Oberst, J.C.S. Lai, Chaos in brake squeal noise, J. Sound Vib. 330 (5) (2011) 955−975.
[121] F. Massi, et al., Brake squeal: Linear and nonlinear numerical approaches, Mech. Sys. Sig. Proc. 21 (6) (2007) 2374−2393.

[122] B.A. Wernitz, N.P. Hoffmann, Recurrence analysis and phase space reconstruction of irregular vibration in friction brakes: Signatures of chaos in steady sliding, J. Sound Vib. 331 (16) (2012) 3887–3896.
[123] N. Coudeyras, S. Nacivet, J. Sinou, Periodic and quasi-periodic solutions for multi-instabilities involved in brake squeal, J. Sound Vib. 328 (4) (2009) 520–540.
[124] J.J. Sinou, Transient non-linear dynamic analysis of automotive disc brake squeal—On the need to consider both stability and non-linear analysis, Mech. Res. Comm. 37 (1) (2010) 96–105.
[125] O. Giannini, F. Massi, Characterization of the high-frequency squeal on a laboratory brake setup, J. Sound Vib. 310 (1) (2008) 394–408.
[126] S. Oberst, J.C.S. Lai, S. Marburg, Guidelines for numerical vibration and acoustic analysis of disc brake squeal using simple models of brake systems, J. Sound Vib. 332 (9) (2013) 2284–2299.
[127] T. Butlin, J. Woodhouse, Friction-induced vibration: Model development and comparison with large-scale experimental tests, J. Sound Vib. 332 (21) (2013) 5302–5321.
[128] F. Renaud, et al., Motion capture of a pad measured with accelerometers during squeal noise in a real brake system, Mech. Sys. Sig. Proc. 33 (2012) 155–166.
[129] Proceedings of SAE (2011–12) Annual Brake Colloquium Detroit, Michigan.
[130] M. Pecht, Prognostics and Health Management of Electronics, Wiley, 2008.
[131] D. Balageas, C.P. Fritzen, A. Güemes, Structural Heath Monitoring, Wiley, New York, 2010.

CHAPTER 5

Friction Dynamics of Oil-Well Drill Strings and Sucker Rods

5.1 INTRODUCTION

Oil-well drilling and oil lifting are key technologies in the exploration and extraction of oil. Drilling strings and sucker rod strings are the primary tools for exploring and extracting petroleum from rock in the subsurface. Improvements in drilling technology and oil lifting technology that lower costs and increase the rate of success in finding and extracting petroleum will have a direct benefit to the world in terms of higher energy reserves, stable energy costs, and improved economic competitiveness in the drilling and service industries [1−12]. The friction dynamics of drill strings and sucker rods have had lots of critical problems associated with the quality of drilling and lifting performance, which limits tool life and productivity. This chapter presents the fundamental friction dynamics of drill string and sucker rod string. Concepts, modeling, and analysis connected with friction dynamics are presented, including advances in knowledge of the tool−rock interaction, couple axial and torsional friction dynamics, stick-slip whirl, and fully coupled dynamics of drill strings, as well as friction dynamics of sucker rods.

5.2 DRILLING SYSTEM

A rotating drill string driven by a motor on a drilling rig is a drill pipe, as shown schematically in Figure 5.1, which transmits torque to the drill bit for rock cutting and transmits drilling mud to the bottom for evacuating cut rock.

A drill string is typically made up of bottom hole assembly (BHA) and drill pipe. BHA consists of several components: drill collars, which are heavy, thick-walled tubes that apply weight to the drill bit; drilling stabilizers, which keep the assembly centered in the hole; and a drill bit, which is used to cut the rock formations. The BHA may also contain other components such as a downhole motor and rotary steerable system, measurement while drilling, and logging while drilling. Drill pipe makes up the majority of the drill string that goes back to the surface.

Friction Dynamics
ISBN 978-0-08-100285-8
http://dx.doi.org/10.1016/B978-0-08-100285-8.00005-5

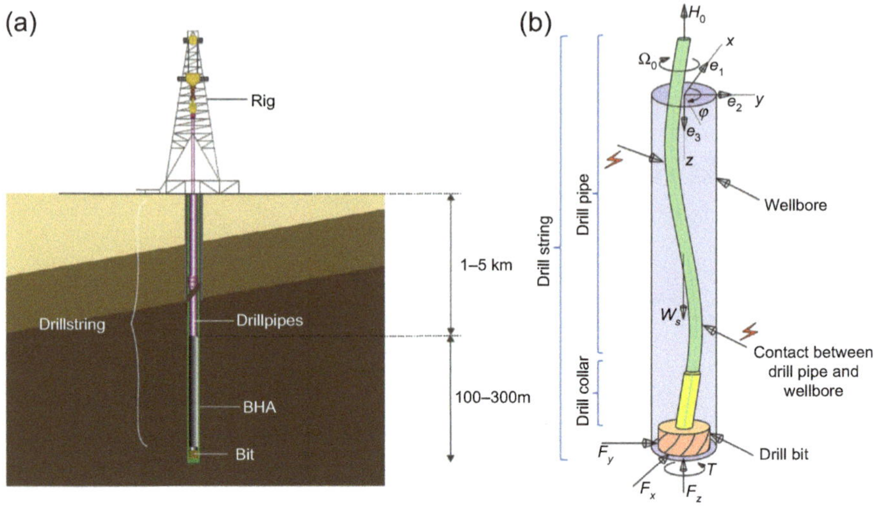

Figure 5.1 Schematic of a drilling system (a), and a drill string (b) [13—14]

5.3 DYNAMICS OF DRILL STRINGS

In oil and gas engineering, drill string dynamics plays a major role in determining the performance of a drill string. There are detrimental motions to the bit and the drill string that cause different failures of equipment and inefficiencies in drilling operations. The presence of detrimental motions is the main cause for shortening tool life and reducing productivity. The dynamics of a drill string system has features of the friction between drill bit and well bottom rock, and the friction between drill string and wellbore, as shown in Figure 5.1b. Most of the faults related to drill strings are attributed to friction vibrations of drill strings and the interactions with rock [1—20]. These motions include the following:

1. Axial motions featured by bit-bouncing, due to the interaction between drill bit and rock, which forces the bit to repeatedly lift off and hit rock. It can occur as a secondary mechanism resulting from whirl or stick-slip. This motion mode consists of irregular movements of the drilling components along its longitudinal axis, causing bit-bounce and rough drilling behavior that destroys the drill bit, damages BHA, and increases total drilling time. Additionally, this motion could be due to downhole coupling mechanisms, which also excite lateral motions of the drilling string.
2. Lateral vibrations due to drill-pipe eccentricity, leading to a rotational motion known as *drill string whirl*. When a rotating BHA makes contact with the borehole wall, it generates impacts from lateral motions. The

impacts with the borehole wall will produce eccentric holes and can damage components of the BHA. The lateral oscillations of the drill string may cause severe damage to the borehole wall and affect the overall drilling direction. Forward and backward whirling may strongly be triggered due to the combined effect of fluid damping, stabilizer clearance, and friction of the drilling assembly against the borehole wall [11–13].

3. Torsion vibrations associated with the stick-slip behavior of the drill bit caused by nonlinear bit–rock interactions, the interactions between drill string and the borehole wall, or both. The stick-slip is self-excited rotational motion induced by the nonlinear relationship between the torque and the angular velocity at the bit. The torsional oscillatory leads to rotational speeds as high as 10 times the nominal rotary table speed or a total standstill of the bit. Torsional vibrations cause fatigues to drill collar connections, damage the drill bit, and slow the drilling operation, thereby prolonging the overall drilling. The bit alternately rotates faster and slower than the table, and the rotational speed and torque on the bit may be several times their value at the surface. Torsional resonance is mainly associated with drill collars, which may likely create resonance of the collars themselves. Bit chatter is a forced torsional vibration, which can be more accurately described as "regenerative waviness." It can be noted as a high-frequency, high-amplitude cyclic acceleration change in the torsional mode.

The major dynamics modes and frequency range for practical drilling systems are briefly summarized in Table 5.1. The dynamics of drill strings

Table 5.1 Dynamics modes of drill-string [21]

Mechanism	Mode	Frequency
Bit bounce	Axial	1-10 Hz
Bitwhirl	Lateral/torsional	3-50 Hz
BHA whirl	Lateral/torsional	5-20 Hz
Stick-slip	Torsional	0.001-5 Hz
Lateral shocks	Lateral/Axial	
Torsional resonance	Torsional	50-250 Hz
Bit chatter	Torsional	50-250 Hz
Modal coupling	Lateral/axial/torsional	1-50 Hz

has been modeled either using multiple-degree-of-freedom systems associated with ordinary differential equations, or using continuous system models characterized by partial differential equations [22–31].

There are many possible excitations acting on drill strings, including imbalance, misalignment, effects of stabilizer blades, mud motors wobbling, the contact and friction between drill string and borehole wall, and particularly the cutting action of the drill bit on rock, which includes specific friction effects and time-delay effects. The bit–rock interaction has been modeled on the basis of varied friction laws. Friction effects lead to stick-slip phenomena. In certain instances, these effects lead to negative damping that can cause self-excited oscillations of the drill string.

Various friction laws have been utilized, including exponential power law, stiction plus Coulomb friction, dry friction plus Karnopp's model, Karnopp's model with a decaying friction term, and Karnopp's model with an exponential decaying friction term. In Karnopp model, the friction is set to be equal to external force for a small velocity zone close to zero, and the friction is a function of velocity outside the zone. Torsional vibrations are characterized by both stick and slip phases, during which the angular velocity of the tool increases up to more than twice the nominal angular velocity. This phenomenon occurs when a section of the rotating drill string is instantly held by friction against the borehole and then released. The bit might stick and then slip in turn, the collar rotation speeds vary dramatically, and large centrifugal accelerations occur. It is usually assumed that the growth of instability eventually leads to stick-slip oscillations arising from specific friction mechanisms governing bit–rock interaction [32–45].

In addition to nonlinear frictional forces acting on the drill bit, time-delay effects in bit-rock cutting are critical. These effects can be another cause of self-excited vibrations in drill bits [46] and have been investigated in a series of studies [47–54].

Uncertainties in drill string systems have also been included in the analysis [55–57]. Because of larger uncertainties in the modeling of a long drill string, including bit–rock interaction, tortuosity, mud properties, fluid viscous effects, wellbore temperature, pressure, and nonuniform geometrical interference between the drill string and wellbore, the coefficient of friction is considered as a composite factor including the effects of uncertainties Some system parameters are modeled using different stochastic processes that yield stochastic differential equations, and the Monte Carlo method is used to find the solution. A number of studies [25,46,58] examined the mechanics of a single cutter and developed a phenomenological model that captures the drag bit–rock interaction through decomposition of the cutting reactions into two independent components: the cutting force and the frictional force. The cutting force is created from the interaction between the cutting face of blade and the rock. Next, we present the bit–rock interaction laws developed by some researchers [46,54,58–61], which represents the drag bit drilling vertically with the bit velocity parallel to its axis of revolution, rotating around its axis of symmetry and representing the action of a single cutter on the bit profile.

Figure 5.2. shows a lumped model of the drilling system, which has a section of the bottom-hole profile located between two successive blades of a drill bit.

Both the total torque and the weight on the bit consist of a cutting term acting on the cutting faces of a cutter and a friction term acting at the all wear-flats/rock interface:

$$T = T_c + T_f, \quad W = W_c + W_f, \qquad [5.1]$$

where T and W denote the torque and force on the drill bit, respectively, as a result of the bit–rock interaction. This consists of cutting and friction components, denoted by the subscripts c and f, respectively. The cutting process takes place on the cutting face of the blades on the bit and describes the removal of the rock, whereas the friction component is related to the contact between the underside of the blades called the *wear-flat* and the *oil-well bottom*. The cutting term of the torque T_c and weight W_c can be expressed as

$$T_c = \frac{1}{2}\varepsilon a^2 d, \quad W_c = \zeta \varepsilon a d, \qquad [5.2]$$

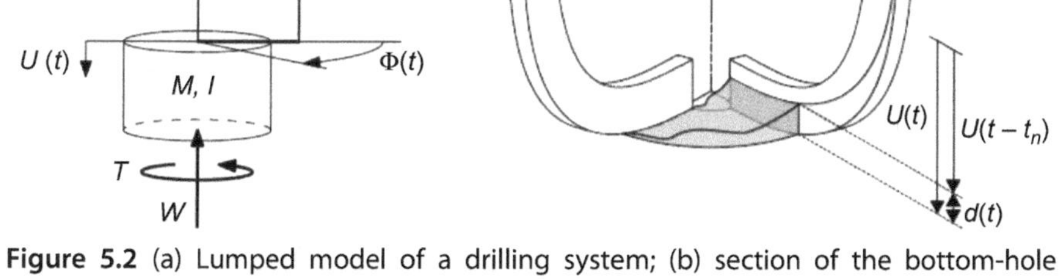

Figure 5.2 (a) Lumped model of a drilling system; (b) section of the bottom-hole profile located between two successive blades of a drill bit [61]

The friction terms of torque T_f and weight W_f are related:

$$2T_f = \mu a \gamma W_f. \tag{5.3}$$

In Eqs. [5.2] and [5.3], a denotes the bit radius and d is a bit geomtry number greater than 1, which characterizes the orientation and spatial distribution of the frictional contact surfaces associated with the cutters. In Figure 5.2, n is the number of blades on the drill bit with radius a. The intrinsic specific energy ε is used to characterize the cutting process, which is the required energy to remove a unit volume of rock. The orientation of the cutting face is represented by ξ. The frictional process takes place on the bit-rock interface with length l at the underside of the blades, which is called the *wear-flat*. The bit-rock contact at the wear-flats is described by the contact stress, which is constant when the bit moves downward into the rock. Each blade is characterized by a subvertical cutting surface and a chamfer of constant width n orthogonal to the cutting face. When such a bit is drilling rock vertically, the depth of cut per blade d_n is constant along the blade and identical for each blade. Here, d_n is related to the vertical position of the bit U as

$$d_n(t) = U(t) - U(t - t_n), \tag{5.4}$$

where t_n is the time required for the bit to rotate at an angle to its current position at time t, as schematically illustrated in Figure 5.2. The delay t_n is the solution of the following:

$$\Phi(t) - \Phi(t - t_n) = 2\pi/n. \tag{5.5}$$

The equivalent wear-flat length for the bit is $n/4$, and the combined depth of cut of the bit is expressed simply:

$$d = n[U(t) - U(t - t_n)]. \tag{5.6}$$

The depth of cut d also represents the depth of cut per revolution of the bit, under steady conditions or when it is averaged over several revolutions. One researcher [50] validated the model through a comprehensive experimental program and showed that there are three distinct operation regimes in the drag bit response, Phase I, II and III, which are directly associated with the variation of the contact forces with the depth of cut per revolution. The friction components are given by

$$W_f = \begin{cases} 0 & \text{if} \quad d \le 0 \\ K_c d & \text{if} \quad d \le d^* \\ R_b \sigma \ell f(\dot{x}_a) & \text{if} \quad d \ge d^* \end{cases} \tag{5.7a}$$

$$T_f = \gamma W_f \frac{R_b}{2} \mu f(\dot{\theta}), \quad [5.7b]$$

where $d \leq 0$ indicates bit bounce or bit loss of contact with formation and all the reaction forces will vanish. K_c and K_b are constants. Once the bit starts to penetrate, the formation is initially assumed to deform elastically, such that the axial reaction force, W_f, can be calculated by multiplying the rock contact stiffness, k_c, and the instantaneous depth of cut, d. This is considered Phase I in this bit—rock interaction model. As the depth of cut increases beyond the critical value d^*, the interaction will be developed into Phase II, where the rock fails similarly to yielding and the contact forces will be fully mobilized. In Phase II, the change in bit axial velocity sign is assumed to represent the bit unloading path or indentation variation, which in turn affects the frictional forces. This boundary condition captured by using a Heaviside function to avoid singularity is given as follows:

$$f(\dot{x}_a) = \frac{1}{1 + e^{(a_1 \dot{x}_a + b_1)}}, \quad [5.8]$$

where a_1 and b_1 are set to be constants, a_1 is used to minimize the transition period, and b_1 is used to reduce its value below 1 at zero, which defines the boundary condition. When the bit rotates backward, which may occur under extreme conditions, the cutting components W_c and T_c vanish due to the cutting faces losing contact with the ridge of the rock, and the friction component T_f changes sign.

5.4 FRICTION DYNAMICS OF COUPLED AXIAL AND TORSIONAL MODES

Rotary drilling systems with drag bits usually exhibit torsional vibration, which may lead to self-excited vibrations due to torsional stick-slip. The severe stick-slip vibrations reduce drilling efficiency, accelerate the wear of drag bits, and lead to drill string failure due to fatigue. Many studies have been conducted to investigate the behavior of the stick-slip vibration in drill strings. For instance, [16] studied the drill string dynamics of stick-slip type by using a simple one-degree-of-freedom torsional pendulum model, with the BHA being treated as a rigid flywheel, which gives rise to a mathematical stick-slip model, including the simple friction law without viscous effects. By using a similar simple pendulum model, [17] studied the stick-slip vibration of drill string by investigating the dry friction effect on the drill string vibrations, in which a continuous exponential form of dry friction is used:

$$f(\dot{\varphi}) = f_2 + (f_1 + f_2)\exp(-\gamma\dot{\varphi})^{\delta}, \qquad [5.9]$$

where f_1 and f_2 are the maximum and minimum values of $f(\dot{\varphi})$, and γ and δ are constants. Based on this model, stick-slip cycles are obtained by numerical solutions. The beating phenomenon exists due to stick-slip vibration, which is elaborated using a phase plane. The effects of rotational speed, viscous damping, and the natural frequency of the drilling system on stick-slip vibrations have been quantified.

Advanced analysis was conducted in [16,46], which are based on modeling coupled axial and torsional dynamics of a drill string. This coupling dynamics are studied by using a rate-independent bit—rock interaction law associated with a regenerative effect. Another study [46] shows that axial dynamics exhibit a stick-slip limit cycle, which is responsible for the generation of an apparent velocity-weakening effect (i.e., a negative slope of the friction-velocity curve) in the torque-on-bit. This velocity weakening effect, in turn, leads to the onset of torsional vibrations. In [46], the proposed bit—rock interaction laws consist of both rate-independent cutting process and frictional contact at the bit rock interface. A number of studies have been dedicated to a similar model [47—54]. Based on [46], [59—71] considered additional factors including the axial flexibility of the drill string and the dissipation due to friction along the bottom hole assembly. In the research, dimensionless formulation is used to describe the system. The researchers used an approximate model of the decoupled axial dynamics to derive the solution of the axial equation and the expression of the velocity-weakening friction law.

In the analysis of the drill string dynamics model, the separation in time scales between fast axial dynamics and slow torsional dynamics is exploited. Fast axial dynamics, exhibiting a stick-slip limit cycle, is analyzed separately. A semianalytical approach is developed to calculate the exact axial limit cycle and compute the parametrized parts of the axial limit cycle. In [59—62], a drill string model is depicted as shown in Figure 5.2. The BHA with axial and angular positions U and Φ, respectively, is modeled as a discrete mass M with inertia I. The drill string is modeled as a spring with torsional stiffness C and axial stiffness K. At the top, both the axial and angular displacements are prescribed. This represents the rotary table, which is assumed to exhibit the constant rotational and vertical speeds, Ω_0 and V_0, respectively. The viscous friction parameter D is used to characterize viscous friction along the BHA. The equations of motion for the BHA are given as

$$M\frac{d^2U}{dt^2} + D\frac{dU}{dt} + K(U - V_0 t) = -W_c - W_f \qquad [5.10]$$

$$I\frac{d^2\Phi}{dt^2} + C(\Phi - \Omega_0 t) = -T_c - T_f. \qquad [5.11]$$

Eqs. [5.2]–[5.6] and Eqs. [5.10] and [5.11] constitute a drill string dynamics model for nonnegative depth of cut and positive angular velocity, which is a set of nonlinear delay-differential equations.

These vibrations are coupled through the bit–rock interaction law. At the bit-rock interface, the cutting process is associated with state-dependent delay. The discontinuous terms on the right side of these equations reflect the frictional process of bit. This complex system is characterized by fast axial dynamics compared to slow torsional dynamics. Even though torsional stick is not included, the model can be used to predict the onset of torsional vibrations. The equations of motion can be scaled to reduce the number of parameters by introducing the following parameters:

$$t_* = \sqrt{I/C}, \quad \tau = t/t_*, \quad \eta = \sqrt{\frac{KI}{MC}}, \quad \bar{\tau} = \tau\sqrt{n\psi}, \quad \bar{\eta} = \frac{\eta}{\sqrt{\eta\psi}}, \quad \psi = \frac{\zeta\varepsilon a I}{MC},$$

which enables a two-time-scale analysis that uses a combination of averaging methods and a singular perturbation approach.

Note that the axial drill string stiffness K is much greater than the torsional stiffness C, such that the axial dynamics is fast compared to the torsional dynamics. This allows for the individual analysis of the axial dynamics, in which the angular velocity is assumed to be constant as its rotational speed slowly varies. An approximate model of the decoupled axial dynamics allows for the derivation of a pseudoanalytical expression of the solution of the axial equation. Its averaged behavior influences the slow torsional dynamics by generating an apparent velocity-weakening friction law that has been proposed empirically in previous studies. The analytical expression of the solution of the axial dynamics can be used to derive an approximate analytical expression of the velocity-weakening friction law related to the physical parameters of the system.

In steady-state drilling, the average axial velocity over multiple revolutions of the drill bit should equal the imposed axial velocity at the surface table in order for the drill string length to be constant on average. Axial stability analysis for calculating the axial limit cycles can be done by using the coordinate transformation method or using periodic solvers such as the shooting method or numerical integration. The results of a stability analysis for varying dimensionless stiffness parameter $\overline{\eta}$ and delay $\overline{\tau}_n$, which is inversely proportional to the rotational speed of the BHA, are given in Figure 5.3 [61]. Figure 5.3a shows that for no axial damping case and a small $\overline{\eta}$ case, the range in the delay for which the equilibrium point is locally asymptotically stable and decreases with increasing stiffness. For higher stiffness, multiple stability regions exist due to the complex interaction between the dynamics and the delay. Figure 5.3b (for $\overline{\gamma} = 0.5$) shows that the parameter region for which the dynamics is locally asymptotically stable increases with increasing delay as the axial damping increases. For realistic values of the stiffness, damping, and delay, the axial equilibrium point is unstable. As such, the axial dynamics is analyzed under the assumption that the parameters related to the slow torsional dynamics are constant.

For the torsional dynamics, the parameters related to the fast axial dynamics can be approximated by their averaged values, in which these parameters are approximated by the averaged value over one axial limit cycle for the current delay. The fast axial vibrations initiates the onset of the torsional vibrations. The onset of torsional vibrations is analyzed by averaging the influence of the fast axial dynamics on the slow torsional

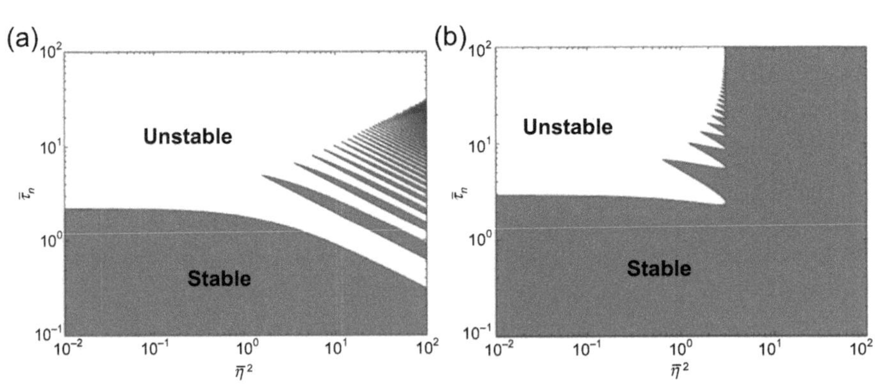

Figure 5.3 Stability diagram in the $(\overline{\eta}^2, \overline{\tau}_n)$ space for $\overline{\gamma} = 0$ (a) and $\overline{\gamma} = 0.5$ (b) [61]

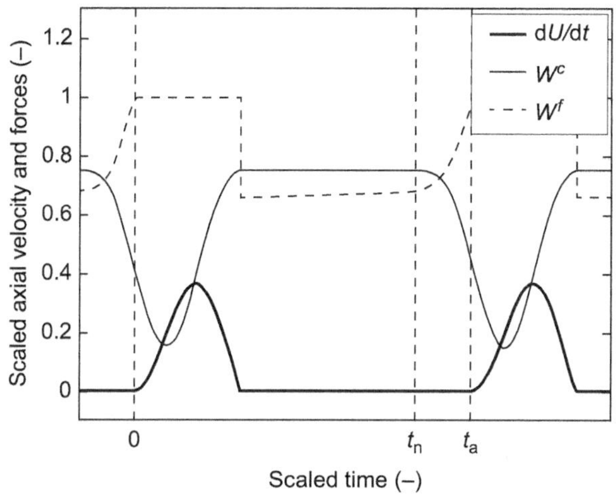

Figure 5.4 Axial limit cycle example in scaled coordinates [61]

dynamics. The axial dynamics results in a velocity-weakening effect in the torque-on-bit, which leads to torsional vibration.

It is shown that the axial equilibrium, which corresponds to a constant axial velocity, is unstable for realistic parameter values. The small perturbations around the axial equilibrium grow and lead to an axial stick-slip limit cycle, as shown in Figure 5.4 for a certain angular velocity. The limit cycle is a consequence of the regenerative effect introduced by the delay-dependency of the depth of cut, in which the stick phase occurs due to the discontinuity in the contact forces. Because of this discontinuity, the friction force W_f is able to compensate for the cutting forces W_c and the forces exerted by the drill string at zero axial velocity, which leads to the axial stick phase. Due to the positive angular velocity, the bit still removes rock during this phase. A change in the depth of cut leads to the onset of a new slip phase. In the analysis, the axial cutting force W_c, and axial friction force W_f are averaged over one axial limit cycle. The bit—rock interaction law then gives rise to the corresponding torque-on-bit. It is noted that the axial limit cycle and the averaged torque-on-bit are dependent on the value of the delay t_n, as shown in Figure 5.5. It shows that the cutting torque exhibits a velocity-weakening effect, even if the friction torque is velocity-strengthening in this situation. For realistic bit geometry, the total torque-on-bit is velocity-weakening, which initiates the onset of torsional vibration.

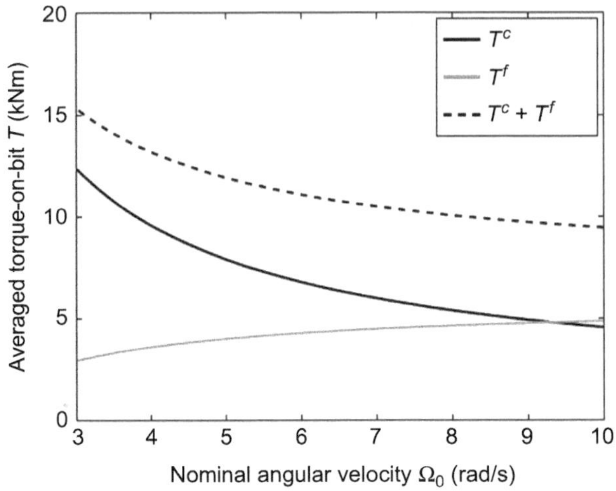

Figure 5.5 Averaged torque-on-bit due to axial vibrations

Hence, the averaged axial dynamics gives rise to a velocity-weakening effect on the torque acting on the bit, which results in unstable torsional dynamics. Some studies [46–54] have demonstrated that although axial and torsional dynamics are coupled, the underlying mechanisms for respective vibrations and instabilities are different. The driving mechanism of the axial vibrations is the regenerative effect. The driving mechanisms of the torsional vibrations include (i) the cutting action of the bit and (ii) wear-flat–rock interaction. In the case of torsional vibration, the generative effect plays only a secondary role. Axial compliance can play a stabilizing role, and this role increases as the ratio of the torsional to the axial natural frequency of the drill string vibration decreases. Varied stability charts in the plane of drilling rates and rotary speeds are obtained by solving a fully coupled, two-degrees-of-freedom model using linear stability analysis without making any asymptotic assumptions. Although the steady-state motion of the bit is always unstable, the nature of the instability depends on the nominal angular velocity Ω_0 of the drill string imposed at the rig. If Ω_0 is larger than a critical velocity Ω_c, the angular dynamics is responsible for the instability. If Ω_0 is smaller than Ω_c, the instability manifests itself on the time scale of the bit motion due to dominating unstable axial dynamics, and the perturbations to the steady-state motion degenerate rapidly into stick-slip limit cycles or bit-bouncing.

5.5 FRICTION WHIRL

Coupled torsional and bending vibrations of drill strings were investigated in [18] by combining simulation results with laboratory and field observations. It was shown from the study results that controlling surface torque rather than rotary table speed could eliminate stick-slip vibration. In [19], the drill string dynamics were studied using a lumped parameter model, which considered the coupling between torsional and bending vibrations, as well as the interactions between the drill string and the wellbore. In other research, [11−13] developed a stick-slip whirl model to explain the drill string dynamics when both stick-slip and whirl vibrations were involved, using bifurcation theory to explain the occurrence of stick-slip vibration. It explains the complicated behavior of oil-well drill string motion when both torsional stick-slip and lateral whirl vibration are involved. It is demonstrated that the observed phenomena in experimental drill string motion could be due to the fluid forces of the drilling mud. A stick-slip whirl model is presented that consists of a whirling motion portion and a stick-slip motion portion. The stick-slip whirl model assumes that a drill string is confined in a borehole wall with drilling mud. Here, [11−13] studied stick-slip effects associated with the whirling motions of a drill-pipe section and illustrated the presence of nonstandard bifurcations of the system response due to nonsmoothness of the system from nonlinear dry friction. The derived bifurcation diagrams of this discontinuous model for varying rotation speeds reveals discontinuous bifurcations. The disappearance of stick-slip vibration when whirl vibration appears is explained by bifurcation theory.

The numerical results are compared with experimental data from a full-scale drilling. The measurements on a real drill string show that torsional stick-slip and lateral whirl can be coexisting attractors. The time history of the rotation speed of the torsionally rigid lower part of the drill string is shown in Figure 5.6a. The drill string is in a torsional stick-slip vibration for the first 30 s, during which the BHA acts like a torsional pendulum in which one can see the drill-pipes as a torsional spring. After 30 s, the stick-slip vibration suddenly stops and the rotation speed of the BHA becomes constant. Whirl vibration is initiated, which is a lateral type of vibration. The downhole measurement device also measured the bending moment that was exerted on it.

The mean bending moment as a function of the downhole rotation speed is shown in Figure 5.6b. To some extent, the mean bending moment

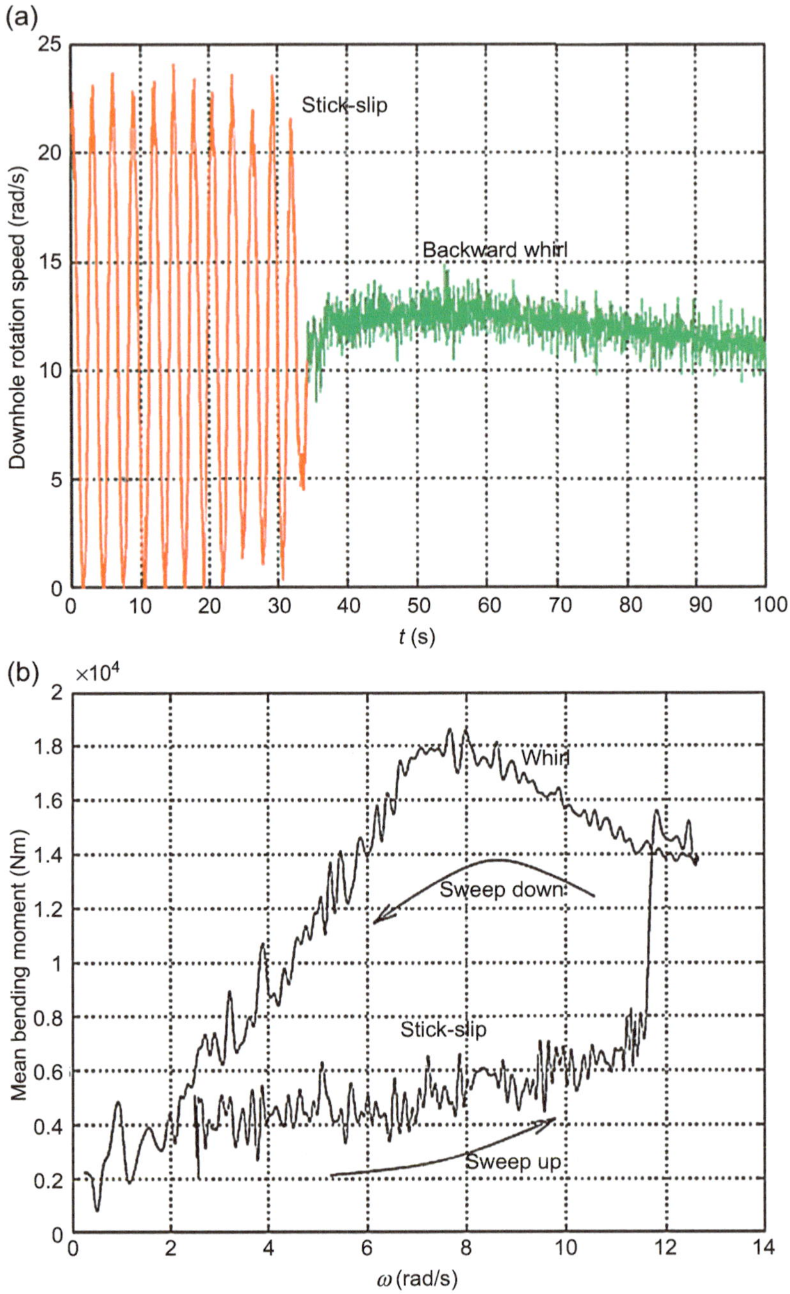

Figure 5.6 Measured downhole angular velocity versus time (a); measured downhole bending moment versus surface angular velocity, with sweep-up followed by sweep-down (b) [13]

is a measure for the radial deflection of the drill string. During whirl, the mean bending moment is considerably larger than that during stick-slip vibration. This indicates that the vibration denoted by "Whirl" in Figure 5.6b has a much larger lateral deflection and is indeed a whirl type of vibration.

The measured frictional torque as a function of the rotational speed exhibits interesting features. During stick-slip vibration, the friction curve shows a Stribeck effect, which explains the instability of steady rotation. During whirl, the friction torque is higher and has a positive slope. The additional torque is due to the lateral deflection during whirl. The larger lateral deflection will not only increase the contact between the BHA and the borehole wall, but also increase fluid drag forces on the drill string that give a kind of viscous friction torque. Therefore, the whirl phase is not only characterized by a higher level of the friction curve, but also by a positive slope of the friction curve due to the viscous friction of the fluid effect. The fluid forces, therefore, annihilate the Stribeck effect and constant rotation becomes stable, and the whirl vibration excludes the stick-slip motion: the vibration is either lateral whirl or torsional stick-slip. Stick-slip occurs for low angular velocities and whirl for high angular velocities. There is a hysteresis phenomenon in between, where the two types of vibration are coexisting stable attractors.

These researchers [11—13] give a detailed analysis with a model of three-degrees-of-freedom: one twist and two lateral displacements, as shown in Figure 5.7.

In this model, it is assumed that fluid forces form the interaction mechanism between torsional motion and lateral motion and the fluid forces are modeled as

$$F_{fr} = -m_f\left(\ddot{r} - \dot{\alpha}^2 r - \frac{\dot{\phi}^2}{4} r + \dot{\phi}\dot{\alpha} r\right) - (D + \psi_2(r))\dot{r} - \psi_1(r) r \quad [5.12a]$$

$$F_{f\alpha} = -m_f\left(\ddot{\alpha} r + 2\dot{r}\dot{\alpha} - \dot{\phi}\dot{r}\right) - \left(\dot{\alpha} - \frac{\dot{\phi}}{2}\right)(D + \psi_2(r)) r. \quad [5.12b]$$

The fluid motion results in a lift force F_{fr} in the radial direction on the rotor and a drag force $F_{f\alpha}$ in the tangential direction. The constant m_f is the added mass of the fluid and D is a damping parameter. The nonlinear functions $\psi_1(r)$ and $\psi_2(r)$ constitute the higher-order terms. The whirl speed is denoted by $\dot{\alpha}$ and the rotation speed of the disk by $\dot{\phi}$. Contact between the rotor and stator induces the contact forces F_{bN} and

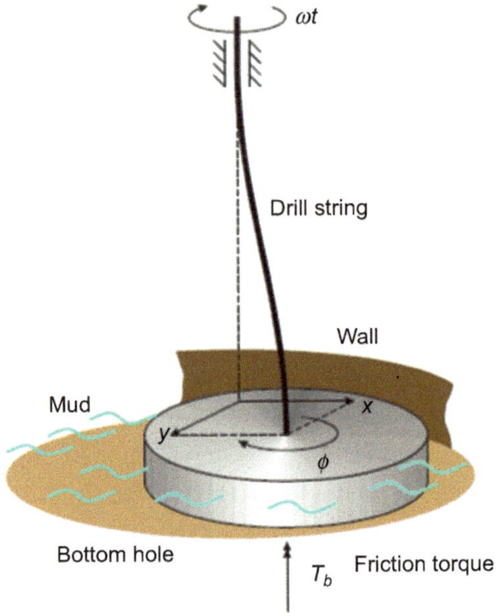

Figure 5.7 Stick-slip whirl model [12]

F_{bT}. The normal contact force is modeled as a linear spring with spring stiffness k_b:

$$F_{bN} = \begin{cases} k_b(r - R_c) & r > R_c \\ 0 & r \leq R_c \end{cases}. \quad [5.13]$$

By assuming a constant friction coefficient μ_b, the friction is given by

$$F_{bT} = -\mu_b \, \mathrm{sign}(v_{\mathrm{rel}}) F_{bN}, \quad v_{\mathrm{rel}} \neq 0. \quad [5.14]$$

The equations of motion for the whirling rotor with fluid and contact forces in polar coordinates are

$$m_a\left(\ddot{r} - \Omega^2 r\right) + (c + D + B_2 r^2)\dot{r} + \left(k - \frac{\dot{\phi}^2}{4} m_f + m_f \dot{\phi}\Omega + B_1 r^2\right) r = -F_{bN}$$

$$m_a\left(\dot{\Omega} r + 2\dot{r}\Omega\right) + (c + D + B_2 r^2)\Omega r = m_f \dot{\phi}\dot{r} + (D + B_2 r^2)\frac{\dot{\phi}}{2} r + F_{bT}$$

$$J\ddot{\varphi} = -k_\varphi \varphi + T_f + T_b + T_d.$$

[5.15]

In the full stick-slip whirl model, the disk represents a very simple model of the BHA acting by a dry friction torque T_b of bit—rock interaction (as well as the stabilizer wall contact), due to tangential contact force $T_b = F_{bT}R$. T_d is torque due to fluid forces $T_d = -F_{f\alpha}r$. Dry friction torque T_f is acting on the rotor, which only depends on the angular velocity $\dot{\phi}$:

$$T_f = -\text{sgn}\dot{\phi}\frac{T_0}{1+\delta|\dot{\phi}|} \quad [5.16]$$

Contact between the rotor and stator induces the contact forces F_{bN} and F_{bT}. The nonlinear dynamics of the stick-slip whirl model was investigated numerically. The pure whirl model containing only the two lateral degrees of freedom of the full stick-slip whirl model is investigated, and a bifurcation diagram is obtained with the rotational speed as the bifurcation parameter and the lateral deflection on the vertical axis. Then the full model is solved to obtain the full bifurcation diagram of the model (as shown in Figure 5.8a), which indicates that the equilibrium branches in a corotating frame as well as the periodic ones.

The Stribeck effect in the friction curve destabilizes the trivial equilibrium (curve E1) of the pure whirl model. Part of the forward whirl branch is unstable: branch E2 contains a Hopf bifurcation. For small rotational speeds, the full stick-slip whirl model undergoes a stable torsional stick-slip oscillation. If the rotational speed is increased, then the fluid forces destabilize the undeflected trivial position and the drill string will deflect in the lateral direction. When the drill string is laterally deflected, the torsional stick-slip vibration will be damped due to the viscous fluid friction. The branch E4 of stable backward whirling solution is unaffected by the torsional model. For low rotational speeds, the model is in torsional stick-slip oscillation, whereas for high rotation speeds, it is in lateral backward whirl motion. A very complicated hysteresis region exists between these conditions, where both types of motion can be present, with 10 other mixed and period-doubled types of oscillation. This region, with very complicated dynamic behavior, is localized. Figure 5.8b plots a zoom of the hysteresis region. Periodic branches are created at points where the equilibrium branches lose stability. From the Hopf bifurcation at the equilibrium, branch E2 emanates a periodic branch P6. Branches P2—P6 connect the solutions found from the pure whirl model with the solutions found from the pure stick-slip model. The

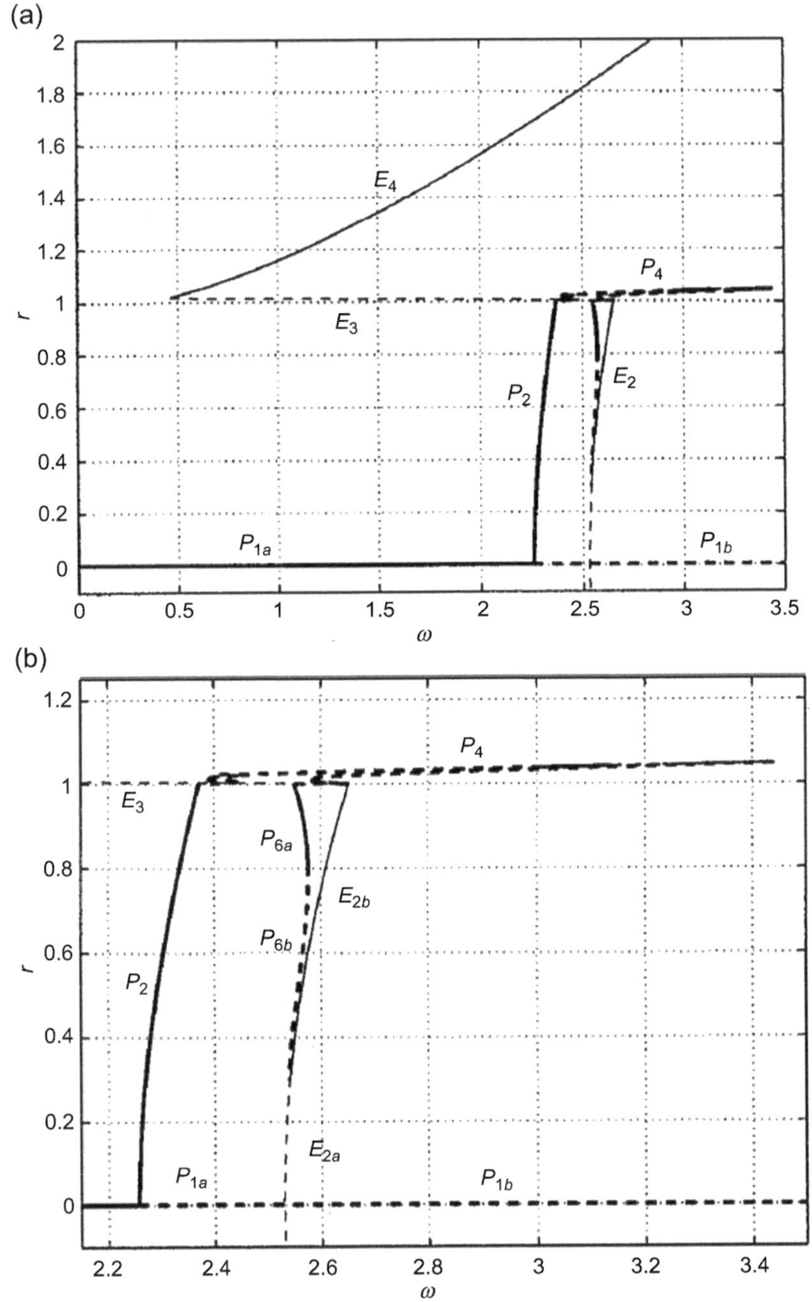

Figure 5.8 Bifurcation diagram of the stick-slip whirl model: (a) periodic branches (bold); (b) zoom of (a) [13]

bifurcation structure is apparently very complicated. Branch P4 undergoes a number of period-doubling bifurcations. A number of bifurcation points are smooth, while other bifurcation points are nonsmooth. These nonsmooth bifurcation points sometimes behave like classical smooth bifurcation points, but they can also show a nonstandard behavior.

5.6 FULLY COUPLED DYNAMICS WITH FRICTION

Instabilities caused by whirling, lateral, and torsional-axial modes of drill strings lead to dimensional inaccuracies of the holes and may damage the drills. A fully coupled model for axial, lateral, and torsional vibrations of drill strings has been desired. In several studies, [69—74] developed and investigated a dynamic model, including the gyroscopic effect, coupling between torsional and bending vibrations, and the effect of the gravitational force field. A fully coupled model for axial, lateral, and torsional vibrations of actively controlled drill strings is presented in [72]. The proposed model includes the mutual dependence of these vibrations, which arises due to bit/formation and drill string—borehole wall interactions, as well as other geometric and dynamic nonlinearities. In [73], two fully coupled models were developed with four and five degrees of freedom, and the model predictions showed phenomena consistent with experimental observations. In these studies on lateral vibrations, which have focused on the whirling of a drill pipe or drill collar, drill bit dynamics has been disregarded. In another study, [14] developed a discrete model of drill string with eight degrees of freedom, considering both the stick-slip vibration and time-delay effect. Nonlinearities that arise due to dry friction, loss of contact, and collisions are considered in the development. State variable-dependent time delays associated with axial and lateral cutting actions of the drill bit are introduced in the model. Numerical studies with different drilling operations showed the self-excited characteristics and the effect of driving speed on system stability. The model considers the regeneration of chip thickness due to the tool deflection in the lateral, torsional, and axial directions. The tool wear is considered on the damping of high-frequency torsional vibrations during the drilling process. The stability for whirling, lateral, and torsional axial chatter is predicted simultaneously. The fully coupled model with axial, torsional, and lateral motions of the drill string given in [14] is shown in Figure 5.9. The model has two disks connected over one constant length

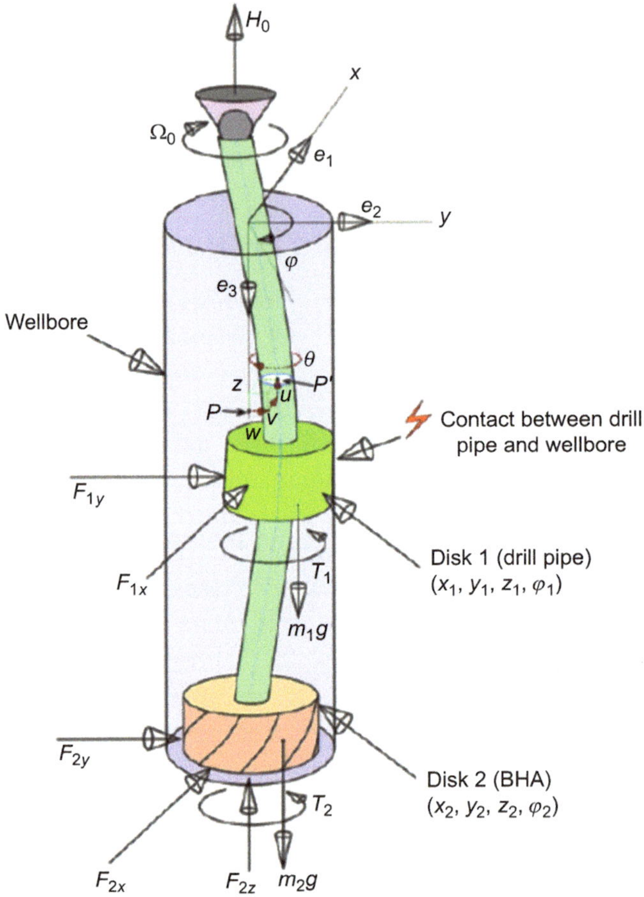

Figure 5.9 Schematic of the discrete system model along with the forces and torques acting on the drill string system [14]

span, with each disk having different translational inertia and rotary inertia properties.

The rotational moments of inertia of the drill pipes and BHA are respectively represented by the first disk and the second one. The dynamics of each disk consists of two lateral motions, an axial motion, and a torsional motion, specified by the lateral, longitudinal, and torsional coordinates (x_1, y_1, z_1, ϕ_1) and (x_2, y_2, z_2, ϕ_2). The system is driven by the top table with a constant angular speed Ω_0, and the two disks are connected by two elastic elements, which reflect the elastic properties in axial, torsional, and bending deformations. The governing equations of motion for the eight-degree-of-freedom system are given by [14]:

$$\begin{cases} m_1\ddot{x}_1 + c_l\dot{x}_1 + 2k_l\left(x_1 - \dfrac{x_2}{2}\right) = m_1 e\dot{\varphi}_1^2 \cos\varphi_1 + F_1x \\ m_1\ddot{y}_1 + c_l\dot{y}_1 + 2k_l\left(y_1 - \dfrac{y_2}{2}\right) = m_1 e\dot{\varphi}_1^2 \sin\varphi_1 + F_1y \\ m_1\ddot{z}_1 + c_a\dot{z}_1 + k_a(z_1 - z_2) = m_1 g - H_0 \\ J_1\ddot{\varphi}_1 + c_t\dot{\varphi}_1 + k_t(2\varphi_1 - \varphi_2) = -T_1 + k_t\Omega_0 t \\ m_2\ddot{x}_2 + c_l\dot{x}_2 + k_l\left(\dfrac{x_2}{2} - x_1\right) = m_2 e\dot{\varphi}_2^2 \cos\varphi_2 + F_2x \\ m_2\ddot{y}_2 + c_l\dot{y}_2 + k_l\left(\dfrac{y_2}{2} - y_1\right) = m_2 e\dot{\varphi}_2^2 \sin\varphi_1 + F_2y \\ m_2\ddot{z}_2 + c_a\dot{z}_2 + k_a(z_2 - z_1) = m_2 g - F_{2z} \\ J_2\ddot{\varphi}_2 + c_t\dot{\varphi}_2 + k_t(\varphi_2 - \varphi_1) = -T_2 \end{cases}, \qquad [5.17]$$

in which m_1, m_2, and J_1, J_2 are the mass and the rotary inertia of the disks; e is the eccentricity, which is assumed to be the same; the influence of torsional acceleration on the lateral centrifugal force has been neglected. The subscripts 1 and 2 donate the top and bottom disks, respectively; and the stiffness and damping are denoted by k and c, respectively. The subscripts a, t, and l correspond to the directions on which these properties have influence: axial, torsional, and lateral. The quantities F_{1x}, F_{1y}, and T_1 are the respective force components and the torque that arises due to interactions between the drill pipe and wellbore. Similarly, the quantities F_{2x}, F_{2y}, F_{2z}, and T_2 are the respective force components and torque that arise due to interactions between the bit and rock. The axial, torsional, and lateral vibrations of the discrete system are coupled through these interaction forces and torques.

Contact interactions between drill pipe and wellbore are modeled by using Eq. [5.13] plus certain damping. The tangent force is obtained by using a specific function of coefficient of friction. Contact interactions between drill bit and rock during axial cutting follow the model of Eqs. [5.1]−[5.7], with a velocity-weakening function $\eta(\omega_2)$ used to model the friction components of cutting forces:

$$\eta(\omega_2) = \frac{2}{\pi}\arctan(\varepsilon\omega_2)\left(\frac{f_1 - f_0}{1+\delta|\omega_2|} + f_0\right), \qquad [5.18]$$

where ε, δ, f_1, and f_2 tailor the properties of the friction model. Contact interactions between drill bit and rock during lateral cutting are also considered in [14]. The time-delay effect in the lateral cutting process is more complex than that in the axial cutting due to the presence of multiple regenerative effects. In [14], a model based on the multiple regenerative effects [75—76] of lateral cutting is used to understand the backward whirl of the drill bit. Bit whirl is expected to introduce cutting along the lateral directions and result in lateral forces. Following [14], the hatched area of Figure 5.10a represents the rock to be removed by the drill bit. The quantity *ri* is the radius vector from the wellbore center *o* to the cutting edge of the drill bit. The location O_2 marks the center of the drill bit, and Θ is the angle between point 0 and O_2 with respect to the point of contact. In order to simplify the geometry of the lateral-cutting process, a couple of assumptions have been made.

Numerical studies were conducted with this model. It shows that the varying time-delay effects in conjunction with the stick-slip effects play a key role in determining the nonlinear motions of the system. In terms of external parameters, the system stability is largely dependent on the motor's drive speed. Parametric studies revealed that the drill string dynamics is sensitive to the driving speed [73]. As a typical case shown in [14], when the

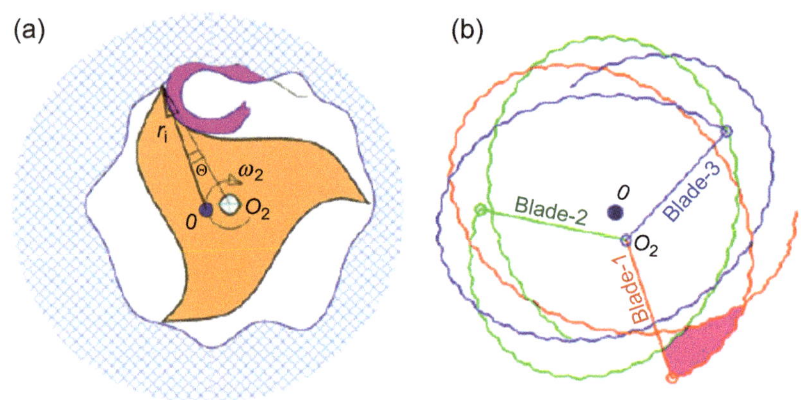

Figure 5.10 Drill bit—rock interaction during lateral cutting: (a) schematic of lateral cutting; (b) multiple regenerative cutting effects [14]

driving speed is 50.0 rad/s, the phase-plane plot for torsional vibration of the drill bit is periodic. Upon increasing the driving speed further to 56.2 rad/s, a qualitative change (period-2 motion) is observed. Ultimately, the aperiodic state is reached through a period-doubling route. To illustrate the local bifurcations, Poincaré sections were constructed in the 16-dimensional state space in [14]. A Poincaré section is chosen as $\dot{\varphi}_2 = \Omega_0$, to ensure that it is a global section. When the drive speed is varied in a quasi-static way, the resulting bifurcation diagram on a Poincaré section is shown in Figure 5.11a. Periodic and chaotic windows can be observed, and a period-doubling route can be noted from the expanded diagram presented in Figure 5.11b.

One study [77] analyzed the chaotic properties of drill strings based on experimental data. It developed a real-time measurement system near the drill bit and extracted the time series of lateral vibration and longitudinal vibration of a bottom-rotating drill string. Then the phase space was reconstructed by estimating the delay time with mutual information and

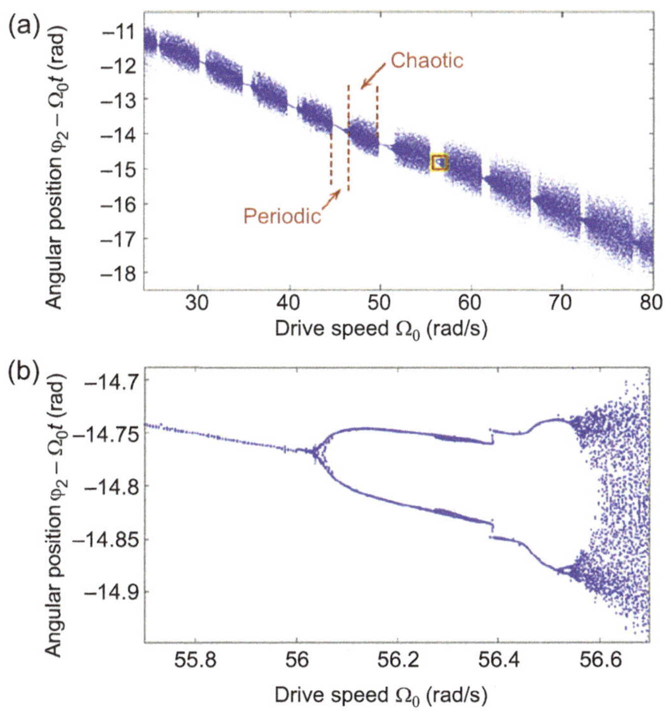

Figure 5.11 Bifurcation diagram on a Poincaré section for $\dot{\varphi}_2 = \Omega_0$: (a) Ω_0 range of 22.0–80.0 rad/s; (b) Ω_0 range of 55.7–56.7 rad/s [14]

calculated the embedding dimension. The chaotic characterization of the system was analyzed by calculating the correlation dimension and the largest Lyapunov exponent. The results showed that such a system can exhibit positive finite-time Lyapunov exponents and a clear convergence toward the correlation dimension, which is a strong indicator of the existence of chaotic behavior of the system.

A nonlinear analysis is critical for processing the test data of a nonlinear system, where traditional tools such as Fourier transform may fail to detect substantial intrinsic quantities in the nonlinear system. A complex dynamics system could be quantified by the concept of an attractor associated with chaos properties. It has been proven that the topology of the attractor can be preserved by using time delays [78–80]. The embedding dimension must be large enough to fully unfold the attractor structure; and in principle, $2D+l$ dimensions are required, where D is the dimension of the attractor. Usually an embedding dimension larger than D is enough. If we have a time series, $x(1), x(2), \ldots$, we can construct an n-dimensional vector such as

$$X(i) = [x(i), x(i+\tau), \ldots, x(i+(m-1)\tau)] \qquad [5.19]$$

where τ is the time delay chosen by one of the standard methods from a set of observations, and multivariate vectors in n-dimensional space are used to trace the orbit of the system. Let $X(i)$ be the ith point on the orbit describing a dynamics system in an n-dimensional space for $i = 1, \ldots, N$.

Complex systems may exhibit dynamic chaotic behavior. The Lyapunov exponents of a dynamic system can be used as a measure of a system's sensitivity to initial conditions associated with chaos. When a chaotic system evolves from a set of initial conditions within radius d in the phase plane, after time t, the divergence of the trajectories is characterized by

$$d = d_0 2^{\lambda t}, \qquad [5.20]$$

in which λ is a Lyapunov exponent corresponding to the average rate of the divergence of the trajectories. When the largest Lyapunov exponent is positive or the system contains at least one positive Lyapunov exponent, the system exhibits chaotic behavior. The direct method has been used to estimate the largest Lyapunov exponents of the reconstructed state space from a time series [81–82]. It is supposed that the chaotic time series is $\{x_1, x_2, \ldots, x_n\}$, embedding dimension is m, and time delay is τ. Then reconstruction phase space is as follows:

$$Y(t_i) = (x(t_i), x(t_i+\tau), \ldots, x(t_i+(m-1)\tau)), \quad i = 1, 2, \ldots, N. \qquad [5.21]$$

Take the initial point of phase space as $Y(t_0)$; suppose that its most neighboring point is $Y_0(t_0)$; and then suppose that the distance between these two points is $L(t_0)$. From time t_0, start tracking the time evolution of these two points until the spacing between the two points exceeds a predetermined value ε on the moment t_1:

$$L'(t_1) = |Y(t_1) - Y_0(t_1)| > \varepsilon, \qquad \varepsilon > 0. \tag{5.22}$$

Retaining point $Y(t_1)$, and supposing that its most neighboring point is $Y_1(t_1)$, then

$$L(t_1) = |Y(t_1) - Y_1(t_1)| < \varepsilon, \ \ \varepsilon > 0. \tag{5.23}$$

Also, make the angle θ between (t_1) and $L'(t_1)$ as small as possible. Repeat this procedure until (t) reaches the end of the time series. The total number of iterations is M; and the largest Lyapunov exponent will be given as follows:

$$\lambda_1 = \frac{1}{t_M - t_0} \sum_{k=1}^{M} \ln \frac{L'(t_k)}{L(t_{k-1})}. \tag{5.24}$$

[77] considered the representation of the time series data as a trajectory in the embedding space. Assuming that (n) exhibits a linear increase with identical slope for all m larger than a certain m_0 and for a reasonable range of ε, then this slope can be taken as an estimate of the largest exponent. The results obtained for $\varepsilon = 0.02$ and $m = 3-7$ are presented in Figure 5.12.

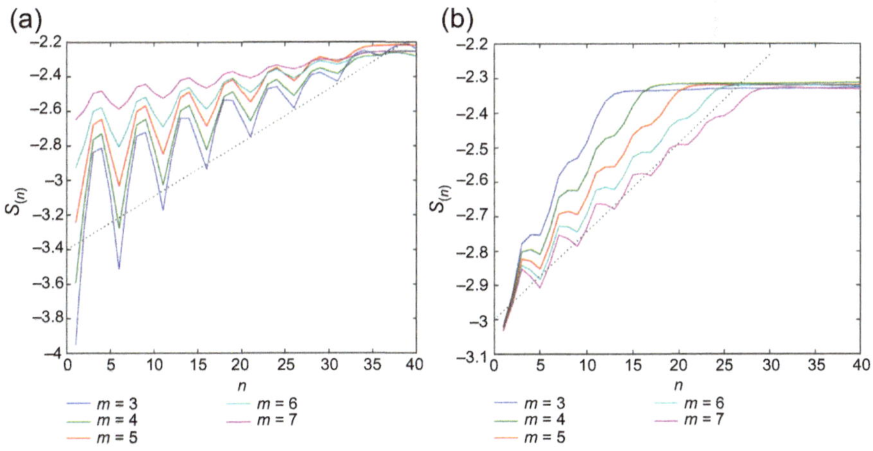

Figure 5.12 Calculation of the largest Lyapunov exponent of lateral vibration and longitudinal vibration and the effective expansion rate (n), of which the linear slope in dependence on n is a robust estimate for the largest Lyapunov exponent [77]

The function $S_{(n)}$ shows a rather robust linear increase for all the embedding dimensions. Thus, the slope of $S_{(n)}$, indicated by the dashed straight line in the figure, is a good estimate for the largest Lyapunov exponent of the system. The largest Lyapunov exponent of the lateral vibration is calculated through a least-squares line fit for the time series and is found to be around 0.032. The largest Lyapunov exponent of the longitudinal vibration is calculated to be around 0.028. This positive value indicates an exponential divergence of trajectories, and hence evidence of the low-dimensional chaos that occurs during the entire drilling process [73]. This is consistent with conventional analysis of [83–84] and [14].

In field measurement data of drill string vibration, finding chaotic characteristics is helpful to rediscover the drill string vibration phenomenon and to improve the control algorithm of drill string vibration in oil and gas engineering.

Many efforts have been made to mitigate and control the friction dynamics of drill string systems [85–99]. This type of vibration can be suppressed by means of many approaches, such as varying parameters and use of control tools. Although tremendous improvements have been made in overcoming this dysfunction, stick-slip vibration suppression remains a large problem in the drilling industry. Passive control includes the optimization of BHA configurations, bit selection and bit redesign, and use of downhole tools. The active control approaches for stick-slip mitigation include the optimization of drilling parameters based on real-time measurement and use of varied active control systems.

5.7 FRICTION DYNAMICS OF SUCKER RODS

Sucker rod pumping systems are the oldest and most widely used type of artificial lifting system for oil wells. Figure 5.13 shows a schematic of a rod pumping system. Sucker-rod pumping is the most widely used artificial lift method for onshore oil wells. Sucker-rod strings transfer the motion from the driving machine on the wellhead to the downhole oil pump.

Pump diagnoses are generally conducted through an interpretation of the pump dynamometer graph (pump card), which is a plot of the calculated loads at various positions of a pump stroke. In deviated wells, mechanical friction becomes nonnegligible since there is considerable contact between rods, rod guides, and tubing. Hence, the dynamic behavior of the rod string is different for deviated wells and vertical wells.

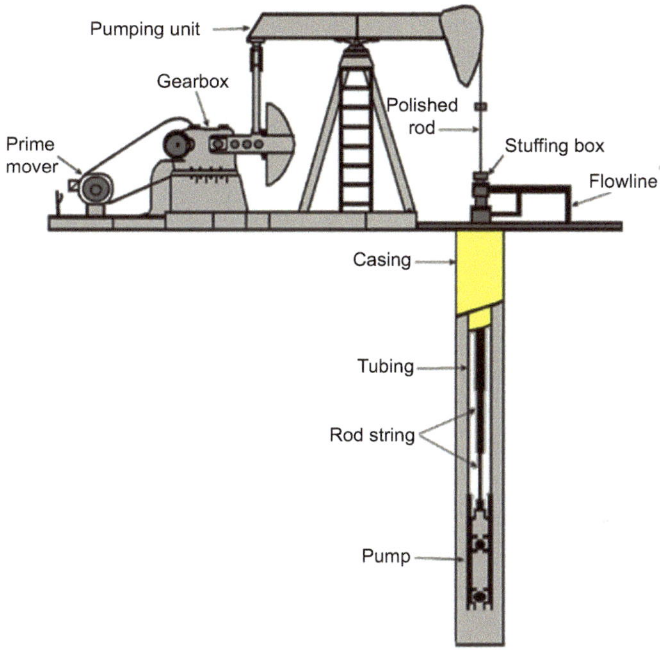

Figure 5.13 Schematic of a simplistic sucker pump system

Correct modeling of contact friction and correct incorporation of contact friction into calculation are important to solve diagnostic problems of deviated wells. Also, since wells are deviated, some sections of the rod string can be bent between two couplings in the middle of a dogleg turn, which introduces the concept of curvature of the rod string. It is essential to capture the behavior of the longitudinal stress waves, as well as the lateral stress waves of a rod element, by understanding the rod-liquid-tubing tubular rod dynamics of sucker-rod pumping systems involved in friction.

In order to optimize oil production, it is important to obtain a pump dynamometer graph that can accurately represent the pump conditions. Pump conditions such as plunger or valve leakage, worn or split barrel, gas interference, bent barrel, sticking pump, pump hitting, tubing movement, and fluid pound can be effectively pinpointed from pump cards. A pump can also be slowed down or shut off if the pump card shape indicates incomplete pump fillage. Direct measurement of the downhole operational data is not applicable for most cases since the downhole conditions make the placement of position or load sensory equipment at the pump costly and difficult. Consequently, an approach to creating a relevant, sufficiently accurate mathematical model for sucker-rod pumping systems and estimating the downhole dynamometer based on measured polished rod displacement and load is increasingly attracting the attention of experts in both industry and academia.

A typical three-dimensional (3D) wellbore trajectory for S-curved wells is shown in Figure 5.14. For this type of well, the inclination angle describing the wellbore geometry varies linearly from 0 at the top of the wellbore to dozens of degrees at certain depth. The angles remain constant to a further depth and then decrease linearly to 0 at the maximum depth. Interestingly, several existing research algorithms for diagnosing S-curved wells in North America generated a common phenomenon, in that the pump around discontinuous stroke transition time points experiences excessive load or displacement backlash. This kind of low-speed, nonsteady-state motion phenomena characterized by stop-slip or stop-reversal of displacement and excessive force are classified as *stick-slips*. In [100], the coupled longitudinal and lateral motions, classic Coulomb friction model, and finite element method are considered. The simulation results show the excessive pump load and pump displacement oscillation at the end of the upstroke. In [101–102], the finite difference method and classic Coulomb friction generated from the 3D wellbore trajectories are used to solve the nonlinear wave equation for deviated wells. In [102], the end of the upstroke demonstrates excessive pump load. The pump load increases as the Coulomb friction increases. The stick

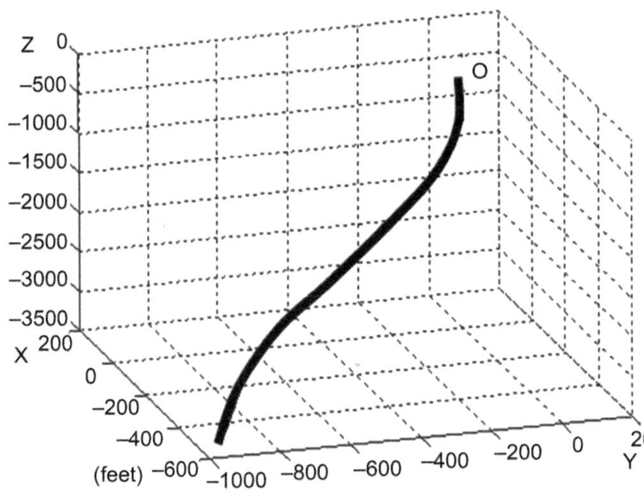

Figure 5.14 3D rod trajectory

slips occur at the ends of both the upstroke and the downstroke. The stick slip is more severe at the end of the upstroke than at the end of the downstroke.

Extremely costly repairs are requried to fix damaged sucker-rod pumping systems. Calculations that consider more factors related to the real systems may provide better diagnostics and control of sucker-rod pumping systems. Many studies have been undertaken to build hand-calculated or computer-aided models that can reproduce the dynamic behavior of sucker-rod pumping systems [100–115]. In the 1990s, [100,102] investigated the dynamics of sucker-rod pumping systems for deviated wells by taking into account the effects of contact, friction, and fluid effects. [104–105] proposed two-dimensional (2D) and 3D models for rod, tubing, and fluid systems. [107] extended the model for deviated wells by consiering longitudinal and transversal vibrations with friction in the interface between rod and tubing; [106] proposed tubular dynamics of rod-liquid-tubing and precisely yielded the effects of viscous damping; [108] used the reduced model of the lumped system and simplified the analysis; and [109] used a Fourier series to solve the wave equation on a semi-infinite domain and utilized the measured surface position and load as the two necessary boundary conditions to model a rod string of finite length. However, confined to linear systems or weak nonlinear systems inherently, the Fourier series method cannot describe the discontinuous point of the wave accurately and may introduce oscillations. [113] studied the dynamic stick-slip behavior of sucker-rod pumping systems with variational inequality and multivalued force-velocity relations, and [114–115] investigated the identification of the system parameters using tested results. Their approach seemed to be able to detect the stroke transitional time point of each node along the rod string and thus obtained reliable predictions.

In these papers, no explanation or analysis was given about why or how these stick-slips are generated. In reality, oil producers are upset when they see these stick-slips since these phenomena are very similar to motion oscillations that occur when a plunger accidentally overtravels to hit the upper and bottom structures of the pump. Therefore, the controllers for diagnostic and controlling purposes based on these algorithms for deviated wells cannot be accepted by domestic or foreign oil producers. These stick-slips may not exist in reality. They may be induced by insufficient rod/friction models or insufficient computational methods.

A thorough understanding of these stick-slip phenomena and the dynamics of deviated wells is desired. Fault diagnosis of sucker rod pumping system is an important research subject of oil extraction engineering. The shape of the plot reflects the actual pumping conditions, so different shaped features of dynamometer cards represent different downhole conditions.

In Figure 5.15, ρ, E are the mass density and elastic modulus of the rod, respectively; and A_r is the cross-section area of the rod.

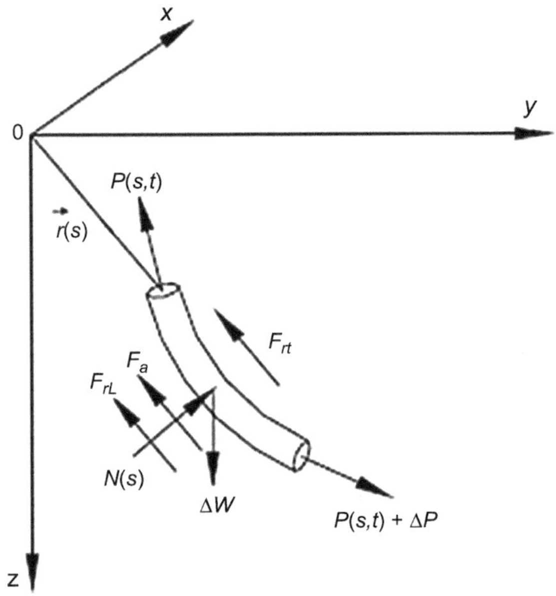

Figure 5.15 Schematic of forces acting on the infinitesimal rod element

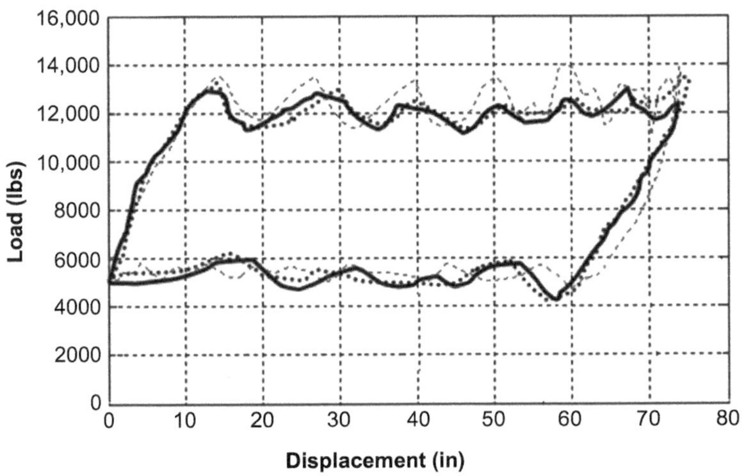

Figure 5.16 Measured and simulated surface dynamometer graph (dashed line: measured; solid line: friction model I; dotted line: friction model II)

Taking an infinitesimal rod element as shown in Figure 5.15, which illustrates the axial forces, viscous force, friction force, and initial force, the viscous force of liquid is

$$F_{rL} = \nu_e \left(\frac{\partial u}{\partial t} \right) \Delta s, \qquad [5.25]$$

where ν_e is the viscous resistance coefficient per unit length. The contact force between tube and rod is

$$N = \sqrt{\left[P \frac{\partial \theta}{\partial s} - q_r' \sin \theta \right]^2 + \left(P \frac{\partial \phi}{\partial s} \sin \theta \right)^2}, \qquad [5.26]$$

where Δs is the rod segment, θ is the inclination angle, φ is the azimuth angle, and P is the equivalent axial force. If the friction coefficient is μ, the sliding friction force per unit length on Δs is

$$F_{rt} = \mu \Delta s N, \qquad [5.27]$$

in which μ is the coefficient of friction. $q_r' = (\rho - \rho_L)Ag$, and ρ_L is the mass density of liquid.

Acceding to the d'Alembert principle, the equation of motion can be derived:

$$\frac{\partial^2 u}{\partial t^2} = (E/\rho) \frac{\partial^2 u}{\partial s^2} - (\nu_e/\rho A) \frac{\partial u}{\partial t} - \mu N/\rho A + (\rho_L - \rho)g/\rho. \qquad [5.28]$$

This is a nonlinear partial differential equation. By incorporating initial and boundary conditions, Eq. [5.28] can be solved, which leads to a closed curve of the rod hanging end load versus the displacement of stroke. The dominant effects are the viscous-damping coefficients and friction coefficients. The shape and corners of the curve change with the variation of the friction model used.

In order to describe the nonlinear friction characteristics of the sucker-rod pumping system, certain advanced friction model such as LuGre friction model can be used. Fig. 5.16 shows the measured and simulated surface dynamometer graphs.

REFERENCES

[1] Committee on Advanced Drilling Technologies, Drilling and Excavation Technologies for the Future, National Research Council, National Academies Press, Washington, D.C., 1994.

[2] A. Kyllingstad, G.W. Halsey, A study of slip/stick motion of the bit, SPE Drill. Eng. 3 (4) (1988) 369–373.

[3] B. Saldivar, et al., An overview on the modeling of oilwell drilling vibrations, in: Preprints of the 19th World Congress. International Federation of Automatic Control, 2014, pp. 24–29. Cape Town, South Africa, August.
[4] X. Zhu, L. Tang, Q. Yang, A literature review of approaches for stick-slip vibration suppression in oilwell drillstring, Adv. Mech. Eng. (Article ID 967952) (2014) 1–17.
[5] K.A. Womer, et al., Results of July 15, 2010, IADC stick-slip mitigation workshop, in: Proceedings of the SPE/IADC Drilling Conference and Exhibition, 2011. Amsterdam, The Netherlands, March, 2011.
[6] A.P. Christoforou, A.S. Yigit, Active control of stick-slip vibrations: The role of fully coupled dynamics, SPE Middle East Oil Show, Proc. SPE, Dallas, March 2001.
[7] R. Dawson, Y.Q. Lin, P.D. Spanos, Drill-string stickslip oscillations, in: Proceedings of the Spring Conference on Experimental Mechanics (SEM '87), 1987. Houston, TX.
[8] G. Takacs, Sucker-Rod Pumping Handbook: Production Engineering Fundamentals and Long-Stroke Rod Pumping, Elsevier, Amsterdam, 2015.
[9] G. Chen, Friction Dynamics and Diagnosis of Sucker Rod, Report submitted to Lufkin company, 2012.
[10] S.G. Gibbs, Rod Pumping: Modern Methods of Design, Diagnosis and Surveillance, BookMasters Inc., Ashland, OH, 2012.
[11] R.I. Leine, D.H. Van Campen, W.J.G. Keultjes, Stick-slip whirl interaction in drill-string dynamics, J. Vib. Acous. 124 (2) (2002) 209–220.
[12] R.I. Leine, D.H. Van Campen, Stick-slip whirl interaction in drillstring dynamics, in: G. Rega, F. Vestroni (Eds.), IUTAM Symposium on Chaotic Dynamics and Control of Systems and Processes in Mechanics, Springer, Dordrecht, the Netherlands, 2005, pp. 287–296.
[13] R.I. Leine, Bifurcations in Discontinuous Mechanical Systems of Filippov-Type, Ph.D. thesis, Mechanical Engineering, Eindhoven University of Technology, Eindhoven, the Netherlands, 2000.
[14] X. Liu, et al., Coupled axial-torsional dynamics in rotary drilling with state-dependent delay: Stability and control, Nonlin. Dyn. 78 (3) (2014) 1891–1906.
[15] D.W. Dareing, J. Tlusty, C. Zamudio, Self-excited vibrations induced by drag bits, J. Ener. Res. Tech. 112 (1) (1990) 54–61 (1990).
[16] A. Kyllingstad, G.W. Halsey, A study of slip/stick motion of the bit, SPE Drill. Eng. 3 (4) (1988) 369–373.
[17] Y.Q. Lin, Y.H. Wang, Stick-slip vibration of drill strings, J. Eng. Indus. 113 (1) (1991) 38–43.
[18] A.S. Yigit, A.P. Christoforou, Coupled torsional and bending vibrations of drillstrings subject to impact with friction, J. Sound Vib. 215 (1) (1998) 167–181.
[19] A.S. Yigit, A.P. Christoforou, Coupled torsional and bending vibrations of actively controlled drillstrings, J. Sound Vib. 234 (1) (2000) 67–83.
[20] T. Richard, C. Germay, E. Detournay, Self-excited stickslip oscillations of drill bits, Comp. Rend. Mécan. 332 (8) (2004) 619–626.
[21] R. Allen, Severe downhole drilling dynamics, IADD Mini Forum, Halliburton, 2013.
[22] K.A. Womer, et al., Results of July 15, 2010 IADC stick-slip mitigation workshop, in: Proceedings of the SPE/IADC Drilling Conference and Exhibition, 2011. Amsterdam.
[23] A.P. Christoforou, A.S. Yigit, Active control of stick slip vibrations: The role of fully coupled dynamics, SPE Middle East Oil Show (2001). Vol. 68093 of Proc. SPE, March.
[24] L.W. Ledgerwood, et al., Downhole measurement and monitoring lead to an enhanced understanding of drilling vibrations and polycrystalline diamond compact bit damage, SPE Drill. Complet. 28 (3) (2013) 254–262.

[25] T. Richard, C. Germay, E. Detournay, Self-excited stickslip oscillations of drill bits, Comp. Rend. Mécan. 332 (8) (2004) 619–626.
[26] X. Wu, et al., Decoupling stick-slip and whirl to achieve breakthrough in drilling performance, in: Proceedings of the IADC/SPE Drilling Conference and Exhibition, 2010, pp. 966–978. New Orleans, LA, February.
[27] P.C. Kriesels, et al., Cost saving through an integrated approach to drillstring vibration control, in: Proceedings of the SPE/IADCMiddle East Drilling Technology Conference, 1999. Vol. 57555 of SPE, Abu Dhabi, UAE, November.
[28] N. Challamel, H. Sellami, E. Chenevez, A stick-slip analysis based on rock/bit interaction: theoretical and experimental contribution, in: Proceedings of the IADC/SPE Drilling Conference, SPE 59230, NewOrleans, LA, 2000, pp. 23–25.
[29] R. Dawson, Y.Q. Lin, P.D. Spanos, Drill-string stickslip oscillations, in: Proceedings of the Spring Conference on Experimental Mechanics, (SEM '87), Houston, TX, 1987, pp. 590–595.
[30] D.W. Dareing, J. Tlusty, C. Zamudio, Self-excited vibrations induced by drag bits, J. Ener. Res. Tech. 112 (1) (1990) 54–61.
[31] Y.Q. Lin, Y.H. Wang, Stick-slip vibration of drill strings, J. Eng. Indus. 113 (1) (1991) 38–43.
[32] H. Melakhessou, A. Berlioz, G. Ferraris, A nonlinear well-drillstring interaction model, J. Vib. Acous. 125 (1) (2003) 46–52.
[33] P.D. Spanos, et al., Oil and gas well drilling: A vibrations perspective, Shock Vib. Dig. 35 (2) (2003) 85–103.
[34] N. Mihajlovi, et al., Analysis of friction-induced limit cycling in an experimental drill-string system, J. Dyn. Sys. Meas. Ctrl. 126 (4) (2004) 709–720.
[35] J.F. Brett, The genesis of torsional drillstring vibrations, SPE Drill. Eng. 7 (3) (1992) 168–174.
[36] N. Challamel, Rock destruction effect on the stability of a drilling structure, J. Sound Vib. 233 (2) (2000) 235–254.
[37] V.A. Dunayevsky, F. Abbassian, A. Judzis, Dynamic stability of drillstrings under uctuating weight on bit, SPE Drill Complet 8 (2) (1993) 84–92.
[38] Y.Q. Lin, Y.H. Wang, Stick-slip vibration of drill strings, J. Eng. Indus. 113 (1) (1991) 38–43.
[39] J.D. Macpherson, J.S. Mason, J.E.E. Kingman, Surface measurement and analysis of drill-string vibrations while drilling, SPE/IADC 25777, Society of Petroleum Engineers, 1993.
[40] E. Detournay, P. Defourny, A phenomenological model for the drilling action of drag bits, Intl. J. Rock Mech. Min. Sci. Geomech. Abs. 29 (1) (1992) 13–23.
[41] E. Detournay, C.P. Tan, Dependence of drilling specific energy on bottom-hole pressure in Shales, SPE/ISRM 78221, Rock Mechanics Conference, Irving, Texas, October 2002, pp. 20–23.
[42] A.S. Yigit, A.P. Christoforou, Stick-slip and bit-bounce interaction in oil well drill-strings, J. Ener. Res. Tech. 128 (4) (2006) 268–274.
[43] G.H.M. Van der Heijden, Nonlinear drillstring dynamics, Ph.D thesis, Mechanical Engineering, University of Utrechtthe, the Netherlands, 1994.
[44] Y.A. Khulief, F.A. Al-Sulaiman, S. Bashmal, Vibration analysis of drillstrings with self-excited stick-slip oscillations, J. Sound Vib. 299 (3) (2007) 540–558.
[45] T. Richard, C. Germay, E. Detournay, A simplified model to explore the root cause of stick-slip vibrations in drilling systems with drag bits, J. Sound Vib. 305 (3) (2007) 432–456.
[46] I. Boussaada, et al., Analysis of drilling vibrations: A time-delay system approach, in: 20th Mediterranean Conference on Control and Automation MED, 2012. Barcelona, Spain.

[47] I. Boussaada, et al., Control of drilling vibrations: A time-delay system-based approach, in: 11th Workshop on Time-Delay Systems Part of 2013 IFAC Joint Conference SSSC, 2013. Grenoble, France.
[48] A. Depouhon, E. Detournay, Instability regimes and self-excited vibrations in deep drilling systems, J. Sound Vib. 333 (7) (2014) 2019–2039.
[49] E. Detournay, T. Richard, M. Shepherd, Drilling response of drag bits: Theory and experiment, Intl. J. Rock Mech. Min. Sci. 45 (8) (2008) 1347–1360.
[50] C. Germay, V. Deno, E. Detournay, Multiple mode analysis of the self-excited vibrations of rotary drilling systems, J. Sound Vib. 325 (1) (2009) 362–381.
[51] Y. Kovalyshen, Understanding root cause of stick-slip vibrations in deep drilling with drag bits, Intl. J. Nonlin. Mech. 67 (2014) 331–341.
[52] K. Nandakumar, M. Wiercigroch, Stability analysis of a state-dependent delayed, coupled two DOF model of drill-string vibration, J. Sound Vib. 332 (10) (2013) 2575–2592.
[53] M.K. Jasem, A.S. Yigit, Modeling and analysis of stick-slip and bit bounce in oilwell drillstrings equipped with drag bits, J. Sound Vib. 333 (25) (2014) 6885–6899.
[54] T.G. Ritto, R. Sampaio, Measuring the efficiency of vertical drill-strings: A vibration perspective, Mech. Res. Comm. 52 (2013) 32–39.
[55] T.G. Ritto, C. Soize, R. Sampaio, Nonlinear dynamics of a drill-string with uncertain model of the bit-rock interaction, Int. J. Nonlin. Mech. 44 (8) (2009) 865–876.
[56] R. Samuel, Friction factors: What are they for torque, drag, vibration, bottom hole assembly, and transient surge/swab analyses? J. Petrol. Sci. Eng. 73 (3) (2010) 258–266.
[57] L. Perneder, E. Detournay, G. Downton, Bit/rock interface laws in directional drilling, Intl. J. Rock Mech. Min. Sci. 51 (2012) 81–90.
[58] B. Besselink, N. Van de Wouw, H. Nijmeijer, Model-based analysis and control of axial and torsional stick-slip vibrations in drilling systems, in: N. van de Wouw, E. Detournay (Eds.), Second International Colloquium on Nonlinear Dynamics and Control of Deep Drilling Systems Proceedings, 2012. May 15–16.
[59] B. Besselink, N. Van de Wouw, H. Nijmeijer, A semi-analytical study of stick-slip oscillations in drilling systems, J. Comp. Nonlin. Dyn. 6 (2) (2011), 021006.
[60] C. Germay, et al., Nonlinear drillstring dynamics analysis, J. Appl. Dyn. Sys. 8 (2) (2009) 527–553.
[61] B. Besselink, N. Van de Wouw, H. Nijmeijer, Model-based analysis and control of axial and torsional stick-slip oscillations in drilling systems, in: IEEE Multi-Conference on Systems and Control, 2011. Denver, CO.
[62] C.M. Liao, et al., Parametric studies on drill-string motions, Int. J. Mech. Sci. 54 (1) (2012) 260–268.
[63] Y. Kovalyshen, Understanding root cause of stick-slip vibrations in deep drilling with drag bits, Int. J. Nonlin. Mech. 67 (2014) 331–341.
[64] E.M. Navarro-López, D. Cortés, Avoiding harmful oscillations in a drillstring through dynamical analysis, J. Sound Vib. 307 (1) (2007) 152–171.
[65] E. Detournay, T. Richard, M. Shepherd, Drilling response of drag bits: Theory and experiment, Intl. J. Rock Mech. Min. Sci. 45 (8) (2008) 1347–1360.
[66] A. Depouhon, E. Detournay, Instability regimes and self-excited vibrations in deep drilling systems, J. Sound Vib. 333 (7) (2014) 2019–2039.
[67] E. Fridman, S. Mondie, S. Saldivar, Bounds on the response of a drilling pipe model, Special Issue on Time-Delay Systems in IMA J. Math. Ctrl Inform. (2010).
[68] Y.A. Khulief, F.A. Al-Sulaiman, S. Bashmal, Vibration analysis of drillstrings with self-excited stick-slip oscillations, J. Sound Vib. 299 (3) (2007) 540–558.
[69] Y.A. Khulief, F.A. Al-Sulaiman, S. Bashmal, Vibration analysis of drillstrings with string-borehole interaction, Proc. Inst. Mech. Eng., Part C: J. Mech. Eng. Sci. 222 (11) (2008) 2099–2110.

[70] Y.A. Khulief, F.A. Al-Sulaiman, Laboratory investigation of drillstring vibrations, Proc. Inst. Mech. Eng., Part C: J. Mech. Eng. Sci. 223 (10) (2009) 2249–2262.
[71] A.P. Christoforou, A.S. Yigit, Fully coupled vibrations of actively controlled drill-strings, J. Sound Vib. 267 (5) (2003) 1029–1045.
[72] C.M. Liao, et al., Drill-string dynamics: Reduced-order models and experimental studies, J. Vib. Acous. 133 (4) (2011), 041008.
[73] K. Ahmadi, Y. Altintas, Stability of lateral, torsional, and axial vibrations in drilling, Intl. J. Mach. Tools Manuf. 68 (2013) (2013) 63–74.
[74] M. Wan, et al., A unified stability prediction method for milling process with multiple delays, Intl. J. Mach. Tools Manuf. 50 (1) (2010) 29–41.
[75] M. Wan, et al., Prediction of chatter stability for multiple-delay milling system under different cutting force models, Intl. J. Mach. Tools Manuf. 51 (4) (2011) 281–295.
[76] Q. Xue, et al., Chaotic vibration analysis of the bottom rotating drill string, Shock Vib. Dig. (2014). Article ID 429164.
[77] H. Kantz, T. Schreiber, Nonlinear Time Series Analysis, Cambridge University Press, Cambridge, UK, 1997.
[78] F. Takens, Detecting strange attractors in turbulence, Lect. Notes Math 898 (1982) 366–381.
[79] L. Cao, Practical method for determining the minimum embedding dimension of a scalar time series, Phys. D: Nonlin. Phenom. 110 (1) (1997) 43–50.
[80] R. Brown, P. Bryant, H.D.I. Abarbanel, Computing the Lyapunov spectrum of a dynamical system from observed time series, Phys. Rev. A 43 (6) (1991) 2787–2806.
[81] M.T. Rossenstein, J.J. Collins, C.J. Deluca, A practical method for calculating largest Lyapunov exponents from small data sets, Phys. D 65 (1–2) (1993) 117–134.
[82] T. Richard, et al., Influence of bit-rock interaction on stick-slip vibrations of PDC bits, in: Proceedings of the 2002 SPE Annual Technical Conference and Exhibition, MS77616, 2002, pp. 2407–2418. San Antonio, TX, September.
[83] K. Mongkolcheep, et al., A modal approach for chaotic vibrations of a drillstring, in: Proceedings of the ASME International Design Engineering Technical Conferences and Computers and Information in Engineering Conference (DETC '09), 2009, pp. 1305–1314. San Diego, CA, August.
[84] G.W. Halsey, A. Kyllingstad, A. Kylling, Torque feedback used to cure slip-stick motion, in: Proceedings of the 63rd Society of Petroleum Engineers Drilling Engineering, 1988, pp. 277–282. Houston, TX.
[85] J.D. Jansen, Nonlinear dynamics of oilwell drill-strings, Ph.D thesis, Mechanical Engineering, Delft University of Technology, the Netherlands, 1993.
[86] J.D. Jansen, L. van den Steen, Active damping of self-excited torsional vibrations in oil well drill-strings, J. Sound Vib. 179 (4) (1995) 647–668.
[87] J.S. Mason, B.M. Sprawls, Addressing BHA whirl: The culprit in mobile bay, SPE Drill. Complet. 13 (4) (1998) 231–236.
[88] R.F. Mitchell, M.B. Allen, Case studies of BHA vibration failure, in: SPE Annual Technical Conference and Exihibition, 1987. Dallas, September.
[89] E. Navarro-Lopez, R. Suarez, Practical approach to modelling and controlling stick-slip oscillations in oilwell drill-strings, in: Proceedings of the 2004 IEEE International Conference on Control Applications, 2, 2004, pp. 1454–1460.
[90] E. Navarro-Lopez, D. Cortes, Sliding-mode of a multi-DOF oilwell drill-string with stick-slip oscillations, in: Proceedings of the 2007 American Control Conference, 2007, pp. 3837–3842. New York.
[91] M.B. Saldivar, S. Mondie, J.J. Loiseau, Reducing stick-slip oscillations in oilwell drill-strings, in: 6th International Conference on Electrical Engineering, Computing Science and Automatic Control (CCE), 2009, pp. 1–6. Toluca.
[92] M.B. Saldivar, et al., Stick-slip oscillations in oillwell drilstrings: Distributed parameter and neutral type retarded model approaches, in: 18th IFAC World Congress, 2011, pp. 284–289. Milano, Italy.

[93] M.B. Saldivar, S. Mondie, Drilling vibration reduction via attractive ellipsoid method, J. Frank. Inst. 350 (3) (2013) 485–502.

[94] R.J. Shyu, Bending vibration of rotating drill strings, Ph.D thesis, Mechanical Engineering Massachusetts Institute of Technology, Cambridge, MA, 1989.

[95] E. Skaugen, The effects of quasi-random drill bit vibrations upon drill-string dynamic behavior, in: SPE Annual Techncal Conference and Exihibition, 1987. Dallas, September.

[96] A.F.A. Serrarens, et al., H1 control for suppressing stick-slip in oil well drill-strings IEEE Cntrl. Sys. 18 (2) (1998) 19–30.

[97] R.W. Tucker, C. Wang, On the effective control of torsional vibrations in drilling systems, J. Sound Vib. 224 (1) (1999) 101–122.

[98] J.K. Vandiver, J.W. Nicholson, R.J. Shyu, Case studies of the bending vibration and whirling motion of drill collars, SPE Drill. Eng. 5 (4) (1990) 282–290.

[99] S.A. Luiksiewicz, Computer model evaluates oil pumping units in inclined wells, J. Canad. Petrol. Tech. 29 (6) (1990) 76–79.

[100] J. Xu, Diagnostic analysis of deviated rod-pumped wells, Southwestern Petroleum Short Course (2001) 133–140.

[101] S.G. Gibbs, Design and diagnosis of deviated rod-pump wells, J. Petrol. Tech. 44 (7) (1992) 774–781.

[102] I.N. Shardakov, I.N. Wasserman, Formulation and solution of dynamic problems of elastic rod systems subjected to boundary conditions described, J. Appl. Mech. Tech. Phys. 44 (3) (2003) 406–414.

[103] I.N. Shardakov, I.N. Wasserman, Numerical modeling of longitudinal vibrations of as sucker rod string, J. Sound Vib. 329 (3) (2010) 317–327.

[104] D.R. Doty, Z. Schmidt, An improved model for sucker-rod pumping, SPE J. 23 (1) (1983) 33–41.

[105] G.A. Yu, Y.J. Wu, G.Y. Wang, Three-dimensional vibration in a sucker rod beam pumping systems, Acta Pet. Sin. 10 (2) (1989) 76–83.

[106] S.D. Lekia, R.D. Evans, A coupled rod and fluid dynamic model for predicting the behavior of sucker-rod pumping systems, SPE Prod. Fac. 10 (1) (1995) 26–33.

[107] S.A. Lukasiewicz, Dynamic behavior of the sucker rod string in the inclined well, in: SPE Production Operations Symposium, 1991, pp. 313–321. Oklahoma City, OK, April.

[108] J.E. Chacin, J.C. Purcupile, A new model for studying oil well pumping installations, in: SPE Annual Technical Conference and Exhibition, 1987. Dallas.

[109] J.J. DaCunha, S.G. Gibbs, Modeling of a finite-length sucker rod using the semi-infinite wave equation and a proof to Gibbs' conjecture, SPE J. 14 (1) (2009) 112–119.

[110] L. Liu, et al., A uniform and reduced mathematical model for sucker rod pumping, in: 4th International Conference on Computational Science, 2004, pp. 372–379. Krakow.

[111] D.N. Yuan, F. Kang, Multi-parameter friction model and its application to dynamics analysis of a mechanism, J. Mech. Sci. Tech. Aerosp. Eng. 28 (7) (2009) 333–342.

[112] S. Goyal, N.C. Perkinsb, C.L. Lee, Non-linear dynamic intertwining of rods with self-contact, Intl. J. Nonlin. Mech. 43 (1) (2008) 65–73.

[113] S. Gavrilov, Non-stationary problems in dynamics of a string on an elastic foundation subjected to a moving load, J. Sound Vib. 222 (3) (1999) 345–361.

[114] P. Xu, S. Xu, H. Yin, Application of self-organizing competitive neural network in fault diagnosis of suck rod pumping system, J. Petrol. Sci. Eng. 58 (1) (2007) 43–48.

[115] B.X. Liu, H.Z. Liu, Prediction of dynamic parameters of sucker rod pumping system in directional well based on LuGre friction model, Acta Pet. Sin. 29 (6) (2008) 938–941.

CHAPTER 6

Friction Dynamics and Diagnosis of Rotor Systems

6.1 INTRODUCTION

Rotor systems are widely used as important components in complex machinery such as airplane engines and air compressors (as shown in Figure 6.1). The rub-impact between the rotor and the stator in operation is one of the most common problems with rotating machinery. Rub-impact mainly takes place between the blade tip and the casing or between the seal and the shaft. Rub-impact between the blade tip and the casing can enlarge the operating clearance, thus reducing the efficiency or even breaking the blade. Rub-impact between the seal and the shaft may result in seal failure and a deformed shaft due to local frictional heat. Also, rub-impact may cause nonconforming precession and alternating stress, speeding failure damage of the rotor [1].

The main root causes of rub-impact phenomena in rotor systems are as follows:

1. The clearance between the rotor and the stator is inadequate. This is mainly attributed to structure design, minor mounting clearance, installation and maintenance error, and thermal expansion or contraction of the component.

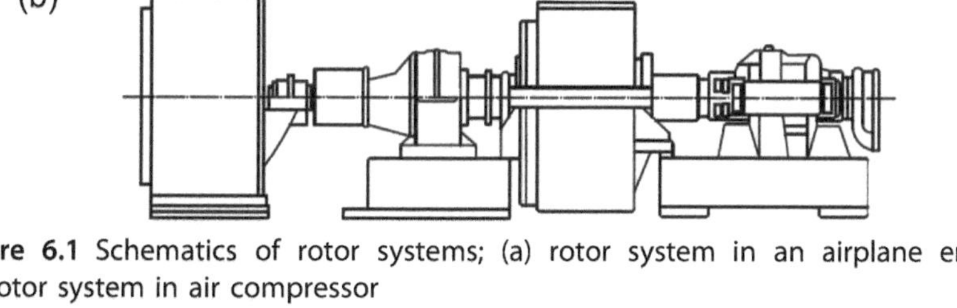

Figure 6.1 Schematics of rotor systems; (a) rotor system in an airplane engine; (b) rotor system in air compressor

Friction Dynamics
ISBN 978-0-08-100285-8
http://dx.doi.org/10.1016/B978-0-08-100285-8.00006-7

2. The vibration amplitude of rotor is larger than the clearance due to oil whipping, shaft bending, unstable shaft motion, or imbalance mass.
3. The clearance between the rotor and the stator decreases or disappears due to the axis misalignment caused by poor installation.

6.2 MECHANISM AND MODELS OF RUB-IMPACT BETWEEN ROTOR AND STATOR

6.2.1 Physical Phenomena and Processes of Rub-Impact

The rub-impact phenomena between rotor and stator may be divided into three cases: single point, partial, and full annular rub-impact. The term *single-point rub-impact* means that the rotor only collides with a specific point of the stator. It usually occurs when the stator has local deformation or when the contact stiffness between rotor and stator is so large that the rotor bounces off the stator by the impact force at the proper rotational speed. The term *partial rub-impact* means that contact occurs occasionally during a part of the period of precession. It is the most common and complex problem, and it results from a variety of reasons. *Annular rub-impact* means that the rotor maintains contact with the stator during the complete cycle of its precession motion. Strictly speaking, annular rub-impact only contains the friction effect, without impulsive impact. In general, rub-impact starts from one point and then develops to the partial, and finally to the annular one. Annular rub-impact is a severe issue that could lead to shaft deformation due to local frictional heat, which aggravates the vibration or causes even worse results, such as backward precession of the rotor. The backward precession of the rotor makes stress more severe and easily causes fatigue damage.

The rub-impact process is usually divided into three stages [2]: collision/impact stage, maintenance stage, and separation stage. In the first stage, the rotor collides with the stator, and great changes of contact force and speed of the rotor center take place. The time duration is very short, so the displacements of rotor and stator are not obvious. However, there is one exceptional case where the impact force is very small when the direction of rotor velocity is almost parallel to the tangent direction of the contact point on stator.

The maintenance stage is also called the *stable contact stage*. At this stage, the contact force between the rotor and the stator varies in stability, and the rotor gets obvious displacement. Therefore, the effect of friction is significant. This stage does not exist when the contact stiffness between rotor and stator is so large that the rotor bounces off the stator by the action of impact during the first stage.

During the last stage, as the relative velocity of rotor and stator on contact position decreases, the contact force gradually declines to zero where the rotor fully separates from the stator. The duration of this process depends on the relative velocity of the rotor and the stator at the beginning of rub-impact and their stiffness.

In the process of rub-impact, the rotor is subjected to the normal force and tangential friction force. Partial rub-impact is the main fault pattern in the rotor system. Three physical phenomena occur in partial rub-impact: impact, friction, and modification of stiffness.

6.2.1.1 Impact

When a collision between the rotor and the stator takes place, the impact force leads to local compression deformation at the contact position and rebounding lateral motion of the rotor. After impact, the rotor responds with complex transient lateral and torsional motions. The rebounding motion may follow the original direction of the precession or reverse it to backward precession.

6.2.1.2 Friction

Friction is the tangential force acting on the rotor at the contact position when rub-impact occurs. It has a backward direction relative to the rotating speed. Its magnitude mostly depends on the normal force between rotor and stator, material and surface properties, and the direction of their relative velocity. The friction force can decrease the rotational speed of the rotor and lead to transient torsional vibrations of shaft. In the case of full annular impact-rub, high friction force may cause a dramatic change of the precession direction from forward to backward whirl [3]. The rotor in the backward precession mode is subjected to high-frequency alternating stresses and deformations, which may eventually lead to damage of the shaft or even the whole machine. Due to the friction effect, the kinetic energy of motion is partly transformed into heat and partly spent on wearing the contact surfaces. Locally generated heat may lead to the shaft deformation or seal destruction.

6.2.1.3 Modification of System Stiffness

The stiffness of a rotor system, defining its natural frequencies, is determined by the stiffness of the shaft, the stiffness of the pedestal, and the stiffness of the bearings and seals. When the rotor occasionally makes contact with the stator, the new boundary conditions modify its stiffness. The modification depends on the contact position along the rotor and the stiffness of the stator.

This analysis indicates that the rub-impact between rotor and stator, involving impact, friction, and parameter modification, is a complex dynamic process. Its dynamic behaviors and diagnosis methods have been widely examined.

6.2.2 Analytical Model of Rub-Impact

The rub-impact between rotor and stator is characterized by typical nonlinearities. Therefore, the research is much more complicated than that of the ordinary linear system. The behavior of nonlinear systems is highly dependent on the initial conditions and environment. Experimentation, as one of the methods used to study rub-impact, is easily affected by external conditions. So it is necessary to investigate rub-impact theoretically. Obviously, it is critical to establish the rub-impact model with typical rubbing and impact characteristics for theory research. Therefore, several popular rub-impact models based on different cases are introduced in this section.

6.2.2.1 Model 1

The elastic model, shown in Figure 6.2, has been widely used in research. This model assumes that the normal force is generated by the elastic deformation of the stator during the rub-impact process and only includes the Coulomb friction force in the tangential direction. When the rotor system is in standby mode, the center of the rotor is assumed to coincide with one of the stators, and the radial clearance is defined as r_0. In running the process, the distance between the two centers is defined as

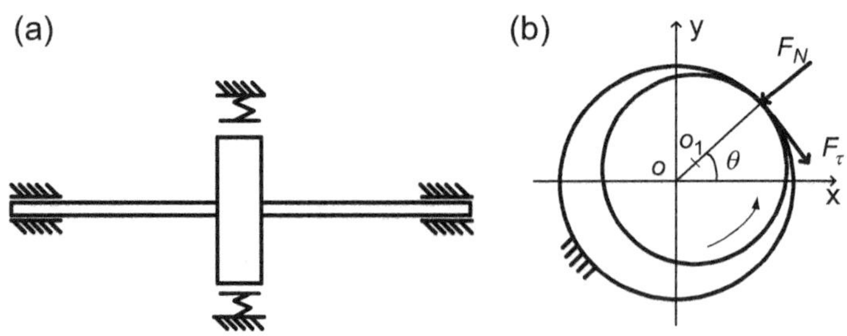

Figure 6.2 Schematic of the rub-impact between rotor and stator; (a) Jeffcott rotor system; (b) forces between rotor and stator

$r = \sqrt{x^2 + y^2}$, where x and y are rotor horizontal and vertical displacements. When $r > r_0$, rub-impact occurs. The normal impact force and tangential friction force can be expressed as

$$F_N = k_c(r - r_0) \quad [6.1]$$

$$F_t = \mu F_N, \quad [6.2]$$

where μ is the friction coefficient and k_c indicates the contact stiffness coefficient. Since the elastic model can describe some rub-impact phenomena very well using simple functions, it has been adopted in many studies. In the elastic model, the contact stiffness k_c is usually considered as a constant.

The direction of rotor precession needs to be determined before calculating the friction force by Eqs. [6.1] and [6.2]. When the rub-impact is not very severe, the rotor precession has a forward direction. Therefore, the direction of the friction force acting on the rotor is the opposite of its rotating direction [just like Figure 6.2 (b)]. However when the rub-impact fault is severe, it is possible for the rotor precession to go in a backward direction. This needs to be analyzed deliberately [2,4].

Assuming that $\vec{i}$, $\vec{j}$, and $\vec{k}$ are the three unit vectors in the x-, y-, and z-directions, respectively, and $\vec{r}$ and $\vec{v}$ indicate the displacement and velocity vectors of the rotor center, respectively, the relationship is as follows:

$$\vec{r} = x\vec{i} + y\vec{j} \quad [6.3]$$

$$\vec{v} = \dot{x}\vec{i} + \dot{y}\vec{j}. \quad [6.4]$$

If the rotor center moves in a circle around its static position, the phase angle of its velocity is $\pi/2$ greater than that of its displacement. The velocity vector $\vec{r}_1$ of the rotor center could be expressed as $\vec{r}_1 = \vec{k} \times \vec{r} = x\vec{j} - y\vec{i}$ and is normalized as

$$\vec{r}_r = \frac{\vec{r}_1}{\left|\vec{r}_1\right|} = \frac{x\vec{j} - y\vec{i}}{\sqrt{x^2 + y^2}}. \quad [6.5]$$

The projection of the real velocity $\vec{v}$ of the rotor center along vector $\vec{r}_r$ can be written as

$$\vec{v} \bullet \vec{r}_r = \left[\dot{x}\vec{i} + \dot{y}\vec{j}\right] \bullet \frac{x\vec{j} - y\vec{i}}{\sqrt{x^2 + y^2}} = \frac{\dot{y}x - \dot{x}y}{\sqrt{x^2 + y^2}}. \quad [6.6]$$

If $\vec{v} \bullet \vec{r}_r > 0$ (i.e., the velocity of the rotor center is in the same direction with the rotating velocity), the rotor precession takes a forward direction. The direction of friction force is the opposite of the rotating direction.

If $\vec{v} \bullet \vec{r}_r < 0$ (i.e., the velocity of the rotor center is in the opposite direction with the rotation speed), the rotor precession takes a backward direction. Furthermore, when $|\vec{v} \bullet \vec{r}_r| < \omega R_r$ (where ω and R_r are angular velocity and radius of the rotor, respectively), the direction of friction force is still the opposite of the rotation direction. But when $|\vec{v} \bullet \vec{r}_r| > \Omega R_r$, the friction force goes in the same direction as the rotation speed and the friction coefficient in Eq.[6.2] should be negative.

6.2.2.2 Model 2

In 1984, Muszynska [5] proposed a mathematical model based on the partial rub-impact phenomenon of the Jeffcott rotor as follows:

$$M\ddot{z} + C\dot{z} + Kz = Fe^{j\omega t} - F(\Omega t)[k_0|z| + N(1 + j\mu)]e^{j\beta}, \qquad [6.7]$$

where M, C, and K are rotor mass, damping, and stiffness coefficients, respectively; $z = x + jy$ is the displacement vector of the rotor center; $Fe^{j\omega t}, \mu, \beta$ are imbalance forces acting on the rotor, friction coefficient, and angular position of the rub location, respectively; and $F(\Omega t)$ is a function indicating periodicity of contact. The frequency Ω is equal to the frequency of the precession motion. Different contact regions between the rotor and the stator could be obtained by adjusting $F(\Omega t)$. This model assumes that the normal force is a constant and the rotor stiffness increases with a modification of k_0 when the partial rub-impact occurs.

6.2.2.3 Model 3

In 1979, Childs [6] modeled the rub-impact force by additional elastic and damping forces, and established the formulation as

$$M\ddot{z} + C\dot{z} + Kz = Fe^{j\omega t} - c_r\dot{\delta} - k_r(\delta - c)(1 + j\mu)e^{j\beta}, \qquad [6.8]$$

where c_r and k_r are the additional damping and stiffness coefficients, respectively; and δ and c are the relative displacement and static clearance between the rotor and the stator.

6.2.2.4 Model 4

These two models [Eqs. [6.7] and [6.8]] are applicable to single-point or partial rub-impact. They emphasized the collision effect in different ways. In fact, with the increase of the displacement of the rotor center after contact, the contact area grows larger, leading to an increase in the contact stiffness. By taking this into account, Yan and Wang [7,8] proposed a model based on Hertz contact theory, and the rub-impact force is formulated as

$$F = K_r(\delta - c)^{\frac{3}{2}}(1 + j\mu)e^{j\beta}. \tag{6.9}$$

The collision effect is not involved in this model, but the nonlinear relation of rub-impact force and displacement is introduced. This model can be applied in the case of small contact stiffness. The exponent of $(\delta - c)$ in Eq. [6.9] has significant influence on the motion of the rotor when a simulation is performed. Therefore, how to determine a proper value of this exponent needs further exploration. The exponential term of the rub-impact force is highly dependent on the deformation. In the case of small deformation, a linear function can accurately reflect the actual situation and has good agreement with the testing results. However, in the case of larger deformations, nonlinear properties must be considered. For this case, the finite element method (FEM) can be used to establish a more accurate model for a specific rotor system.

6.2.3 Finite Element Model of Rub-Impact Between Rotor and Stator

The contact stiffness increases with the growing contact area (relative displacement) after the rotor collides with the stator. For a specific structure, a more accurate model can be established through FEM. A specific example of rub-impact analysis on a finite element model of the rotor system [9] is given next.

The schematic of the rotor system is shown in Figure 6.2, but here, the elasticity of the rotor disc and stator (casing) is considered. For the finite element model of the rotor system, some assumptions are made: (i) the stiffness of the rotor disc is much larger than the stiffness of the casing; (ii) the collision period of time is short, and only the small region near the contact location deforms; and (iii) the collision effect on the area far from contact location is ignored.

Figure 6.3 and Table 6.1 show the models and material properties of the rotor and casing, respectively.

A further displacement of 2 mm is assumed after the rotor disc makes contact with the casing. The contact pair is established between the outer surface of the rotor and the inner surface of the casing using the

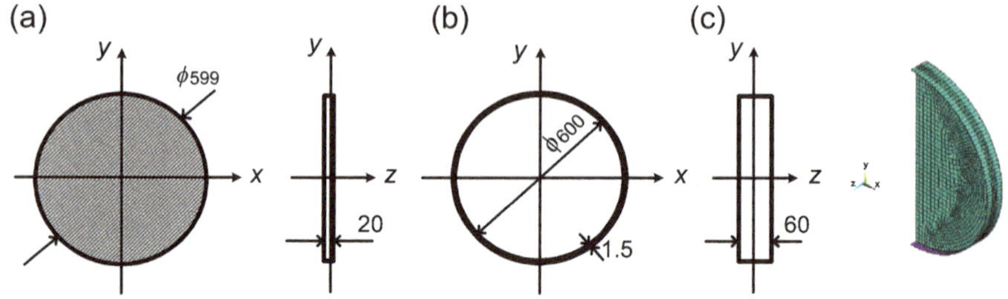

Figure 6.3 Models of the rotor disc and casing; (a) rotor disc; (b) casing; (c) finite element model of disc

Table 6.1 Material properties of the rotor disc and casing

	Material	Elastic Module (GPa)	Poisson's Ratio
Casing	1Cr18Ni9Ti	159	0.3
Rotor disc	K24	191	0.25

surface-to-surface contact analysis function of the ANSYS software. Nonlinear simulation is conducted involving the effect of large contact surface deformation. The relationship between the normal contact force and the displacement of the rotor relative to the casing is obtained and displayed in Table 6.2.

Table 6.2 Relationship between the normal contact force and the relative displacement

Normal force (F/N)	164.2	331.6	588.8	989.8	1630.1	2691.6	4556.2	8103	10737.2
Displacement (Δs/mm)	0.05	0.1	0.175	0.2875	0.4562	0.709	1.089	1.659	2

The data in Table 6.2 can be fitted with a six-order polynominal function:

$$F = -24.3921\Delta s^6 + 138.6772\Delta s^5 - 298.4002\Delta s^4 + 521.1111\Delta s^3 + 482.2159\Delta s^2 + 3268.9\Delta s - 0.5496. \quad [6.10]$$

It can be observed that the contact force is a nonlinear function of the relative displacement, which could be more obvious when the shell is thicker or the rotor moves with more displacement. In the simulation of the rub-impact phenomenon of a rotor system, the normal contact force can be determined directly from normal relative displacement based on the fitting function like Eq. [6.10].

6.3 COMPLEX DYNAMIC RESPONSE OF A ROTOR SYSTEM WITH RUB-IMPACT

A rotor system with a rub-impact fault is usually considered as a nonlinear vibration system with piecewise linear stiffness. The system may behave as various forms of periodic motions or chaotic motions under different conditions, and it may show complex bifurcation phenomena [10–14]. In this section, the governing equations of motion of the rotor system with imbalance mass and misalignment are established and then solved numerically. The influence of the nonlinear properties and key parameters on the dynamic behaviors of the system is also analyzed.

6.3.1 Nonlinearities in Rub-Impact Rotor System

The partial rub model of a Jeffcott rotor system [see Figure 6.2 (a)] is shown in Figure 6.4. The nonlinearities that occur during rub-impact are illustrated through the analysis of a Jeffcott rotor system by Muszyska [5].

The governing equation of motion of the system can be written as

$$M\ddot{z} + C\dot{z} + Kz + F(\Omega t)[K_0|z| + N(1 + j\mu)]e^{j\beta} = mr_e\omega^2 e^{j(\omega t + \delta)}, \quad [6.11]$$

where $z = x + jy$, $|z| = \sqrt{x^2 + y^2}$, $j = \sqrt{-1}$; x and y are rotor horizontal and vertical deflections, respectively; M, C, and K are rotor mass, damping, and stiffness coefficients, respectively; m, r_e, and δ are imbalance mass, radius, and angular position, respectively; ω is the rotational speed; β determines the angular position of the stator; $K_0|z|$, N, and μN describe the

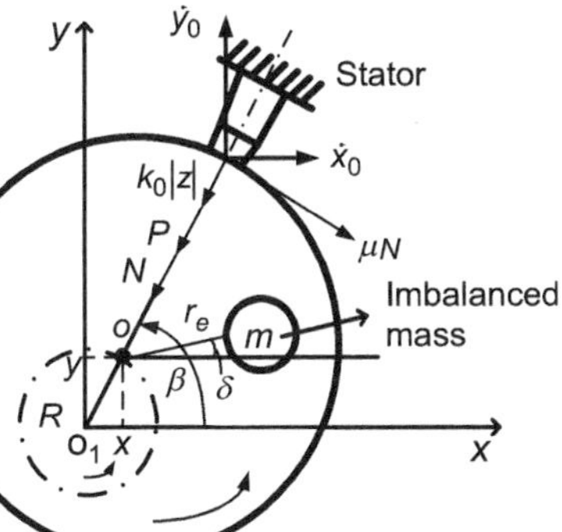

Figure 6.4 Forces acting on rotor with rub-impact

effects of stiffness modification, impacts, and friction, respectively; $F(\Omega t)$ is a function indicating periodicity of the effect of rub-related forces; and the angular speed Ω is equal to one of the precession motions. Assuming that the shaft touches the stator during a fraction of the precession period and neglecting the transient process, $F(\Omega t)$ can be modeled by a periodic step function:

$$F(\Omega t) = 1 - \left\{ \Omega t_1/2 - \sum_{i}^{n} \frac{2}{i} \sin(i\Omega t_1/2)\cos\left[i\Omega(t - t_1/2)\right] \right\} / \pi, \quad [6.12]$$

where t_1 is the moment of contact that is smaller than $2\pi/\Omega$. It follows the expression indicating the rotor/stator contact and lack of contact:

$$\frac{1}{2}\left[\text{sign}\left(z - Re^{j\beta}\right) + 1\right] = \begin{cases} 1, & \text{for } z \geq Re^{j\beta} \\ 0, & \text{for } z > Re^{j\beta} \end{cases}, \quad [6.13]$$

where R and β determine the position of the rub between the rotor and the stator.

6.3.1.1 Effect of Stiffness Modification

The term $K_0|z|$ in Eq. [6.11] is an additional elastic force generated by the contact between rotor and stator, which indicates that the stiffness variations during the rub-impact process may lead to parametrically self-excited vibration. Research results indicate that the rotor response may contain many harmonics of the main component with the lowest frequency equal to $\Omega/2$. In the case of initial synchronous precession, the main component will have a frequency equal to half of the rotational speed. Nonlinear factors in real machines may limit the rotor vibration amplitudes, leading to limit cycles of these self-excited vibrations.

The rotor model, with variable stiffness due to the contact between rotor and stator, explains the appearance of the rotor subharmonic vibration of a 1/2 order, as well as its higher harmonics. However, it does not predict subharmonic vibrations of any order lower than 1/2. In many practical cases, 1/2 subharmonic vibrations are more often reported. Several factors, such as oil whirl and transversal crack of rotor, most likely contribute to this effect.

6.3.1.2 Effect of Impact and Friction

In Eq. [6.11], the term related to K_0 is now omitted and the effect of periodic action of the normal force N and a resulting friction force μN will be analyzed. The magnitude of the force N depending on rotor/stator impact conditions is undetermined. The collisions repeat periodically following the frequency of the rotor precession. Each collision causes a short transient lateral motion of the rotor (i.e., excites its free vibrations at the rotor natural frequency). The initial conditions (i.e., displacement and velocity) of the free vibration are determined by the rotor motion before collision as well as by friction at the contacting surfaces.

Before collision, the motion is given by

$$\begin{cases} z(t) = \dfrac{mr_e\omega^2}{\sqrt{(K - M\omega^2)^2 + (C\omega)^2}} e^{j(\omega t + \delta + \gamma)} \\ \gamma = \arctan\left(\dfrac{-C\omega}{K - M\omega^2}\right) \end{cases}, \quad [6.14]$$

which corresponds to the imbalance-excited synchronous precession.

At the moment of time $t = 0$, when the rotor has a constant rotating speed ω, the motion described by Eq. [6.14] is perturbed by a rotor contact with the stator located at the radial position given by the distance R and angle β. At the moment of the collision, the initial position is determined by setting $t = 0$ in Eq. [6.14]:

$$z(0) = Re^{j\beta} = \frac{mr_e\omega^2}{\sqrt{(K - M\omega^2)^2 + (C\omega)^2}} e^{j(\beta + \delta + \gamma)}. \quad [6.15]$$

The initial velocity is $\dot{z}(0) = \dot{z}_0$, and its value depends on the rotor velocity, the impact conditions (material and geometric properties), and friction. The solution of the rotor model with action of the average value of the radial force and the initial conditions determined as described previously is as follows:

$$z(t) = \left(\dot{z}_0 + \zeta\frac{P(1 + j\mu)}{K}\omega_n e^{j\beta} - j\omega Re^{j\beta}\right)\frac{1}{\omega_n}e^{-\zeta\omega_n t}\sin\omega_n t$$
$$+ \frac{P(1 + j\mu)}{K}e^{j\beta}\left(e^{-\zeta\omega_n t}\cos\omega_n t - 1\right) + Re^{j(\omega t + \beta)}, \quad [6.16]$$

where $\omega_n = \sqrt{K(1 - \zeta^2)/M}$ is the rotor natural frequency and $\zeta = C/2\sqrt{KM}$ is a damping factor.

The first term of Eq. [6.16] corresponds to the damped free vibrations with the rotor natural frequency. The amplitude of this vibration depends mainly on the rotating speed, initial velocity, and determined initial position (contact position). The second term represents the action of constant force P, which determines the new equilibrium position of the rotor, around which the periodic motion is performed. The third term gives the forced synchronous response due to the imbalance.

The solution of Eq. [6.16] is valid during the time interval $[0, t_1]$. At $t = t_1$, the rotor hits the stator again and remains in contact during $t_1 < t < 2\pi/\Omega$, and then the cycle repeats. Practical observations prove that the precession frequency Ω is equal either to the rotational speed ω or to a fraction of it—i.e., the following relationship holds true $\Omega = \omega/v, v = 1, 2, 3, \ldots$.

6.3.1.3 Effect of Radial Force and Friction Oscillatory Terms

The simplified Eq. [6.11] with the remaining terms of the function [Eq. [6.12]] is as follows:

$$M\ddot{z} + C\dot{z} + Kz = -N(1 + j\mu)e^{j\beta}\sum_{i=1}^{n}\frac{2}{i\pi}\sin\left(\frac{i\Omega t_1}{2}\right)\cos\left(i\Omega\left(t - \frac{t_1}{2}\right)\right]. \quad [6.17]$$

It yields the solution

$$z(t) = -2N(1 + j\mu)e^{j\beta}\sum_{i=1}^{n}\frac{\sin(i\omega t_1/2v)\cos[i\omega(t - t_1/2)/v - \theta_i]}{i\pi\sqrt{(K - Mi^2\omega^2/v^2)^2 + (Ci\omega/v)^2}}, \quad [6.18]$$

where $\theta_i = \arctan[Ci\omega/(K - Mi^2\omega^2/v^2)v]$, and $\Omega = \omega/v$ has been introduced.

The rotor response to rotor/stator partial rub-impact contains the three parts discussed here, which qualitatively explain the appearance of the rotor subharmonic vibration and are helpful for analyzing the experimental results of rub-impact and identifying the partial rub-impact phenomenon in the rotor system [15].

6.3.2 Complex Dynamic Response of a Jeffcott Rotor System With Rub-Impact

One of the most common reasons for the rub-impact between the rotor and the stator is the coexistence of axial misalignment and an imbalanced mass of the rotor. The schematic of a rotor system with axial misalignment is shown in Figure 6.5.

When the system is at rest, the deviation between the rotor centroid O and the stator centroid O_1 along the horizontal and vertical directions are δ_x and δ_y, respectively. The origin of the inertial coordinate system is located at the centroid of the static rotor, and the eccentricity is indicated by e. Thus, the distance between their centroids can be defined as

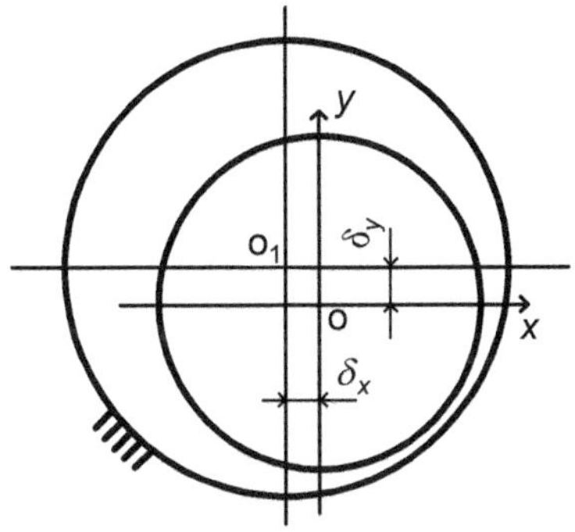

Figure 6.5 Schematic of a rotor system with axial misalignment

$r = \sqrt{(x-\delta_x)^2 + (y-\delta_y)^2}$ (in the example shown in Figure 6.5, $\delta_x < 0$, $\delta_y > 0$). It is assumed that the rotor rotates counterclockwise with a rotating speed ω. The radius of the clearance circle is indicated by r_0; i.e., the difference between the rotor radius and the stator radius. When $r > r_0$, the rotor collides with the stator, and the rub-impact forces [Figure 6.2(b)] acting on the rotor are

$$\begin{cases} F_N = k_c(r - r_0) \\ F_\tau = \mu k_c(r - r_0) \end{cases}. \quad [6.19]$$

The projections of F_N and F_τ in the x- and y-directions are obtained as

$$\begin{Bmatrix} F_x \\ F_y \end{Bmatrix} = -\frac{(r-r_0)}{r} k_c \begin{bmatrix} 1 & -\mu \\ \mu & 1 \end{bmatrix} \begin{Bmatrix} x - \delta_x \\ y - \delta_y \end{Bmatrix}. \quad [6.20]$$

Therefore, Newton's second law leads to the equations of motion of the rotor system with rub-impact:

$$\begin{cases} m\ddot{x} + c\dot{x} + kx + k_c\left(1 - \dfrac{r_0}{r}\right)\left[(x-\delta_x) - \mu(y-\delta_y)\right] = me\omega^2 \cos \omega t \\ m\ddot{y} + c\dot{y} + ky + k_c\left(1 - \dfrac{r_0}{r}\right)\left[\mu(x-\delta_x) + (y-\delta_y)\right] = me\omega^2 \sin \omega t \end{cases}. \quad [6.21]$$

Letting $\delta_x = 0$ and $\delta_y = \delta$, we get

$$\begin{cases} m\ddot{x} + c\dot{x} + kx + k_c\left(1 - \dfrac{r_0}{r}\right)\left[x - \mu(y-\delta)\right] = me\omega^2 \cos \omega t \\ m\ddot{y} + c\dot{y} + ky + k_c\left(1 - \dfrac{r_0}{r}\right)\left[\mu x + (y-\delta)\right] = me\omega^2 \sin \omega t \end{cases}. \quad [6.22]$$

Eq. [6.22] can be normalized with the following dimensionless parameters:

$$X = \frac{x}{e}, Y = \frac{y}{e}, R_0 = \frac{r_0}{e}, \Delta = \frac{\delta}{e}, R = \frac{r}{e} = \sqrt{X^2 + (Y - \Delta)^2} \quad [6.23a\text{–}e]$$

$$\omega_1^2 = k/m, \omega_2^2 = (k + k_c)/m, \tau = \omega_2 t, c/m = 2\zeta\omega_2, \varpi = \omega/\omega_2,$$
$$\beta = k/(k + k_c) \quad [6.23f\text{–}k]$$

The governing equations of motion in nondimensional form are obtained as follows:

$$\begin{cases} X'' + 2\zeta X' + \beta X + (1 - \beta)(1 - R_0/R)[X - \mu(Y - \Delta)] = \varpi^2 \cos(\varpi\tau) \\ Y'' + 2\zeta Y' + \beta Y + (1 - \beta)(1 - R_0/R)[\mu X + (Y - \Delta)] = \varpi^2 \sin(\varpi\tau) \end{cases}, \quad [6.24]$$

where $()' = d()/d\tau$.

Letting $Y_1 = X$, $Y_2 = X'$, $Y_3 = Y$, and $Y_4 = Y'$, Eq. [6.24] is expressed in state space as

$$\begin{cases} Y_1' = Y_2 \\ Y_2' = \varpi^2 \cos(\varpi\tau) - 2\zeta Y_2 - \beta Y_1 - (1 - \beta)\left(1 - \dfrac{R_0}{\sqrt{Y_1^2 + (Y_3 - \Delta)^2}}\right) \\ \qquad \times (Y_1 - \mu(Y_3 - \Delta)) \\ Y_3' = Y_4 \\ Y_4' = \varpi^2 \sin(\varpi\tau) - 2\zeta Y_4 - \beta Y_3 - (1 - \beta)\left(1 - \dfrac{R_0}{\sqrt{Y_1^2 + (Y_3 - \Delta)^2}}\right) \\ \qquad \times (\mu Y_1 + (Y_3 - \Delta)) \end{cases} . \quad [6.25a\text{–}d]$$

When $r \leq r_0$, the rotor will not collide with the stator, and the equations of motion of the system in space state can be obtained by omitting the fourth term of Eqs. [6.25b] and [6.25d].

The parameter values in simulation are as follows: $\zeta = 0.05$, $\mu = 0.1$, $\beta = 0.1$, $R_0 = 2.0$, and $\Delta = 1.0$. In order to obtain steady-state motion, the simulation data are recorded and analyzed after 1000 periods.

The influences of the system parameters, such as speed ratio, friction coefficient, and damping ratio, on system behavior may be studied through the bifurcation diagram. The Poincaré map on the phase plane $Y_3 - Y_4$ may be used to identify the state of system motion. Also, the centroid trajectory of the rotor and the clearance circle may be used intuitively to show their relative position and rub-impact status [16].

6.3.2.1 Effect of Speed Ratio ϖ on Dynamic Behavior of the Rrotor

The bifurcation diagram with varying speed ratio ϖ is shown in Figure 6.6, which contains two regions of complex motion ($\varpi = 0.697-0.9119$ and $\varpi = 1.192-1.317$). Based on the simulation data, the following complex motion properties of the rotor may be obtained through analyzing the centroid trajectories of the rotor and the clearance circle and the Poincaré map.

The period-1 motion occurs for $\varpi < 0.688$, where the rub-impact develops from partial rub-impact to full annular rubbing, then to partial rub-impact again with the increase of speed ratio ϖ (note that through drawing the centroid trajectory of the rotor and the clearance circle, the evolution of rub-impact status may be seen easily).

When $\varpi > 0.684$, quasi-periodic motion appears with the increase of the speed ratio until chaotic motion occurs at $\varpi = 0.697$. The chaotic motion is maintained for the speed ratio range $\varpi = 0.697-0.9119$, except the periodic motion for the narrow range $\varpi = 0.743-0.748$. For $\varpi = 0.912$, the chaotic motion disappears where a period-2 motion is

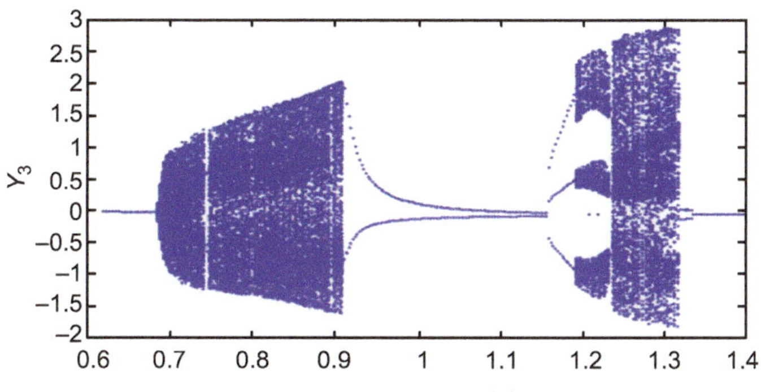

Figure 6.6 Bifurcation diagram with varying speed ratio ϖ

created. This shows that the bifurcation transits from quasi-periodic motion to chaotic motion, and then from chaotic motion to periodic motion in the first region. The period-2 motion lasts for $\varpi = 0.912{-}1.158$, and a period-3 motion lasts for $\varpi = 1.1581{-}1.191$. The chaotic motion appears again for the range $\varpi = 1.192{-}1.317$ and transits to a period-3 motion at $\varpi = 1.318$. Therefore, the bifurcation path in the second region ($\varpi = 1.192{-}1.317$) is similar to that of the first region ($\varpi = 0.697{-}0.9119$). Then the system enters into a period-2 motion for the range of $\varpi = 1.318{-}1.3344$, and again into a period-1 motion for $\varpi = 1.3345{-}1.4$.

6.3.2.2 Effect of Friction Coefficient and Damping Ratio on Dynamic Behavior of the Rotor

The bifurcation diagram with a varying friction coefficient ($\mu \leq 0.25$) for $\varpi = 0.5$ is shown in Figure 6.7. It can be seen that a period-1 motion occurs for $\mu \leq 0.16$, and quasi-periodic motion or motion with many periodic points appears for $\mu \geq 0.161$. Based on the simulation data, the centroid trajectories of the rotor and the clearance circle, as well as the Poincaré map, can be depicted. It may be concluded that there is no chaotic motion for the system with these parameters. In addition, the complex motion appears with a smaller speed ratio for a larger friction coefficient.

The bifurcation diagram with a varying damping ratio ($\zeta \geq 0.02$) for $\varpi = 0.5$ is shown in Figure 6.8. It can be observed that a period-1 motion occurs for $\zeta \geq 0.0319$ and quasi-periodic motion or motion with many periodic points appears for $\zeta \leq 0.0318$. After similar investigations as the ones previously described are performed, the conclusion was reached that no chaotic motion exists for these system parameters. In addition, the

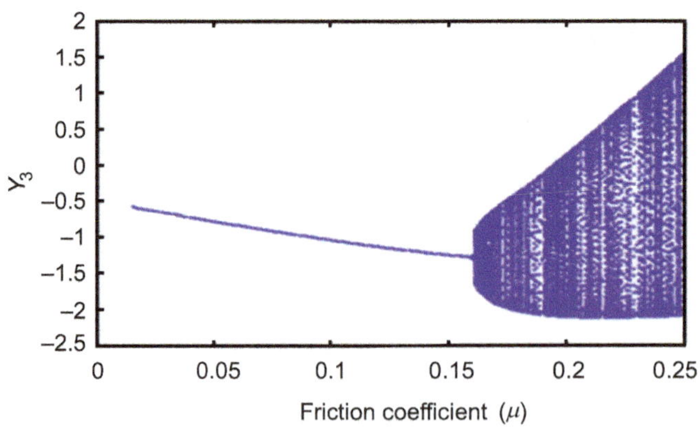

Figure 6.7 Bifurcation diagram with varying friction coefficient

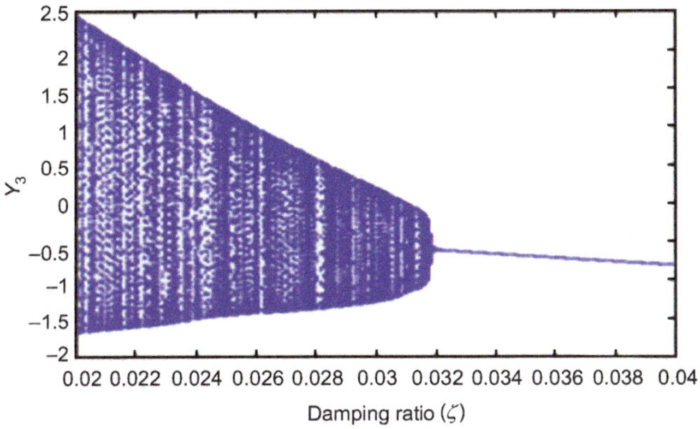

Figure 6.8 Bifurcation diagram with varying damping ratio

complex motion appears with a smaller speed ratio for the lower damping ratio.

6.3.2.3 Summary

According to this analysis, the following conclusions can be obtained. The variation of speed ratio ϖ may lead to various forms of periodic, quasi-periodic, and chaotic motions of the system. With the increase of the friction coefficient, the motion of the rotor with rub-impact evolves from periodic motion to quasi-periodic motion, and then to complex motion. And the complex motion appears with a smaller speed ratio for a larger friction coefficient. With the increase in damping ratio, the motion of the rotor with rub-impact evolves from complex motion to quasi-periodic motion, and then to periodic motion. And the complex motion appears with a smaller speed ratio for the lower damping ratio. There are two kinds of routes to or out of chaos of the rotor system with axial misalignment (i.e., quasi-period bifurcation and intermittent bursting). Many studies show that the various dynamic characteristics of the rotor system, such as chaotic and quasi-periodic motions, are helpful to diagnose and predict the rub-impact fault.

6.3.3 Nonlinear Behavior of Dual-Rrotor System With Rub-Impact

A dual-rotor system is applied for high efficiency in rotary machinery such as airplane engines and air compressors. In the airplane engine, increasing thrust-weight ratio and efficiency is often achieved by tightening the operating clearance between rotor and casing. Reduced clearance and other factors, such as misalignment, mass imbalance, and bend of the axle, lead more frequently to rub-impact fault, especially for high-pressure turbines and casings subjected to high temperatures.

In this section, the dynamic model of dual-rotor system of airplane engines is established, including high- and low-pressure rotors, turbines, gas compressors, and front and rear casings. The governing equations of the system are derived through the Lagrange method and are numerically solved to depict the rub-impact fault between high-pressure turbines and rear casings.

6.3.3.1 Equations of Motion of Dual-Rotor Systems With Rub-Impact

For simplification, some assumptions of the system [7,8] are made as follows:

- The casing has only two translation degrees of freedom in the horizontal and vertical directions.
- Usually, the elastic bending and torsion deformation of both high- and low-pressure rotors are very small. Therefore, the rotor is assumed to be rigid.
- High- and low-pressure rotors work at constant rotating speeds of ω_1 and ω_2, respectively, without axial motion.
- Because the stiffness of the rotor system is large and the influence of the system's gravity is relatively small, the influence of gravity is ignored.
- The rub-impact takes place between the high-pressure turbine and rear casing.

The schematic of a dual-rotor airplane engine system is illustrated in Figure 6.9. For the convenience of dynamic analysis, the dual-rotor system is simplified and divided into four substructures: high-pressure rotor, low-pressure rotor, front casing, and rear casing. The simplified dynamic model is shown in Figure 6.10, where the numbers 1, 2, and 5 represent the supports of low-pressure rotors, 3 and 4 are the bearings of high-pressure rotors, and 6 and 7 are the bearings of casings. The inertia coordinate of the system is o-xyz. It is assumed that the displacements of sections B and E of the low-pressure rotor are y_1, z_1, and y_2, z_2, respectively; the

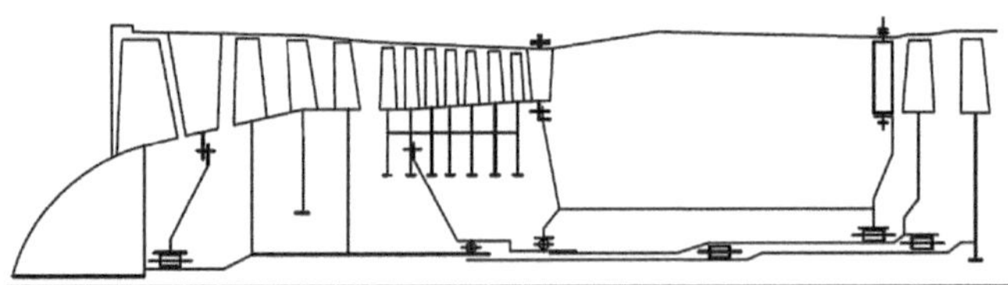

Figure 6.9 Schematic of a dual-rotor airplane engine system

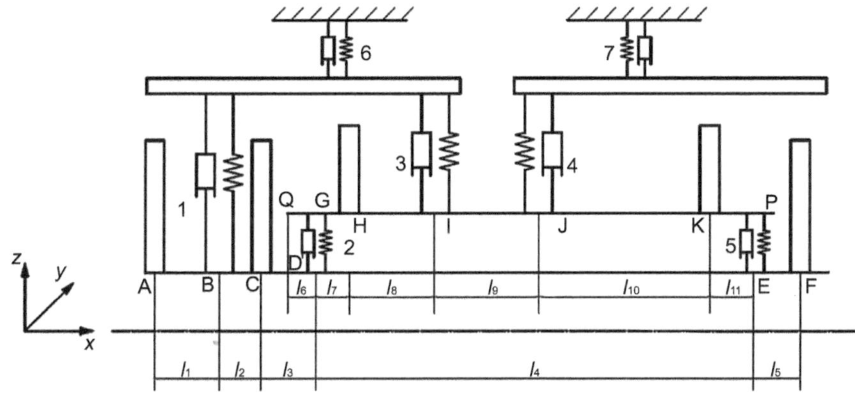

Figure 6.10 Dynamic model of a dual-rotor system

displacements of sections I and J of the high-pressure rotor are y_3, z_3, and y_4, z_4, respectively; the displacements of the front and rear casings are y_5, z_5, and y_6, z_6, respectively; the rotating speed of the low-pressure rotor is ω_1; and the rotation speed of the high-pressure rotor is ω_2. The compressor and turbine on the low-pressure rotor are given simply as disks A, C, and F. Those on the high-pressure rotor are given as disks H and K.

The governing equations of the dual-rotor system obtained by Lagrange's equations may be written in the following matrix form [17]:

$$M\ddot{\mathbf{q}} + C\dot{\mathbf{q}} + G\dot{\mathbf{q}} + K\mathbf{q} = \mathbf{F}, \qquad [6.26]$$

where M, C, G, and K are mass, damping, gyroscopic effect, and stiffness matrices, respectively; $\mathbf{F}$ and $\mathbf{q}$ are generalized force and coordinate vectors, respectively; and $\mathbf{q} = \{ y_1 \quad y_2 \quad y_3 \quad y_4 \quad y_5 \quad y_6 \quad z_1 \quad z_2 \quad z_3 \quad z_4 \quad z_5 \quad z_6 \}^T$.

When the rub-impact occurs between high-pressure turbines and casings, the governing equations remain the same as Eq. [6.26] except for some changes in the generalized force vector. The relative position of the high-pressure turbine and casing and the analysis of rub-impact force are shown in Figure 6.11.

Assume that the radius of the clearance circle (i.e., the difference between radii of the high-pressure turbine and casing) is r_0, and the vertical eccentricity is δ. In the running process, the distance between the two centers is

$$r = \sqrt{(y_k - y_6)^2 + (z_k - \delta - z_6)^2}, \qquad [6.27]$$

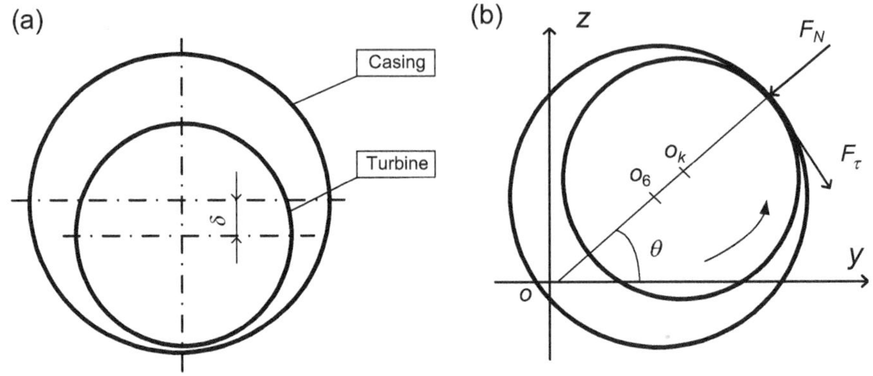

Figure 6.11 Relative position of the high-pressure turbine and rear casing at rest (a) and the rub-impact force (b)

where y_k and z_k are the displacements of the high-pressure turbine center along axes y and z; and y_6 and z_6 are the displacements of the rear casing in the direction of y and z.

It is obvious that the rub-impact does not take place when $r < r_0$, while the turbine collides with the casing with $r > r_0$. Assuming that the high-pressure rotor rotates counterclockwise in the rotating speed ω_2, the normal impact force and its direction can be expressed as

$$F_N = k_c(r - r_0), \cos\theta = \frac{y_k - y_6}{r}, \sin\theta = \frac{z_k - \delta - z_6}{r}. \qquad [6.28a\text{–}c]$$

The tangential friction force is $F_\tau = F_N\mu$, whose direction is perpendicular to that of the normal force, and is opposite to that of the relative speed of the contact surface.

The governing equations of the system with rub-impact can be obtained by adding terms $\Delta F_1, \Delta F_2, \ldots, \Delta F_{12}$ to the corresponding terms of the generalized force vector of Eq. [6.26]. The $\Delta F_1, \Delta F_2, \ldots, \Delta F_{12}$ are the components of the rub-impact forces in generalized coordinates obtained by the virtual work principle, and $\Delta F_1 = \Delta F_2 = \Delta F_5 = \Delta F_7 = \Delta F_8 = \Delta F_{11} = 0$ may also be obtained [17].

In general, the rotor precession has a forward direction when the rub-impact fault is not severe. Therefore, the direction of the friction force acting on the rotor is the opposite of its rotating direction. When the rub-impact is severe, the rotor precession may have a backward direction. The direction of the friction force acting on the rotor remains if the velocity of backward precession is small, but reverses if the velocity of backward precession is over a certain value. Therefore, the direction of the friction force should be determined at each step of the calculation according to the method described in the section entitled "Analytical Model of Rub-Impact," earlier in this chapter.

6.3.3.2 Analysis of Nonlinear Characteristics

It is seen from the preceding discussion that the dual-rotor system with rub-impact is strongly nonlinear and has a dual-frequency excitation (even if the mass imbalance just exists in the high-pressure turbine). In this section, the rub-impact between a high-pressure turbine and rear casing is numerically simulated, and the frequency spectrum characteristics are investigated. More detailed characteristics of the vibration signals will be analyzed by using the holospectrum and wavelet transform methods in the section entitled "Fault Diagnosis of the Rotor System with Rub-Impact," later in this chapter.

The corresponding parameters in this example are obtained from an airplane engine. The simulation results show that when partial rub-impact between rotor and casing takes place and develops to a certain level, the frequency spectrum diagram of the casing or rotor may include the rotating frequencies of the low- and high-pressure rotors, fractional and double terms, and various combinations of these two frequencies (as shown in Figure 6.12), which depend on the combination of the system parameters.

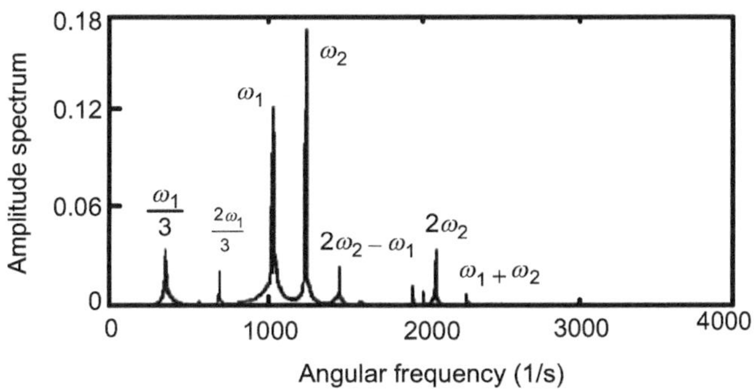

Figure 6.12 Amplitude spectrum of vertical vibration of the rear casing

6.4 STABILITY AND HOPF BIFURCATION OF A RUB-IMPACT JEFFCOTT ROTOR SYSTEM WITH IMBALANCE

In the section entitled "Complex Dynamic Response of a Jeffcott Rotor System with Rub-Impact," earlier in this chapter, the dynamic equations of a Jeffcott rotor system with axial misalignment and mass imbalance were derived via Eq. [6.24]. If the axial misalignment is ignored (i.e., $\Delta = 0$), Eq. [6.24] can be rewritten as

$$\begin{cases} X'' + 2\zeta X' + \beta X + (1-\beta)(1-R_0/R)[X - \mu Y] = \varpi^2 \cos(\varpi\tau) \\ Y'' + 2\zeta Y' + \beta Y + (1-\beta)(1-R_0/R)[\mu X + Y] = \varpi^2 \sin(\varpi\tau) \end{cases}. \tag{6.29}$$

In this section, the influence of friction coefficient μ on the precession frequency and the jump phenomenon, stability, and Hopf bifurcation will be analyzed by numerically simulating Eq. [6.29].

6.4.1 Influence of a Friction Coefficient on Precession Frequency

By means of numerical simulation [2], it is known that for the case of $\mu \le 0.12$, the vibration caused by the mass imbalance plays a major role and the contribution of vibrations with other frequencies is very little. Thus, only one frequency component is obvious in the rotor precession. In this case, the increase of ϖ only causes the change of the phase difference φ between the rotor precession and the centrifugal force. The variation of φ with ϖ is shown in Table 6.3.

For the case of $\mu > 0.12$, with the increase of ϖ, the rotor precession goes through four stages: periodic motion, quasi-periodic motion, unstable motion, and then periodic motion, in that order. Unstable motion is not the final status; if the rotating speed increases rapidly, so that the rotor can cross the speed range of instability, the rotor will revert to the periodic motion. When the quasi-periodic motion of the rotor occurs, in addition to the frequency component caused by unbalanced mass, the precession

Table 6.3 Variation of φ with ϖ ($R_0 = 1.0$, $\mu = 0.1$)

ϖ	0.3	0.4	0.7	0.75	0.79
φ (°)	34.52	35.68	47.43	53.76	60.62

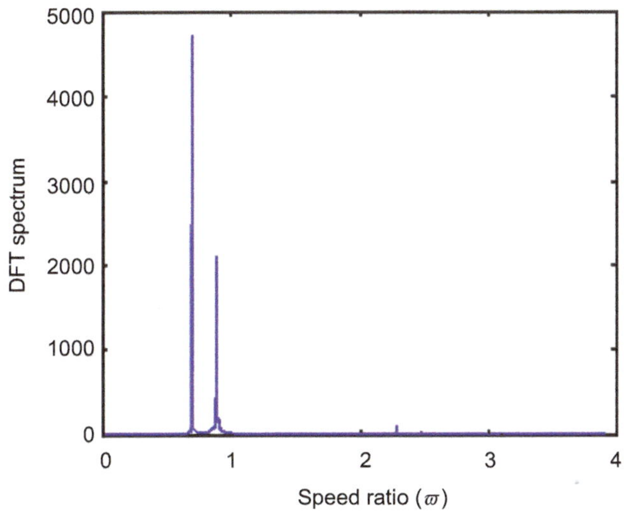

Figure 6.13 DFT spectra of horizontal vibration of the rotor procession

contains other frequency components, as shown in Figure 6.13 ($R_0 = 1.0$, $\mu = 0.13, \varpi = 0.7$). Actually, the corresponding motion of the rotor shown in Figure 6.13 is quasi-periodic motion. Figure 6.14(a) indicates that the rotor behaves as periodic motion for $\varpi = 0.6$. Figures 6.14(b) and (c) are the trajectory of the rotor precession and corresponding Poincaré map for $\varpi = 0.7$, respectively, which clearly illustrate that the rotor behaves as quasi-periodic motion whose discrete Fourier transform (DFT) is displayed in Figure 6.13. Figure 6.14(d) indicates that the rotor precession becomes unstable at $\varpi = 0.9$; i.e. with the precession going, the precession amplitude will increase continuously if the contact stiffness is linear. In reality, due to the nonlinearity of the contact stiffness, the precession amplitude does not continuously increase. In addition, this indicates that the rotor follows a backward precession due to the large normal force and friction force between the rotor and the stator. For $\varpi \geq 1$, the trajectory of the rotor precession restores to a circle and the rotor returns to forward precession gradually, since the rub-impact decreases due to the self-centering function of the rotor over the critical speed. Many calculations [2] also show that the range of speed ratio ϖ corresponding to the stable precession of the rotor narrows with the increase of friction coefficient μ.

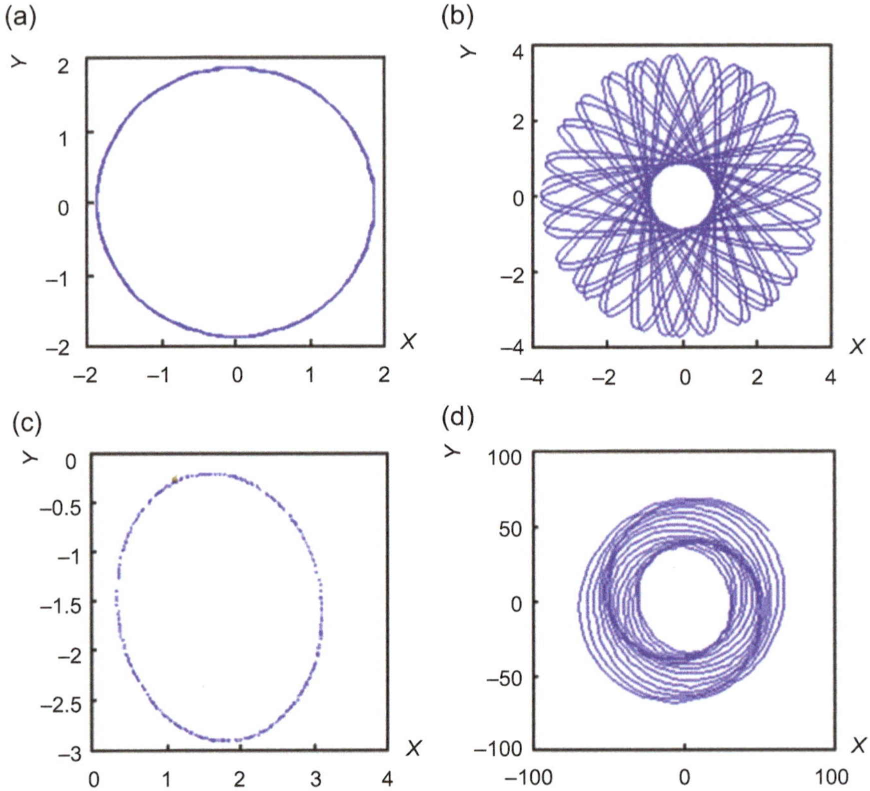

Figure 6.14 Influence of speed ratio on rotor precession characteristic: (a) trajectory of the rotor precession for $\varpi = 0.6$; (b) trajectory of the rotor precession for $\varpi = 0.7$; (c) Poincaré map for $\varpi = 0.7$; (d) trajectory of the rotor precession for $\varpi = 0.9$

6.4.2 Jump Phenomenon of a Rotor System With Full Annular Rub-Impact

The previous section established that, when $\mu \leq 0.12$, the rotor mainly performs as the forward precession caused by the unbalanced force. Therefore, the trajectory of the rotor precession is approximately 1 cycle with a constant radius. One study [18] showed that the jump phenomenon may occur in the case of the full annular rub-impact. In this section, the jump phenomenon and its occurrence conditions are analyzed.

Multiplying the second equation of Eq. [6.29] by the imaginary unit j, adding it to the first equation, and then taking into account the vector of rotor precession $\Re = X + jY$ yields

$$\Re'' + 2\zeta\Re' + \left[1 - (1-\beta)\frac{R_0}{|\Re|}\right]\Re + j(1-\beta)\left(1 - \frac{R_0}{|\Re|}\right)\mu\Re = \varpi^2 e^{j\varpi\tau}. \tag{6.30}$$

Letting $\Re = \Re_A e^{j\omega\tau}, \beta_0 = 1 - \beta$, Eq. [6.30] becomes

$$-\varpi^2\Re_A + 2j\varpi\zeta\Re_A + \left[1 - \beta_0\frac{R_0}{|\Re|}\right]\Re_A + j\beta_0\left(1 - \frac{R_0}{|\Re|}\right)\mu\Re_A = \varpi^2. \quad [6.31]$$

The amplitude $\Re_A$ can be solved as

$$\Re_A = \frac{\varpi^2}{\left[1 - \beta_0\frac{R_0}{|\Re|}\right] - \varpi^2 + j\left[2\zeta\varpi + \beta_0\left(1 - \frac{R_0}{|\Re|}\right)\mu\right]}. \quad [6.32]$$

$\Re_A$ can be expressed as $|\Re|e^{-j\varphi}$ or $Re^{-j\varphi}\left(R = |\Re| = \sqrt{X^2 + Y^2}\right)$, in which R is

$$R = \frac{\varpi^2}{\left\{\left[1 - \beta_0\frac{R_0}{R} - \varpi^2\right]^2 + \left[2\zeta\varpi + \beta_0\left(1 - \frac{R_0}{R}\right)\mu\right]^2\right\}^{\frac{1}{2}}}. \quad [6.33]$$

The relationship can be simplified as

$$P_1R^2 + 2P_2R + P_3 = 0. \quad [6.34]$$

where the functions P_1, P_2, and P_3 are defined as

$$P_1 = \left(1 - \varpi^2\right)^2 + \left(2\zeta\varpi + \beta_0\mu\right)^2 \quad [6.35a]$$

$$P_2 = -\left(1 - \varpi^2\right)\beta_0R_0 + \left(2\zeta\varpi + \beta_0\mu\right)\beta_0R_0\mu \quad [6.35b]$$

$$P_3 = \left(\beta_0R_0\right)^2 + \left(\beta_0R_0\mu\right)^2 - \varpi^4. \quad [6.35c]$$

The solution of Eq. [6.34] is

$$R_{1,2} = \frac{-P_2 \pm \sqrt{P_2^2 - P_1P_3}}{P_1}. \quad [6.36]$$

Figure 6.15(a) illustrates the relationship between R and ϖ for $R_0 = 2.0$ and $\mu = 0.08$. There are three different solutions at a certain speed range, which are R_1, R_2 and the amplitude of rotor precession without rub-impact. Obviously, the jump phenomenon of the system occurs with a certain circumstance. Figures 6.15(b–c) are the trajectories of the rotor precession for $\varpi = 0.71$ and 0.72. For $\varpi = 0.71$, the amplitude of the rotor precession is larger than 4.0, but for $\varpi = 0.72$, the amplitude is about 1.2, which indicates that the jump phenomenon occurs when the speed ratio varies from 0.71 to 0.72, which is different from the speed ratio ($\varpi = 0.82$) of the jump phenomenon shown in Figure 6.15(a). The reason for this difference is that the rotor precession is assumed to include only the forward

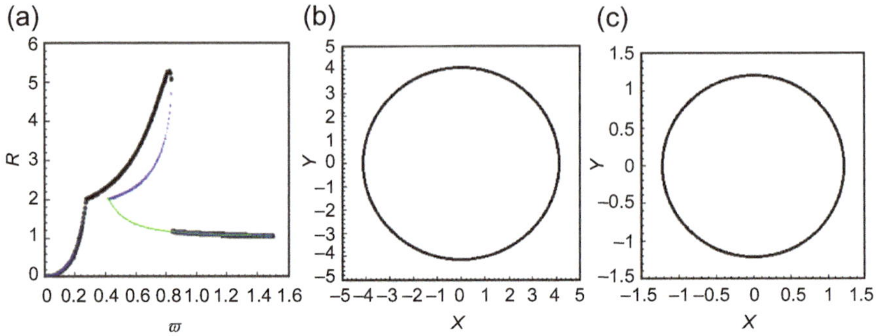

Figure 6.15 Jump phenomenon of the rotor precession: (a) variation of amplitude of the rotor with speed ratio; (b) trajectory of the rotor precession for $\varpi = 0.71$; (c) trajectory of the rotor precession for $\varpi = 0.72$

precession frequency components caused by an unbalanced force for the convenience of theoretical analysis, but that's not the real case.

6.4.3 Stability and Hopf Bifurcation of the Rotor System With Full Annular Rub-Impact

Although the full annular rub-impact of a rotor system can be caused by various reasons, in this section only the case caused by imbalance is discussed. It is known from the previous two sections that when the full annular rub-impact occurs, the trajectory will be circular if the speed ratio is not very large; when the speed ratio exceeds a certain value, the trajectory becomes a petal shape [Figure 6.14(b)]; if the speed continues to increase, the amplitude of the rotor precession increases accordingly. If the rotor precession is depicted in the phase planes of R and $\dot{R}$ [displacement and velocity of the rotor center, as given by Eq. [6.33]], it can be seen that when the speed ratio is small, the stable motion where the precession trajectory is circular corresponds to a stable focus, and the stable motion where the precession trajectory is a petal shape corresponds to a stable limit cycles. Therefore, it can be deduced that the Hopf bifurcation phenomenon may occur in the full annular rub-impact [19].

The bifurcation theory of dynamical systems describes the qualitative change and jump phenomenon of the phase trajectory of the differential equations when the parameters change. In [20,21], the mechanism of instability and the Hopf bifurcation phenomenon of a nonlinear rotor system are investigated. In [22], the Galerkin method, which combines Floquet theory and the numerical integral method, is used to analyze the stability and the bifurcation phenomenon of a rotor-bearing system. In this section, the equations of motion of the rotor system with full annular rub-impact caused by imbalance will be established, and the stability and Hopf bifurcation will also be discussed.

Only periodic solutions are considered, and Eq. [6.29] can be transformed to the following polar form [18]:

$$\begin{cases} X(\tau) = R(\tau)\cos[\varpi\tau + \psi(\tau)] \\ Y(\tau) = R(\tau)\sin[\varpi\tau + \psi(\tau)] \end{cases}. \qquad [6.37]$$

Letting $X_1 = R, X_2 = R', X_3 = \psi, X_4 = \psi'$, and using Eq. [6.37], Eq. [6.29] can be rewritten as

$$\begin{cases} X_1' = X_2 \\ X_2' = (1-\beta)R_0 + \varpi^2\cos(X_3) - 2\zeta X_2 - \left(1 - \varpi^2 - 2\varpi X_4 - X_4^2\right)X_1 \\ X_3' = X_4 \\ X_4' = \dfrac{1}{X_1}[\mu(1-\beta)R_0 - \varpi^2\sin(X_3) - 2(X_2 + \zeta X_1)(\varpi + X_4) \\ \qquad - \mu(1-\beta)X_1] \end{cases}. \qquad [6.38]$$

6.4.3.1 Stability Analysis

Letting $X_1' = 0, X_2' = 0, X_3' = 0, X_4' = 0$, the equilibrium position $(X_{10}, X_{20}, X_{30}, X_{40})$ of the equations of motion of the rotor system with full annular rub-impact can be obtained by solving Eq. [6.38] as follows:

$$X_{10} = R_0 + \frac{-P_2 \pm \sqrt{P_2^2 - P_1P_3}}{P_1}, \quad X_{20} = 0, \quad X_{40} = 0, \qquad [6.39a\text{–}c]$$

where P_1, P_2, and P_3 are defined as

$$P_1 = (1-\varpi^2)^2 + [\mu(1-\beta) + 2\zeta\varpi]^2 \qquad [6.40a]$$

$$P_2 = R_0\{(1-\varpi^2)(\beta-\varpi^2) + 2\zeta\varpi[\mu(1-\beta) + 2\zeta\varpi] \qquad [6.40b]$$

$$P_3 = R_0^2\left[(\beta-\varpi^2)^2 + 4\zeta^2\varpi^2\right] - \varpi^4. \qquad [6.40c]$$

X_{30} is determined by the following equations:

$$\begin{cases} \sin(X_{30}) = [\mu(1-\beta)(R_0 - X_{10}) - 2\zeta X_{10}\varpi]/\varpi^2 \\ \cos(X_{30}) = [(1-\varpi^2)X_{10} - (1-\beta)R_0]/\varpi^2 \end{cases}. \qquad [6.41]$$

If $X_{10} > R_0$, rub-impact between the rotor and the stator takes place. When both of the two solutions of X_{10} are larger than R_0, there are two equilibrium points, $\left(X_{10}^A, 0, X_{30}^A, 0\right)$ and $\left(X_{10}^B, 0, X_{30}^B, 0\right)$, in which the superscript A corresponds to the plus signs in Eq. [6.39a], and the superscript B corresponds to the minus signs. If $X_{10} \leq R_0$ or Eq. [6.39a] has no real solution, rub-impact does not take place. In this situation, the equations of motion of the rotor system can be simplified as

$$\begin{cases} X_1' = X_2 \\ X_2' = \varpi^2 \cos(X_3) - 2\zeta X_2 - \left(\beta - \varpi^2 - 2\varpi X_4 - X_4^2\right)X_1 \\ X_3' = X_4 \\ X_4' = \dfrac{1}{X_1}[-\varpi^2 \sin(X_3) - 2(X_2 + \zeta X_1)(\varpi + X_4)] \end{cases} . \qquad [6.42]$$

Figure 6.16 shows the variation of X_{10} with ϖ for the case of $\zeta = 0.05, \mu = 0.13, \beta = 0.1, R_0 = 1.5$ (in the subsequent discussion, these parameters will be assumed to be the same values). Obviously, Figure 6.16 gives a typical bistable curve. It indicates that the speed ratio is ϖ_a when rub-impact starts, and $\varpi_b = \varpi_d$. The coordinate X_{10} of the equilibrium point has more than one value for $\varpi_b > \varpi > \varpi_c$. The curve branch *ab* shows the variation of X_{10}^A with ϖ, and *cd* shows the variation of X_{10}^B with ϖ. The curve branch below *cd* corresponds to the case without the rub-impact. Here, only the motion after the occurrence of rub-impact ($\varpi > \varpi_a$, $X_{10} > R_0$) is analyzed.

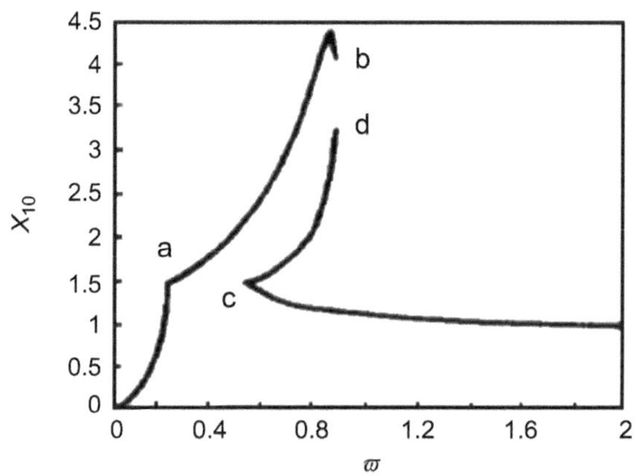

Figure 6.16 Variation of X_{10} with ϖ

Transforming the origin of coordinates corresponding to Eq. [6.38] to the equilibrium point $(X_{10}, X_{20}, X_{30}, X_{40})$—i.e., introducing the transformation of $X_i^* = X_i - X_{i0}$ $(i = 1, 2, 3, 4)$— yields $X^{*'} = AX^* + f(X^*)$. For the sake of convenience, the superscript * is dropped. Then Eq. [6.38], in which the origin of coordinates is moved, may be expressed in the form of a matrix as

$$X' = AX + f(X). \quad [6.43]$$

where AX corresponds to the linear term and $f(X)$ corresponds to the nonlinear one, and their expressions are written as

$$X = \{X_1 \quad X_2 \quad X_3 \quad X_4\}^T, \quad [6.44a]$$

$$A = \begin{bmatrix} 0 & 1 & 0 & 0 \\ a_1 & a_2 & a_3 & a_4 \\ 0 & 0 & 0 & 1 \\ b_1 & b_2 & b_3 & b_4 \end{bmatrix}, \quad [6.44b]$$

$$f(X) = \{0 \quad f_2 \quad 0 \quad f_4\}^T. \quad [6.44c]$$

The parameters in Eq. [6.44] are detailed in [19].

The characteristic polynomial of the coefficient matrix A in Eq. [6.43] is given by

$$\lambda^4 + a\lambda^3 + b\lambda^2 + c\lambda + d = 0. \quad [6.45]$$

where

$a = -(a_2 + b_4), b = a_2b_4 - a_1 - b_3 - a_4b_2, c = a_1b_4 + a_2b_3 - a_3b_2 - a_4b_1,$
$d = a_1b_3 - a_3b_1.$

From Eqs. [6.39] and [6.41], it may be seen that the equilibrium point varies with ϖ. Solving Eq. [6.45] gives four eigenvalues. If the following conditions are met for $\varpi = \varpi_{cr}$:

$$\frac{c}{a}\left(\frac{c}{a} - b\right) + d = 0, \frac{c}{a} - b < 0,\ c > 0, \quad [6.46]$$

the conjugated imaginary eigenvalues $\lambda_{1,2} = \pm i\Omega_c$ and complex eigenvalues $\lambda_{3,4} = -\delta_1 \pm i\delta_2$ $(\delta_1 > 0)$ can be obtained [20]. The critical speed ratio ϖ_{cr} can be solved from Eq. [6.46], and the system is stable for $\varpi < \varpi_{cr}$ and unstable for $\varpi > \varpi_{cr}$.

Figure 6.17 shows the variation of the largest real part of the four eigenvalues $\lambda_1, \lambda_2, \lambda_3$, and λ_4 [the solutions of Eq. [6.45]] with ϖ. Curves I and II correspond to the curve branches *ab* and *cd* in Figure 6.16, respectively. Obviously, the equilibrium point $\left(X_{10}^B, 0, X_{30}^B, 0\right)$ corresponding to the curve branch *cd* is unstable, and the equilibrium point $\left(X_{10}^A, 0, X_{30}^A, 0\right)$ corresponding to *ab* is stable for $\varpi < \varpi_{cr} = 0.715$. Figure 6.18 shows the variation of the maximum real part of eigenvalues with a friction coefficient for $\varpi = 0.715$. It can be observed that the maximum value grows with the increase of the friction coefficient—i.e., the stability worsens.

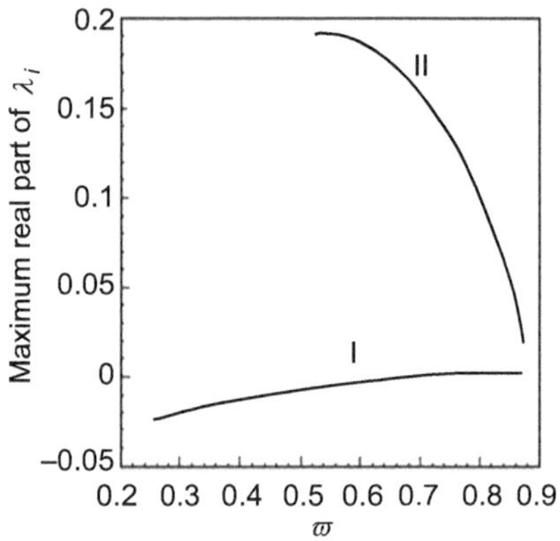

Figure 6.17 Variation of maximum real part of eigenvalues with ϖ

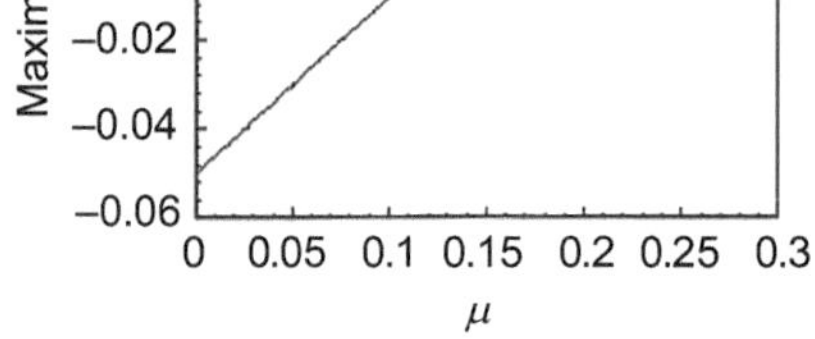

Figure 6.18 Variation of maximum real part of eigenvalues with friction coefficient

6.4.3.2 Hopf Bifurcation of the Rotor System

Hopf bifurcation is the characteristic phenomenon of a nonlinear system. If the phase trajectory of the system converges to a focus gradually under certain configurations of the system parameters (the system is stable), and small changes to the system parameters can make the phase trajectory no longer converge to the focus but rather to a closed curve (called a *limit cycle*), we say that Hopf bifurcation takes place. The occurrence of Hopf bifurcation in a system means that it will behave with periodic motion.

Letting $\sigma = \varpi - \varpi_{cr}$ be the bifurcation parameter and substituting $\varpi = \varpi_{cr} + \sigma$ into the coefficient matrix A in Eq. [6.43] give the eigenvalues of the system in the vicinity of the critical speed ratio ϖ_{cr}: $\lambda_{1,2} = \alpha(\sigma) \pm i\Omega(\sigma)$, and $\lambda_{3,4} = -\delta_1(\sigma) \pm i\delta_2(\sigma)$. Obviously, $\alpha(0) = 0$ and $\delta_1(0) > 0$. If $(d\alpha/d\varpi)_{\varpi=\varpi_{cr}} \neq 0$, Hopf bifurcation of the system occurs at $\varpi = \varpi_{cr}$ according to Hopf theory; i.e., there are periodic solutions in the vicinity of ϖ_c [22,23]. Figure 6.19 shows the variation of $d\alpha/d\varpi$ with ϖ. It can be observed that $(d\alpha/d\varpi)_{\varpi=\varpi_{cr}} \neq 0$ for $\varpi_{cr} = 0.715$. Thus, Hopf bifurcation of the system exists at the critical speed ratio ϖ_{cr}.

Letting $\varpi = \varpi_{cr} + \sigma$, the matrix $A(\varpi)$ in Eq. [6.43] can be represented as a Taylor series:

$$A(\varpi) = A_0(\varpi_{cr}) + \sigma A_1(\varpi_{cr}) + \sigma^2 A_2(\varpi_{cr}) + \sigma^3 A_3(\varpi_{cr}) + \cdots. \qquad [6.47]$$

If σ is a small quantity, the second-order approximation is given by

$$A(\varpi) \approx A_0(\varpi_{cr}) + \sigma A_1(\varpi_{cr}) + \sigma^2 A_2(\varpi_{cr}). \qquad [6.48]$$

Then Eq. [6.43] can be rewritten as

$$X' = A_0(\varpi_c)X + F, \qquad [6.49]$$

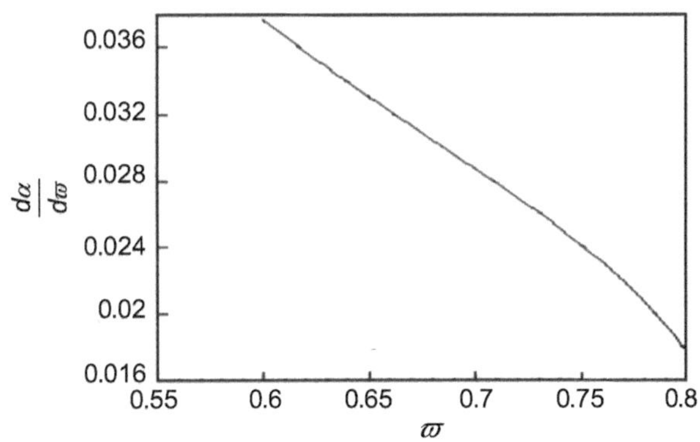

Figure 6.19 Variation of $d\alpha/d\varpi$ with ϖ

where $F = \{0 \quad F_2 \quad 0 \quad F_4\}^T = \{0 \quad f_2 \quad 0 \quad f_4\}^T + \sigma A_1(\varpi_c)X + \sigma^2 A_2(\varpi_c)X$.

Thus, to obtain the Hopf bifurcation solution of the system in the vicinity of the critical speed ratio ϖ_c, the periodic solution of the multiple-degree-of-freedom nonlinear autonomous system [Eq. [6.49]] must be sought.

The matrix $A_0(\varpi_c)$ in Eq. [6.49] has four eigenvalues—a pair of imaginary numbers ($\pm i\Omega_c$) and a pair of complex numbers with negative real parts ($-\delta_1 \pm i\delta_2$). The normalized eigenvectors corresponding to $A_0(\varpi_c)$ and its transposed matrix $A_0^T(\varpi_c)$ for eigenvalues $\pm i\Omega_c$ are assumed to be $\alpha \pm i\beta$ and $p \pm iq$, respectively, where

$$\alpha = \begin{Bmatrix} \alpha_1 \\ \alpha_2 \\ \alpha_3 \\ \alpha_4 \end{Bmatrix}, \beta = \begin{Bmatrix} \beta_1 \\ \beta_2 \\ \beta_3 \\ \beta_4 \end{Bmatrix}, p = \begin{Bmatrix} p_1 \\ p_2 \\ p_3 \\ p_4 \end{Bmatrix}, q = \begin{Bmatrix} q_1 \\ q_2 \\ q_3 \\ q_4 \end{Bmatrix}. \qquad [6.50a\text{–}d]$$

According to [24], the bifurcation solution is given by

$$\begin{cases} X = 2a(\tau)(\alpha \cos\phi - \beta \sin\phi) \\ \phi = \Omega_c\tau + \theta \end{cases}, \qquad [6.51]$$

where

$$\begin{aligned} a' &= (p^T \cos\phi + q^T \sin\phi)\{0 \quad F_2 \quad 0 \quad F_4\}^T \\ &= p_2F_2 \cos\phi + p_4F_4 \cos\phi + q_2F_2 \sin\phi + q_4F_4 \sin\phi \end{aligned} \qquad [6.52a]$$

$$\theta' = \frac{1}{a}(q_2F_2 \cos\phi + q_4F_4 \cos\phi - p_2F_2 \sin\phi - p_4F_4 \sin\phi). \qquad [6.52b]$$

Suppose that a' and θ' are slowly varying functions that can be approximated as the average value in a period of ϕ. Then we have

$$\begin{aligned} a' = \frac{1}{2\pi}[&p_2(A_1a^3 + B_1a) + p_4(A_3a^3 + B_3a) + q_2(A_2a^3 + B_2a) + \\ &q_4(A_4a^3 + B_4a)] \end{aligned} \qquad [6.53a]$$

$$\begin{aligned} \theta' = \frac{1}{2\pi a}[&q_2(A_1a^3 + B_1a) + q_4(A_3a^3 + B_3a) - p_2(A_2a^3 + B_2a) - \\ &p_4(A_4a^3 + B_4a)], \end{aligned} \qquad [6.53b]$$

where A_i $(i = 1, 2, 3, 4)$ and B_i $(i = 1, 2, 3, 4)$ are the integration constants.

Letting $a' = 0$, the stationary solutions of amplitude a are obtained as follows:

$$\begin{cases} a_1^0 = 0 \\ a_2^0 = \left(-\dfrac{p_2B_1 + p_4B_3 + q_2B_2 + q_4B_4}{p_2A_1 + p_4A_3 + q_2A_2 + q_4A_4}\right)^{\frac{1}{2}} = \left(-\dfrac{D}{C}\right)^{\frac{1}{2}} \end{cases} \quad [6.54a\text{–}b]$$

where

$$D = p_2B_1 + p_4B_3 + q_2B_2 + q_4B_4,\ C = p_2A_1 + p_4A_3 + q_2A_2 + q_4A_4. \quad [6.55a\text{–}b]$$

Integration of Eq. [6.53b] yields

$$\theta = \frac{1}{2\pi a}\left[q_2(A_1a^3 + B_1a) + q_4(A_3a^3 + B_3a) - p_2(A_{21}a^3 + B_2a) - p_4(A_4a^3 + B_4a)\right]\tau + \theta_0. \quad [6.56]$$

The condition of the existence of the stationary solution [Eq. [6.54b]] is

$$a_2^0 > 0, \text{i.e.}, \frac{D}{C} < 0. \quad [6.57]$$

By letting $a' = g(a)$, the condition for Eq. [6.53a] to have a stable solution is that the real part of the eigenvalue of $dg(a)/da|_{a^0}$ is negative. Then the condition $D > 0$ is obtained. Therefore, the conditions under which stable stationary solutions of amplitude a exist are $D > 0$ and $C < 0$.

Finally, the bifurcation solutions of the Eq. [6.49] are given by [19]:

$$\begin{cases} X = 2a_2^0(\alpha \cos\phi - \beta \sin\phi) \\ \phi = \Omega_c\tau + \theta \end{cases}. \quad [6.58]$$

Figure 6.20 shows the variation of amplitude a_2^0 with ϖ, which indicates the supercritical bifurcation of this system. According to Eqs. [6.37], [6.38] and [6.43], we know that $R = X_{10} + X_1$. On this basis, the variation of R with τ for $\varpi = 0.74$ is obtained as shown in Figure 6.21. Figures 6.22(a) and (b) show the trajectories of the rotor precession obtained by the numerical solution of Eq. [6.38] for $\varpi = 0.67 < \varpi_{cr}$ and $\varpi = 0.74 > \varpi_{cr}$, respectively. It can be seen that when $\varpi < \varpi_{cr}$, the trajectory is a cycle with constant radius, which corresponds to a stable focus point in the phase

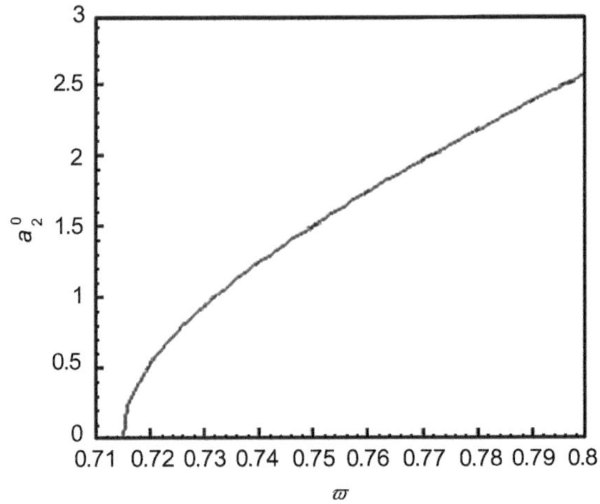

Figure 6.20 Variation of amplitude a_2^0 with ϖ

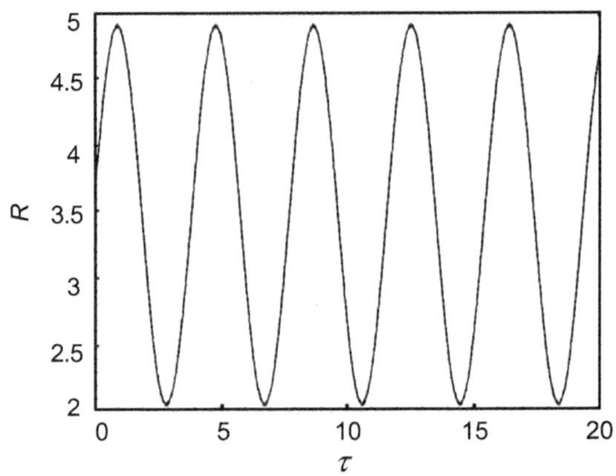

Figure 6.21 Variation of R (from bifurcation analysis) with τ

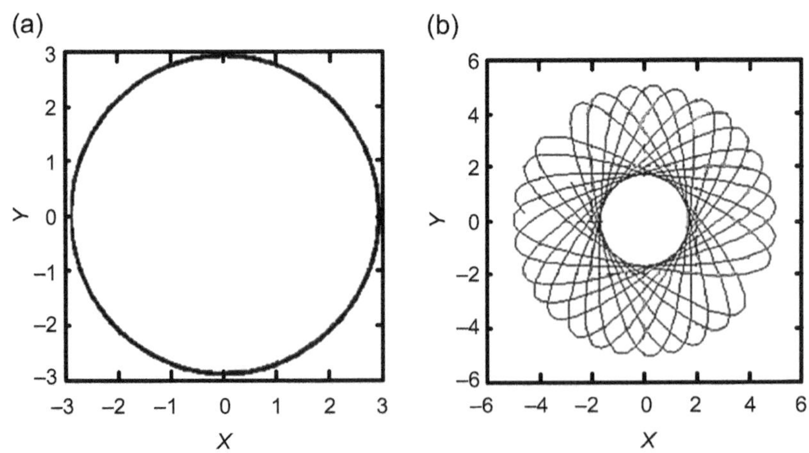

Figure 6.22 Trajectories of the rotor precession for $\varpi = 0.67$ (a) and $\varpi = 0.74$ (b)

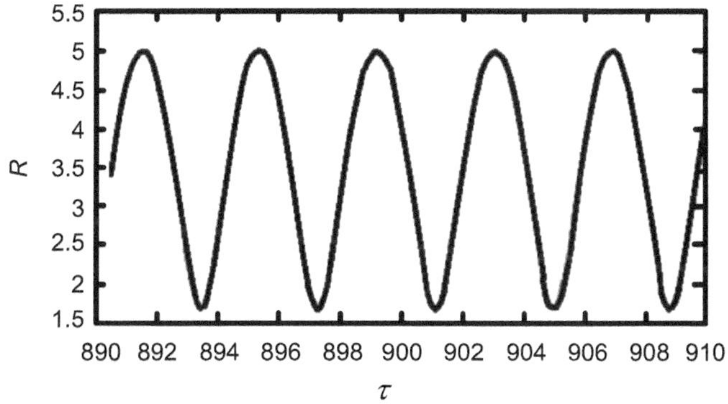

Figure 6.23 Variation of R [from the numerical solution of Eq. [6.38]] with τ

plane $X_1 - X_2$. When $\varpi > \varpi_{cr}$ and σ ($\sigma = \varpi - \varpi_{cr}$) are small values, the trajectory is in a petal shape, which means that the rotor behaves as quasi-periodic motion, and R is not a constant but varies periodically. But the trajectory is a limit cycle when depicted in the phase plane $X_1 - X_2$. Based on the simulation result shown in Figure 6.22(b), the variation of the rotor center deviation R with τ can also be obtained as shown in Figure 6.23. It can be seen that Figure 6.23 coincides with Figure 6.21 very well, although it is obtained through a different method.

6.5 FAULT DIAGNOSIS OF THE ROTOR SYSTEM WITH RUB-IMPACT

It can be seen clearly from the previous sections that nonlinear phenomena appear in the rotor system with rub-impact. Based on these findings, many methods for diagnosing the rub-impact fault of the rotor system have been proposed. In this section, we will discuss the most popular methods.

6.5.1 Frequency Spectrum Method for Diagnosing Impact Characteristics

Vibrations with frequency doubling may be excited when rub-impact fault occurs in a rotor system [1,25]. The underlying mechanism is qualitatively introduced in the following discussion. For simplicity, it is assumed that the rotor collides with the stator once in every circle [15]. Figure 6.24 shows the projection of the rotor precession in a certain axis, and the vibration ranges between A and -1. Thus, the motion can be described by

$$f(\phi) = \begin{cases} A & (-\phi_1 \le \phi \le \phi_1) \\ \cos\phi & (\phi_1 \le \phi \le 2\pi - \phi_1) \end{cases}. \qquad [6.59]$$

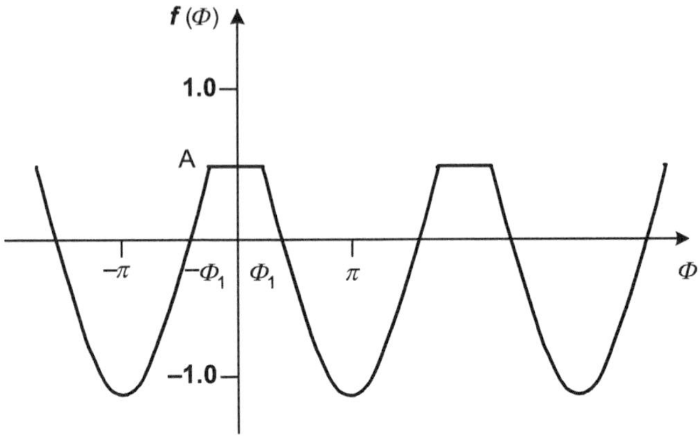

Figure 6.24 Waveform of the rotor precession

The function $f(\phi)$ with period 2π can be expanded in trigonometric form for a Fourier series:

$$f(\phi) = \frac{a_0}{2} + \sum_{n=1}^{\infty} a_n \cos n\omega t, \qquad [6.60]$$

where the static and dynamic components are expressed by

$$\frac{a_0}{2} = \frac{1}{\pi}[A\phi_1 - \sin \phi_1]$$

$$a_1 = \frac{2A}{\pi}\sin \phi_1 + 1 - \frac{\phi_1}{\pi} - \frac{\sin 2\phi_1}{2\pi}$$

$$a_n = \frac{2}{\pi}\left[\frac{A}{n}\sin n\phi_1 + \frac{\cos n\phi_1 \sin \phi_1}{n^2 - 1} - \frac{\sin n\phi_1 \cos \phi_1}{2(n^2 - 1)}\right] \quad (n = 2, 3, \ldots, \infty).$$

[6.61a–c]

It is shown by Eq. [6.60] that the higher-order harmonics with fundamental frequency ω to be the first harmonics occur in the rub-impact event, and their amplitudes decrease gradually. The frequency spectra of a great deal of response signals of the rub-impact rotor system from numerical simulation and experimentation also show that both an operational frequency and its harmonics are included in the signals, and the harmonics increase rapidly as the rub-impact intensifies. The typical waveform of the rub-impact fault and its spectrum are shown in Figure 6.25.

However, not only the rub-impact fault, but also other faults such as bearing looseness, crack of rotor and pitting of bearing, can cause the

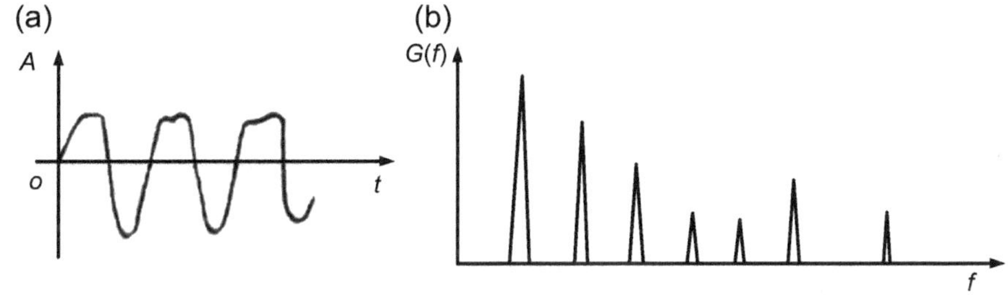

Figure 6.25 Typical waveform of the rub-impact fault (a) and its spectrum (b)

impact characteristic and produce a waveform and spectrum similar to Figure 6.25. Therefore, the frequency spectrum method is unable to diagnose the rub-impact fault of a rotor system accurately.

6.5.2 Holospectrum Diagnosis Method for the Rub-Impact Fault

The traditional spectrum analysis method always plays an important role in rotor fault detection and diagnosis. However, it has many obvious deficiencies. First, the amplitude and phase of the vibration signal are separated, and generally the phase information is ignored. Second, the relationship of the rotor vibrations in the vertical and horizontal directions is difficult to be established. To overcome these problems, the holospectrum diagnosis method is proposed [26,27]. The concept of *holospectrum* includes two-dimensional holospectrum, three-dimensional holospectrum, and holographic waterfall plot. The holospectrum method has been applied successfully to analyze the fault of rotary machinery such as generators and compressors [26—32]. In these studies, the holospectrum method has been explored as a possible way to obtain the shape and rotating direction of the rotor center trajectory and purify the trajectory for various frequency components. For a strong rub-impact fault in a rotor system, the abrupt changes in the vibration signals caused by the rub-impact can be seen clearly. However, for weak rub-impact, this phenomenon is not observable. In [31], it is found that when rub-impact exists, even though the rotor does not run in backward precession, the backward precession can still be found in the trajectories of the rotor for some frequency components obtained by the holospectrum method. This is because the direction of the friction force acting on the rotor is the opposite of the rotating direction and the friction force may lead to backward precession of the rotor.

The key points of the holospectrum diagnosis method lie in accurately obtaining the frequencies, amplitudes, and phases of main components in the frequency spectrum. Thus, methods such as Zoom FFT, diminishing the sampling frequency, and increasing sampling length are employed to obtain accurate data. In [32], a method was proposed to get the precise amplitudes and phases through selecting an appropriate window function and using the dichotomy.

The shape of the rotor center trajectory for each frequency component in a holospectrum depends on the amplitude and phase difference between the horizontal and vertical vibrations, while the rotating direction is only related to the phase difference. The phase difference is more worthy of being investigated for discriminating the backward precession trend of a rub-impact rotor system. Another study [26] indicates that although calculating errors of phases of the horizontal and vertical vibrations on the same section of the rotor by the traditional FFT method exist, the errors for same frequency components are equal. Therefore, the phase difference between the horizontal and vertical vibrations is accurate and the traditional FFT method is qualified for analyzing the rotating direction of the trajectory for each frequency component.

Let $x(i)$ $(i = 0, 1, 2, \ldots, N-1)$ be a real, discrete, zero-mean signal. The DFT of this signal is as follows:

$$X(k) = \sum_{i=0}^{N-1} x(i)e^{-jik\frac{2\pi}{N}} \quad k = 0, 1, 2, \ldots, N-1, \qquad [6.62]$$

where $X(k)$ is a complex value and can be written as

$$X(k) = |X(k)|e^{j\phi_k}. \qquad [6.63]$$

Using the discrete inverse Fourier transform and the characteristics of $X(k)$, the following equation may be obtained:

$$x(i) = \frac{1}{N}\sum_{k=0}^{N-1}|X(k)|e^{j\left(\frac{2\pi}{N}ik+\varphi_k\right)} = \frac{2}{N}\sum_{k=0}^{\frac{N}{2}-1}|X(k)|\cos\left(\frac{2\pi}{N}ik+\phi_k\right). \qquad [6.64]$$

From Eq. [6.64], it can be seen that, for a time-domain signal $x(t)$ with sampling period Δt and total number of discrete points N, the formula of the kth component with frequency $f_k = k \times \Delta f$ $(\Delta f = \frac{1}{N\Delta t})$ is $\frac{2}{N}|X(k)|\cos(2\pi f_k t + \phi_k)$. Similarly, the various components in both the x- and y-directions can be acquired. By combining them, the trajectories and rotating directions of the rotor precessions for various frequency components may easily be constructed. Generally, the trajectories are circles, ovals, or straight lines, and the rotating directions may be clockwise or anticlockwise.

Based on the abovementioned method, holospectrum analysis is performed on the response signals of a dual-rotor system with rub-impact between the low-pressure turbine and the casing. In this example, both the high- and low-pressure rotors rotate anticlockwise. As illustrated in Figure 6.26, when the partial rub-impact fault between the low-pressure turbine and casing becomes severe, the rotational frequencies (f_1,f_2) of the low- and high-pressure rotors and other frequency components appear in the amplitude spectra of the low-pressure turbine precession. Figures 6.26(a) and (b) show the amplitude spectra in the horizontal and vertical directions, respectively. The trajectories and their rotating directions related to the five obvious components shown in the amplitude spectra can be obtained according to the holospectrum method. The rotating directions corresponding to the frequency components $\frac{f_2-f_1}{2}$, $\frac{f_1+f_2}{2}$, and $\frac{3f_2}{2}-\frac{f_1}{2}$ are the opposite of the turbine-rotating direction, which indicates the backward precession, even though the turbine generates the forward precession. Clearly, these three frequency components with backward precession are directly related to the rub-impact. Similar conclusions can be drawn through the holospectrum analysis of other rub-impact patterns.

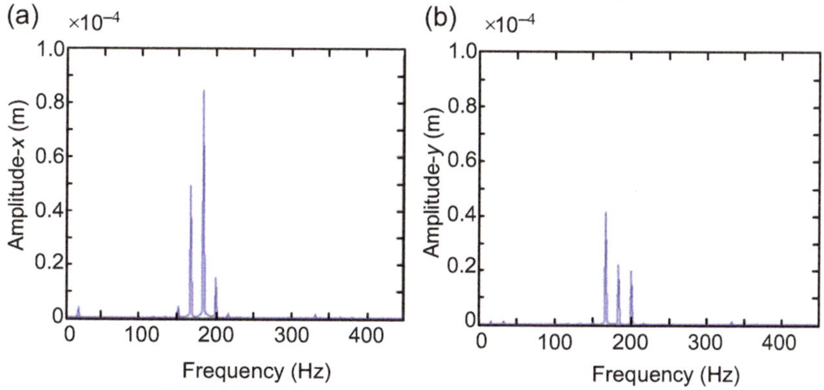

Figure 6.26 Amplitude spectra of a low-pressure turbine: (a) spectrum in horizontal direction; (b) spectrum in vertical direction

6.5.3 Higher-Order Spectrum Diagnosis Method for the Rub-Impact Fault

In the field of signal processing, signals are generally assumed to follow Gaussian distribution. Therefore, the autocorrelation function, autospectral density function, and other second-order statistics can be used for information extraction and parameter identification. For a nonlinear system, although the input signal follows Gaussian distribution, the output is a non-Gaussian signal. In practice, there are many non-Gaussian and nonstationary signals that also need to be processed and analyzed. For these signals, high-order statistics is one effective and important tool to get the detailed characteristics.

6.5.3.1 Brief Introduction of Higher-Order Statistics

The higher-order statistics have been used to describe the higher-order statistical characteristics of a random process. These include the higher-order moment, the higher-order cumulant, and their Fourier transforms known as the higher-order spectra [33]. The higher-order cumulant and the higher-order cumulant spectrum have been widely used in engineering, and they are introduced in this section [34].

The higher-order spectrum can be defined as follows. Let $\{x(n)\}$ be a real, discrete, zero-mean stationary process with a higher-order cumulant $c_{kx} = (\tau_1, \tau_2, \ldots, \tau_{k-1})$ defined as

$$c_{kx}(\tau_1, \tau_2, \ldots, \tau_{k-1}) = cum\{x(n), x(n+\tau_1), \ldots, x(n+\tau_{k-1})\}. \qquad [6.65]$$

The higher-order spectrum is defined as the Fourier transform of the cumulant:

$$S_{kx}(\omega_1, \omega_2, \ldots, \omega_{k-1}) = \sum_{\tau_1=-\infty}^{\infty} \cdots \sum_{\tau_{k-1}=-\infty}^{\infty} c_{kx}(\tau_1, \tau_2, \ldots, \tau_{k-1}) \exp\left[-j\sum_{i=1}^{k-1} \omega_i \tau_i\right] < \infty. \qquad [6.66]$$

The third-order spectrum or bispectrum can be defined as follows. Let $\{x(n)\}$ be a real, discrete, zero-mean stationary process with a third-order cumulant $c_{3x}(\tau_1, \tau_2,)$. Then its Fourier transform is bispectrum:

$$B(\omega_1, \omega_2) = \sum_{\tau_1=-\infty}^{\infty} \sum_{\tau_2=-\infty}^{\infty} c_{3x}(\tau_1, \tau_2,) e^{-j(\omega_1\tau_1+\omega_2\tau_2)}. \qquad [6.67]$$

In the frequency domain, we define

$$X(\omega) = \sum_{k=-\infty}^{\infty} x(k)e^{-j\omega k}. \quad [6.68]$$

Then

$$B(\omega_1, \omega_2) = X(\omega_1)X(\omega_2)X^*(\omega_1 + \omega_2). \quad [6.69]$$

in which * denotes the conjugate operator. The bispectrum is the double Fourier transformation of the second-order moment of a time signal that involves two frequency components (both amplitudes and phases) of the signal.

Higher-order spectra are not sensitive to a Gaussian signal, which can be used to restrain noise, and it can be used to detect a nonlinear signal. The conventional power spectrum density provides information on the second-order properties (i.e., energy) of a signal, whereas the bispectrum can provide information on the signal's third-order properties. In a physical sense, the bispectrum provides insight into the nonlinear coupling between frequencies (as it involves both amplitudes and phases) of a signal compared to the traditional power spectrum density, which gives only the content of different frequencies and their amplitudes in a signal. The higher-order spectra provide the coupling between the frequencies at ω_1, ω_2 and $\omega_1 + \omega_2$ for the time domain signal $x(t)$. In engineering testing, detected signals are not always stationary and may have a low signal-to-noise ratio. Traditional correlation and power spectral analysis based on a Fourier transform could not extract useful information from the nonstationary and nonlinear signals, because in principle, a Fourier transform is based on the assumption that the signals are stationary. Higher-order spectra have been proven to be effective in handling nonstationary and nonlinear signals, which are able to capture the characteristic frequencies, identify phase information, and quantify nonlinear properties. When various frequency components in the signal interact with each other due to nonlinear physical phenomena, new combinations of frequencies are generated in the form of the sum, difference, or fraction of the interacting frequencies. Those frequency components are phase-coupled to the primary interacted frequencies. Higher-order spectra use this phase-coupling signature between frequency components to detect nonlinearities.

6.5.3.2 Application of a Bispectrum to Analyze the Rub-Impact Fault of the Rotor System

The experimental vibration signals of Jeffcott rotor systems without rub-impact, or with weak rub-impact and with severe rub-impact, are processed by the bispectrum method. The results are shown in Figures 6.27–6.32.

The waveform and DFT spectrum of the axle center vibration signal of a rotor system without rub-impact are shown in Figures 6.27(a) and (b), respectively. It can be seen that the signal mainly consists of vibrations caused by weak imbalance and noise. The contour map of the bispectrum is shown in Figure 6.28, which indicates that a wide range of frequency and many peaks exist in the signal. However, the peak values are very small and therefore are supposed to be mainly caused by noise.

The waveform and DFT spectrum of the axle center vibration signal of a rotor system with weak rub-impact are shown in Figures 6.29(a) and (b), respectively. The signal mainly consists of vibrations caused by imbalance, noise, and inadequate rub-impact information. The contour map of a bispectrum with a clearer edge than Figure 6.28 is shown in Figure 6.30. The peak values corresponding to the rotating frequency (40 Hz) can be easily observed, and the regular high-frequency components resulting from the rub-impact occur near 480 Hz.

The waveform and DFT spectrum of the axle center vibration signal of a rotor system with a severe rub-impact fault are shown in Figures 6.31(a)

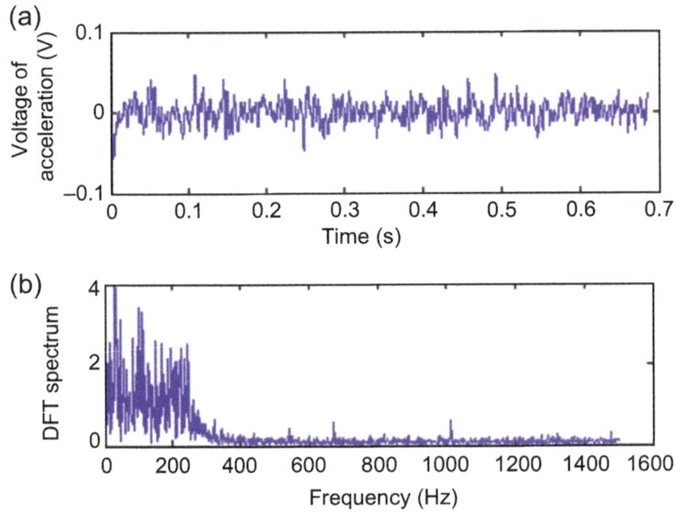

Figure 6.27 Waveform (a) and DFT spectrum (b) of the axle center vibration signal of a rotor system without rub-impact

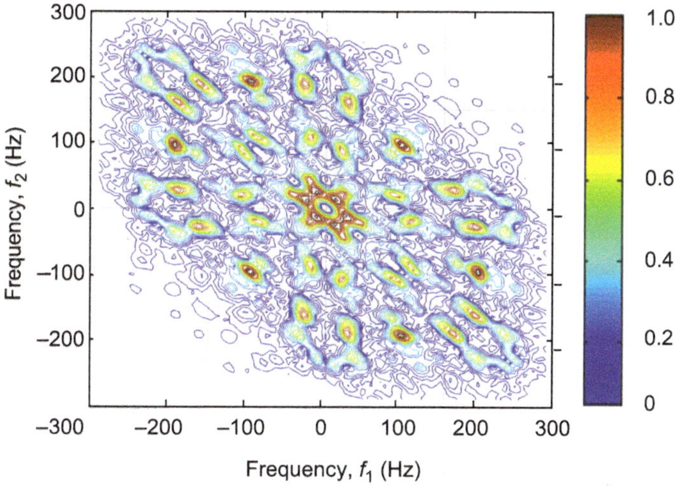

Figure 6.28 Contour map of the bispectrum of the axle center vibration signal of a rotor system without rub-impact

and (b), respectively. The signal mainly consists of vibrations caused by imbalance and noise, as well as obvious rub-impact information such as frequency doubling. The contour map of a bispectrum with a much clearer edge than Figure 6.28 is shown in Figure 6.32. The rotating frequency

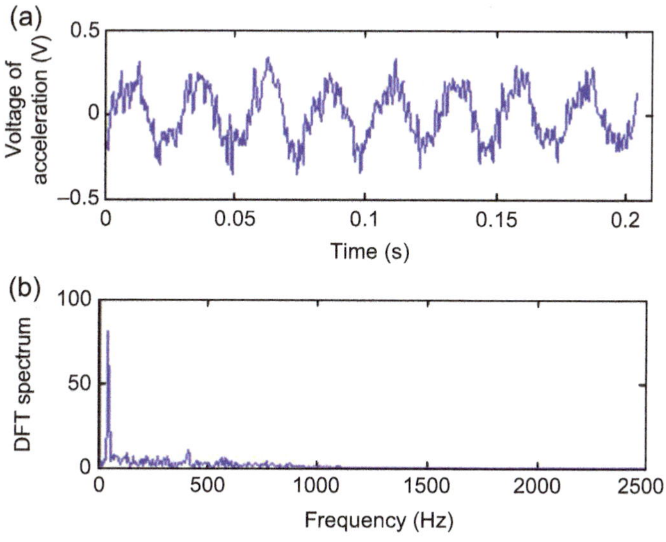

Figure 6.29 Waveform (a) and DFT spectrum (b) of the axle center vibration signal of a rotor system with weak rub-impact

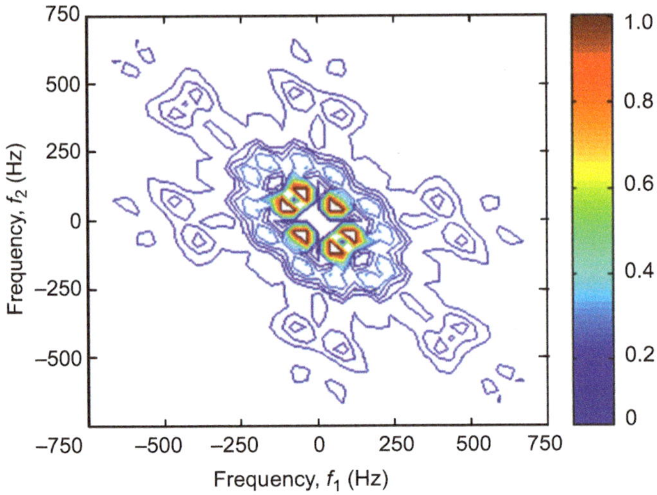

Figure 6.30 Contour map of a bispectrum of the axle center vibration signal of a rotor system with weak rub-impact

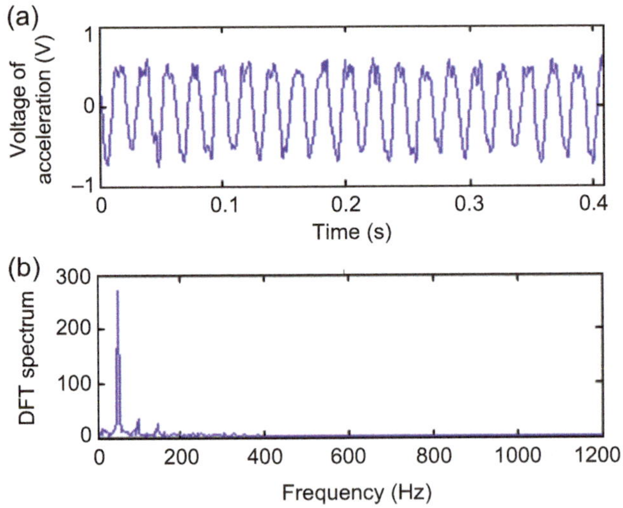

Figure 6.31 Waveform (a) and DFT spectrum (b) of the axle center vibration signal of a rotor system with severe rub-impact

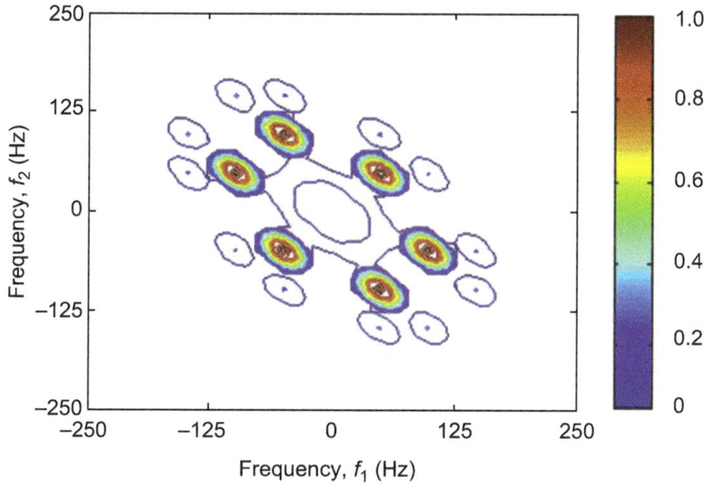

Figure 6.32 Contour map of bispectrum of the axle center vibration signal of a rotor system with severe rub-impact

(50 Hz) and its double frequency (100 Hz) can be easily found, but the other noise components are eliminated.

6.5.4 Application of Wavelet Transform on Feature Extraction of the Rub-Impact Fault

6.5.4.1 Brief Introduction of Wavelet Transform

Many signal processing methods based on the Fourier transform, such as correlation analysis and spectrum analysis, can easily extract different characteristics of signals and have been widely used in signal processing. However, the traditional Fourier transform is limited by the assumption that the signal should be stationary for the whole time domain. Therefore, it is unable to depict the local characteristics of the signal.

A wavelet transform can be used to extract the local characteristics efficiently by adding the scaling window function and dividing the signal into sections. Through wavelet transform of the signal, the abrupt change in the signal may be located accurately, and the low-frequency component in the signal may be identified very well. Therefore, the wavelet transform method has attracted extensive attention from academia and industry since it was proposed in the 1980s, and it has been widely used in the fields of image processing, pattern recognition, and fault diagnosis, among others.

If a function $\psi(t) \in L^1 \cap L^2$ (L^1, L^2 are spaces consisting of absolute integrable and absolute square integrable functions, respectively, in a real number field) satisfies

$$C_\psi = \int_R \left|\hat{\psi}(\omega)\right|^2 \frac{d\omega}{|\omega|} < +\infty, \qquad [6.70]$$

then the function $\psi(t)$ is called the *mother wavelet* or *wavelet basis*. Eq. [6.70] is called the *admissible condition,* where the function $\hat{\psi}(\omega)$ is the Fourier transform of $\psi(t)$. The wavelet family is developed by scaling and translating $\psi(t)$, as follows:

$$\psi_{s,x}(t) = \frac{1}{\sqrt{|s|}}\psi\left(\frac{t-x}{s}\right) \quad s,t \in R,\ s \neq 0. \qquad [6.71]$$

For a signal $f(t)$ ($f(t) \in L^2(R)$), the definition of wavelet transform [35] is

$$Wf(s,x) = <f, \psi_{s,x}> = \frac{1}{\sqrt{|s|}}\int_R f(t)\psi^*\left(\frac{t-x}{s}\right)dt \quad s,t \in R, \qquad [6.72]$$

where the superscript * denotes the conjugate operator. Corresponding to Eq. [6.72], the inverse wavelet transform is

$$f(t) = \frac{1}{C_\psi}\iint_R Wf(s,x)\psi_{s,x}(t)\frac{ds}{s^2}dx. \qquad [6.73]$$

6.5.4.2 Application of a Wavelet Transform to Identifying the Rub-Impact Fault

Wavelet transform and Fourier transform are performed on the simulation signals of a dual-rotor system with rub-impact between the low-pressure turbine and the casing to demonstrate the capability of a wavelet transform in identifying rub-impact fault. Here, a Gaussian wavelet is used, and three cases of no rub-impact, weak rub-impact, and severe rub-impact are simulated numerically [9]. The rotating speeds of the low- and high-pressure rotors are set as 10,000 and 12,000 rpm ($f_1 = 166.7$ Hz, $f_2 = 200$ Hz), respectively.

The amplitude spectrum and contour map obtained by the wavelet transform of the vertical vibration signal of a low-pressure turbine without rub-impact are shown in Figure 6.33. Figure 6.34 shows the results for the case of weak rub-impact between the low-pressure turbine and the casing. Figure 6.35 is for the case of severe rub-impact. From these figures, the following conclusions can be drawn.

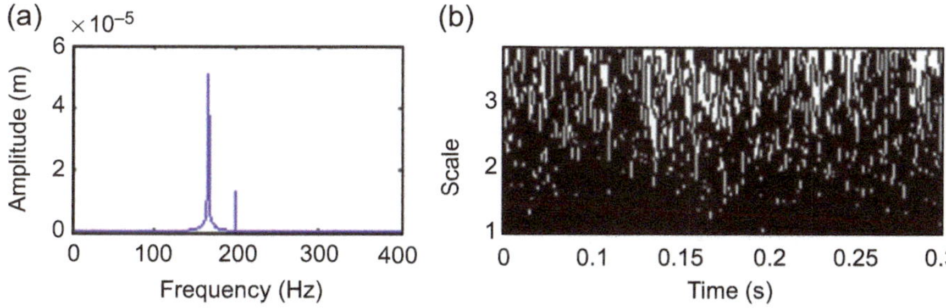

Figure 6.33 Amplitude spectrum (a) and contour map obtained by wavelet transform (b) for the case without rub-impact

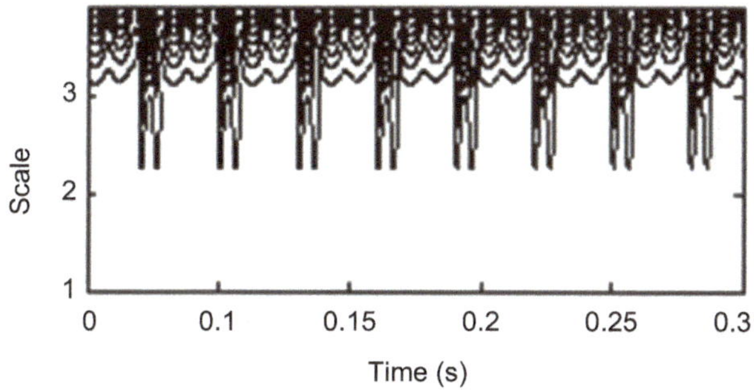

Figure 6.34 Contour map of wavelet transform for a case with weak rub-impact

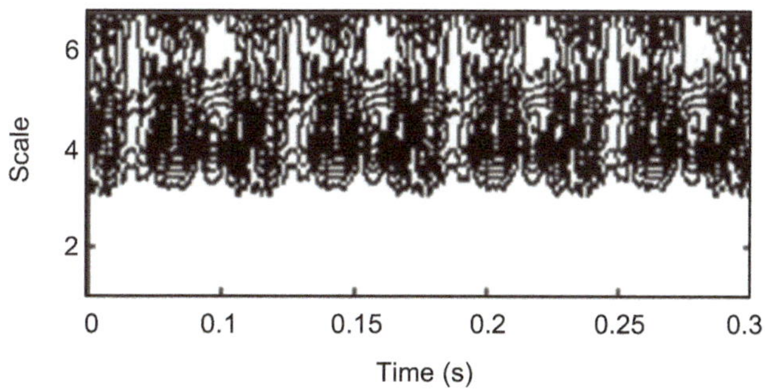

Figure 6.35 Contour map obtained by a wavelet transform for a case with severe rub-impact

The amplitude spectrum of the low-pressure turbine without rub-impact only contains the rotating frequency components ($f_1 = 166.7$ Hz, $f_2 = 200$ Hz) of the low- and high-pressure rotors, while there is not an apparent peak in the contour map obtained by a wavelet transform. That is because the rotors run smoothly, and no local signal abrupt information exists for the case without rub-impact fault. Therefore, the characteristics obtained through frequency spectrum analysis and wavelet transform can be applied to preliminarily identify the rub-impact fault of the rotor system.

When the weak partial rub-impact takes place, the amplitude spectrum of the low-pressure turbine is similar to the case without rub-impact. Although no other frequency components can be seen in the amplitude spectrum diagram except for the rotating frequency components ($f_1 = 166.7$ Hz, $f_2 = 200$ Hz) of low- and high-pressure rotors, the local abrupt information clearly appears on the contour map. Compared with Figure 6.33(b), Figure 6.34 contains many regular peaks, obvious and intermittent lower-frequency components, and higher-energy areas surrounded by several contours. These phenomena are related to the rub-impact characteristics. Based on the intermittent information on the lower scale, the duration of the rub-impact can be defined. In this example, the advantage of using the wavelet transform is illustrated.

When severe rub-impact fault takes place, the subharmonic, superharmonic, and combinations of the rotating frequencies may appear in the amplitude spectrum (as shown in Figure 6.12). The amplitude spectrum can be used to preliminarily diagnose the rub-impact fault; meanwhile, the contour map obtained by wavelet transform also clearly displays the rub-impact characteristics.

6.5.5 Application of the Hilbert-Huang Transform to the Rub-Impact Fault Diagnosis

The Hilbert-Huang transform (HHT) has many advantages in processing nonstationary and nonlinear signals [36]. HHT has similar advantages of multiresolution as the wavelet transform, but it does not involve the selection of a mother wavelet. Through HHT, the signal can be decomposed based on its intrinsic characteristics. Therefore, this method is very promising in the field of fault diagnosis [37]. However, as a new method for processing signals, HHT has some deficiencies as well, such as the fact that, during its time-adaptive decomposition operation named *empirical mode decomposition (EMD),* some undesired intrinsic mode functions (IMFs) at the low-frequency region are generated [38], components with close frequencies are difficult to distinguish, and weak signals are difficult to separate [36]. Therefore, a series of improved HHT methods were proposed to adapt to different applications.

In this section, the vibration signals of the dual-rotor system with rub-impact faults discussed in the section entitled "Application of a Wavelet Transform to Identifying the Rub-Impact Fault," earlier in this chapter, are processed using HHT combined with singular value decomposition (SVD) [39]. Figure 6.36 shows the flowchart of the signal processing.

The instantaneous frequency components of five IMFs of the weak rub-impact signal are shown in Figure 6.37. Not only the rotating frequencies ($f_1 = 166.7$ Hz, $f_2 = 200$ Hz) of high- and low-pressure rotors but also the low-frequency components and periodically abrupt high-frequency components (which are two typical characteristics of rub-impact) are included [40]. At the beginning of the rub-impact process, the instantaneous frequency increases abruptly due to the effect of impact. During the rub-impact process, the instantaneous frequency is restrained to a lower level due to the constraint resulting from the contact between the rotor and the stator. At the end of the rub-impact process, the instantaneous frequency increases abruptly again because of the elimination of contact. Although the

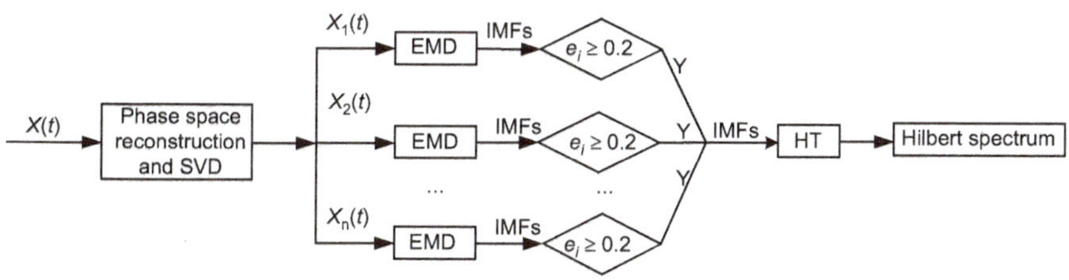

Figure 6.36 Flowchart of signal processing by SVD and HHT

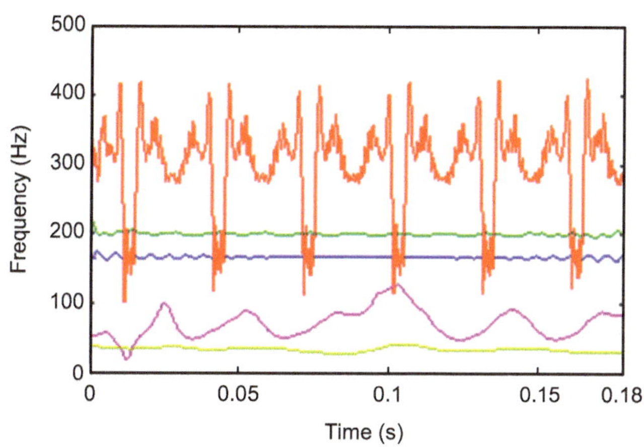

Figure 6.37 Time-frequency diagram of weak rub-impact obtained by HHT

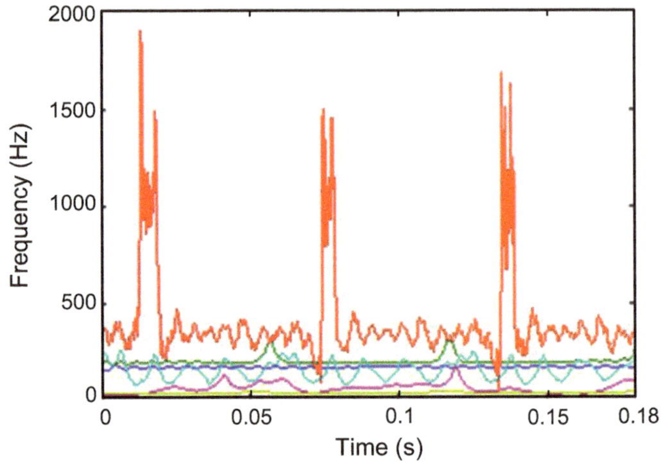

Figure 6.38 Time-frequency diagram of severe rub-impact obtained by HHT

rub-impact signal can be identified by a wavelet transform, the low-frequency components caused by the rub-impact are difficult to distinguish. Moreover, the instantaneous frequency components obtained by HHT are more convenient to understand than that from a wavelet transform.

For a severe rub-impact signal, the instantaneous frequencies obtained by HHT are shown in Figure 6.38. Similarly, not only the rotating frequencies of high- and low-pressure rotors but also the low-frequency components and high-frequency components having periodically abrupt changes are included, and there are some differences between Figures 6.37 and 6.38. When the severe rub-impact takes place, the frequency of the high-frequency components of the system during separation is greater than that of the system with weak rub-impact; meanwhile, the duration of higher frequencies is shorter since this duration corresponds to the separation of rotor and stator. Therefore, the frequency and duration of the high-frequency components having periodically abrupt changes obtained by HHT can be used to determine the severity of the rub-impact between rotor and stator.

REFERENCES

[1] B.C. Wen, et al., Advanced Rotor Dynamics, China Machine Press, Beijing, 2000.

[2] X.D. Liu, Research on rub-impact fault between rotor and stator in rotating machinery and its diagnostic technology, Ph.D. Thesis, Department of Jet Propulsion, Beijing University of Aeronautics and Astronautics, Beijing, 1999.

[3] Z.C. Feng, X.Z. Zhang, Rubbing phenomena in rotor-stator contact, Chaos, Sol. Frac. 14 (2002) 257–267.
[4] J. Jiang, Z.Y. Shang, L. Hong, Characteristics of dry friction backward whirl—A self-excited oscillation in rotor-to-stator contact systems, Sci. China Tech. Sci. 53 (3) (2010) 674–683.
[5] A. Muszyska, Partial lateral rotor to stator rub. 3th International Conference on Vibration in Rotating Machinery, 11–13 September 1984, Heslington, England, 1984, pp. 327–335.
[6] D.W. Childs, Rub-induced parameter excitation in rotors, J. Mech. Des. 101 (4) (1979) 640–644.
[7] L.T. Yan, D.Y. Wang, Vibration features from rubbing between rotor and casing for a dual-shaft aeroengine, J. Aerosp. Power 13 (2) (1998) 173–176.
[8] D.Y. Wang, Theory research and vibration features extracting of rub-impact fault between rotor and stator in aeroengine, Ph.D. Thesis, Department of Jet Propulsion, Beijing University of Aeronautics and Astronautics, Beijing, 1995.
[9] Y.C. Shan, et al., Research on the finite element impact-contact analytical model of dual-rotor system and its diagnosis method, J. Aerosp. Power 20 (5) (2005), 789–194.
[10] F.K. Choy, J. Padovan, Non-linear transient analysis of rotor-casing rub event, J. Sound Vib. 13 (3) (1987) 529–545.
[11] F.F. Ehrich, Some observations of chaotic vibration phenomena in high-speed rotordynamics, J. Vib. Acous. 113 (1) (1991) 50–57.
[12] V. Gangadhar, Rub-impact analysis in rotor dynamic systems, MTech Thesis, Department of Mechanical Engineering, National Institute of Technology—Rourkela, India, 2011.
[13] W.M. Zhang, et al., Nonlinear dynamics of a rub-impact micro-rotor system with scale-dependent friction model, J. Sound Vib. 309 (2008) 756–777.
[14] L. Liu, D.Q. Cao, S.P. Sun, Dynamic characteristics of a disk—drum—shaft rotor system with rub-impact, Nonlin. Dynam. 80 (1) (2015) 1017–1038.
[15] N.Q. Hu, Research on identification of nonlinear behavior and fault of rub-impact in rotors, Ph.D. Thesis, Department of Mechatronics Engineering, National University of Defense Technology, Changsha, China, 2001.
[16] X.D. Liu, Q.H. Li, Impact-rub model and chaotic feature of partial rub of eccentric rotor, J. Aerosp. Power 13 (4) (1998) 361–365.
[17] X.D. Liu, Q.H. Li, D.Y. Wang, Dynamic model of dual-rotor system with rub-impact fault between rotor and stator and its feature of wavelet transform, J. Aerosp. Power 15 (2) (2000) 187–190.
[18] J.L. Isaksson, Dynamics of a rotor with annular rub. Proceedings of the Fourth International Conference on Rotor Dynamics, September 7–9, 1994, Chicago, 1994, pp. 85–90.
[19] X.D. Liu, Q.H. Li, S.P. Yang, The stability and Hopf bifurcation in the annular impact-rub of rotating machinery with imbalance, J. Vib. Eng. 12 (1) (1999) 40–46.
[20] Y.S. Chen, Q. Ding, A study on unstability mechanism and Hopf bifurcation of nonlinear rotor seal system, J. Nonlin. Dyn. Sci. Tech. 3 (3) (1996) 197–205.
[21] M.O.T. Cole, On stability of rotor dynamic systems with rotor-stator contact interaction, Proc. R. Soc. A 464 (2008) 3353–3375.
[22] Y.S. Chen, Q. Meng, Bifurcations of a nonlinear rotor-bearing system, J. Vib. Eng. 9 (3) (1996) 266–275.
[23] Y.S. Chen, Y. Tang, Modern Analysis Method in Nonlinear Dynamics, Science Press, Beijing, 1992.
[24] S.P. Yang, E.L. Chen, The Hopf bifurcation in railway bogie with hysteretic nonlinear hanging, J. China Rail. Soc. 15 (4) (1993) 11–18.
[25] M.A. Abuzaid, M.E. Eleshaky, M.G. Zedan, Effect of partial rotor-to-stator rub on shaft vibration, J. Mech. Sci. Tech. 23 (2009) 170–182.

[26] L.S. Qu, et al., The holospectrum: A new method for rotor surveillance and diagnosis, Mech. Sys. Signal Proc. 3 (3) (1989) 255–267.
[27] L.S. Qu, X. Liu, Y.D. Chen, Discovering the holospectrum, J. Noise Vib. Ctrl World 20 (2) (1989) 58–62.
[28] Y.D. Chen, R. Du, L.S. Qu, Fault feature of large rotating machinery and diagnosis using sensor fusion, J. Sound Vib. 188 (2) (1995) 227–242.
[29] L.S. Qu, D.F. Shi, Holospectrum during the past decade: Review and prospects, J. Vib. Meas. Diag. 18 (4) (1998) 235–242.
[30] X. Liu, Z.Y. Zhao, L.S. Qu, Rotor Monitoring and Diagnosis System, Xi'an Jiaotong University Press, Xi'an, China, 1991.
[31] X.D. Liu, Q.H. Li, Impact-rub fault characteristics and holospectrum analysis of rotating machinery with imbalance, J. Aerosp. Power 13 (4) (1998) 428–430.
[32] Y.L. Cai, L.S. Qu, Y.D. Shen, Accurate determination of amplitude, frequency, and phase of spectral lines of mechanical vibration signals, J. Vib. Meas. Diag. 16 (2) (1998) 13–17.
[33] S.V. Vaseghi, Advanced Digital Signal Processing and Noise Reduction, John Wiley & Sons, West Sussex, UK, 2000.
[34] G. Chen, X.D. Liu, J.Y. Chang, Characterization of nonlinear, non-stationary, nano-scale surfing contact dynamics of TFC slider utilizing high order spectra detection, Microsystem Technologies 21 (2015) 2581–2587.
[35] C.K. Chui, An Introduction to Wavelets, Academic Press, New York, 1992.
[36] N.E. Huang, Z. Shen, S.R. Long, The empirical mode decomposition and the Hilbert spectrum for nonlinear and non-stationary time series analysis, Proc. R. Soc. Lond. A 454 (1998) 903–995.
[37] D.J. Yu, J.S. Cheng, Y. Yang, Application of EMD method and Hilbert spectrum to the fault diagnosis of roller bearings, Mech. Sys. Sig. Proc. 19 (2) (2005) 259–270.
[38] Z.K. Peng, P.W. Tse, F.L. Chu, An improved Hilbert-Huang transform and its application in vibration signal analysis, J. Sound Vib. 286 (1–2) (2005) 187–205.
[39] T. He, et al., Improved Hilbert-Huang transform and its application in rotor-stator rub, J. Aerosp. Power 22 (10) (2007), 1179–1784.
[40] Z.K. Peng, P.W. Tse, F.L. Chu, A comparison study of improved Hilbert–Huang transform and wavelet transform: Application to fault diagnosis for rolling bearing, Mech. Sys. Sig. Proc. 19 (5) (2005) 974–988.

INDEX

'*Note*: Page numbers followed by "f" indicate figures, "t" indicate tables.'

FSC
www.fsc.org
MIX
Papier | Fördert
gute Waldnutzung
FSC® C083411

Kolibri 360 GmbH
Ferdinand-Jühlke-Straße 7
99095 Erfurt, Deutschland
produktsicherheit@kolibri360.de